Springer Series on Cultural Computing

Cultural Computing is an exciting, emerging field of Human Computer Interaction, which covers the cultural impact of computing and the technological influences and requirements for the support of cultural innovation. Using support technologies such as artificial intelligence, machine learning, location-based systems, mixed/virtual/augmented reality, cloud computing, pervasive technologies and human-data interaction, researchers can explore the differences across a variety of cultures and cultural production to provide the knowledge and skills necessary to overcome cultural issues and expand human creativity.

This series presents monographs, edited collections and advanced textbooks on the current research and knowledge of a broad range of topics including creativity support systems, creative computing, digital communities, the interactive arts, cultural heritage, digital culture and intercultural collaboration.

This Series is abstracted/indexed in Scopus.

Craig Vear · Fabrizio Poltronieri
Editors

The Language of Creative AI

Practices, Aesthetics and Structures

Editors
Craig Vear
University of Nottingham
Nottingham, UK

Fabrizio Poltronieri
IOCT
De Montfort University
Leicester, UK

ISSN 2195-9056 ISSN 2195-9064 (electronic)
Springer Series on Cultural Computing
ISBN 978-3-031-10959-1 ISBN 978-3-031-10960-7 (eBook)
https://doi.org/10.1007/978-3-031-10960-7

This Springer imprint is published by the registered company Springer Nature Switzerland AG
The registered company address is: Gewerbestrasse 11, 6330 Cham, Switzerland

Dedicated to those creative AI pioneers whose shoulders we proudly stand on and to all future explorers of creative AI.

Preface

I have a vague recollection of the first time that I encountered a computer in a creative way. In fact, it was my first ever encounter with a computer. I must have been about 7 years old and was selected as part of a national research project that studied the implications of having a computer in the classroom. This, especially given that this was the mid-late 1970s, was a cutting-edge proposal, and, amongst the 'selected' children, an exciting new world. When the Personal Electronic Transducer[1] (or PET) was installed, we were instructed in simple programming. There were not any games for the PET, nor was there Internet, so we programmed in BASIC. I recall making computer art out of letters and numbers and designing glyphs for hours on its green monochrome display and making the PET play simple melodies; it seemed like the logical thing to do. We explored random number generation and created lists of complex looking number sequences because it looked amazing; we printed off abstract generative art made using simple programmes, and we especially loved panicking the staff supervisors as we wrote the classic:

```
10 PRINT "HELLO WORLD"
20 GOTO 10
```

Even then, the personal electronic transducer was something creative, playful, immersive, provocative, disrupting and interactive.

Then in 1982, I received a ZX Spectrum[2] for Christmas. This had games, and colours, and like the PET was programmable in BASIC. I built programmes, made more computer music and played lots of games. The computer signalled and facilitated to me a play space: a world where I could lose time and be immersed in a digital realm. Although the games were not open world, the programming was, and I was able to express myself using the code and create spaces that were borne in my imagination. It was a world that felt free to me as a young boy, and it was exciting to be the kid on the driving seat. I even dreamt of being a systems analyst for a career, and in a way, I am.

[1] See https://en.wikipedia.org/wiki/Commodore_PET for more information.

[2] See https://en.wikipedia.org/wiki/ZX_Spectrum for more information.

A cycle emerged through successive new computers and novel creative software environments: it was new, I explored this newness, and its materiality and functionality would offer and suggest and guide me, while simultaneously transforming my creativity. Over the decades spanning the 1990s to 2010s, this exploration was fuelled by faster and smaller computers. New software specifically designed for media creativity became available, and with it, my language of creativity transformed. The computer, and at its core the new software and operating systems, enhanced my way of being creative: it helped me, offered suggestions, became a partner, teacher and inspiration. It was a free and open space where I could exercise new ideas and reach for seemingly impossible (maybe improbable) solutions to radical ideas.

Skip forward to the recent [re]emergence of AI and the advancement in machine learning, and it felt like a step-change had arrived. I could get a computer to perform music *with me*, I could get it to think about the music we made together and could train it to create and improvise in the moment, *with me*. It was possible to build some creative device that made art and music with others, and perhaps I might be able to find the music equivalent to move 37 like Deep Mind in its game of Go.[3] It seemed that a lot of my colleagues and artist friends also felt the same. But I think the single most exciting aspect of this [re]emergence was that it finally felt that I could hand these decisions and processes over to some other entity and for it to have creative autonomy, something that we started to call creative AI.

Within the last 10 years, I have worked closely with Fabrizio Poltronieri to set up the Creative AI Research Group in the Institute of Creative Technologies at De Montfort University, UK. Interestingly, we have a very similar trajectory from young coder to creative AI researcher and are both interested in the creative relationships between humans and AI. Together, we shared a vision of the language of creative AI and the deep philosophical and conceptual transformations this is having on human creativity. It was this shared vision that led to this book.

Nottingham, UK Craig Vear

[3] Discussed here https://www.wired.com/2016/03/two-moves-alphago-lee-sedol-redefined-future/.

Acknowledgements

We wish to thank all the authors that contributed to this book and those who did not make it through the whole process. COVID was not kind to this project and set up many obstacles which we struggled through. However, it has been a wonderful experience getting to know their work better and to feel part of a global conversation about the nature of creative AI.

Thanks, also, to Helen Desmond and Springer team behind the Cultural Computing series which I am proud to be series editor.

A huge sense of respect and thanks to our good friend Ernest Edmonds whose work and writings have paved the way for our understanding of creative AI and continue to bring sustenance to our practices.

Lastly, thanks to our families who are a constant source of inspiration, fun, play and love.

About This Book

Few technologies have made such an impact upon creativity within such a short space of time. The creative potential of AI was seized immediately by artists and creators from a broad range of disciplines. This led to an exploration through new and innovative practices that seem to have leapt fully formed from their minds. Since the inception of the creative application of artificial intelligence in the 1960s, much has been written about the technical developments of AI in art and approaches to creativity with AI. Analysis and criticism naturally followed, as the academic desire to contain and contextualise the innovative practices through antecedents, influences and technological studies.

Creative AI is an emerging artistic form that defines art and media practices that have AI and, more recently, machine learning embedded into the process of creation, but also encompass novel AI approaches in the realisation and experience of such work, e.g. robotic art, distributed AI artworks across locations, AI performers, artificial musicians, synthetic images generated by neural networks, AI authors and journalist bots. It builds on this previous discourse of AI and creativity (e.g. Dartnell 1994) and extends the notion of embedded and cooperative creativity with intelligent software. It does so through a human-centred approach in which the AI is empowered to make the human experience more creative or join in/cooperate with the creative enterprise in real time.

Creative AI is a rapidly growing area, with many exchanges and connections across disciplines, through shared technology and between the creators. Immediate academization can be risky to the growth of such practices as the critique will extend the previous discourse through analysis of technique and analysis of technical means. A time will come when all creative AI works across all the disciplines are considered worthy of equal analysis, together with the map of who innovated what, and when, and descriptions of technical means. But this kind of academization will inevitably side-step that which is important to the creative technologists and their artistic forms: the aesthetics, philosophies and creative structures. It is also important to such an understanding that these are told through the practitioner's voice from inside the creative relationships between human, computer, code and AI.

This book is one of the first to examine the aesthetic issues central to the rapidly developing field of creative AI. The focus is on the language of creative AI, investigating the diverse ways artists explore the combination and recombination of signs and different syntactic structures using AI techniques, aiming to generate new understandings and new forms of creativity. The chapters examine artistic structures that emerge from the use of computers not only as a technical instrument, but as a creative partner, which poses creative challenges and, in doing so, helps to expand the universe of the artistic and technological language.

The book brings together artists, composers and practice-based researchers from across the world in a study of their approaches, concerns, concepts and aesthetics from inside their creative relationship with AI. This book purposefully avoids the academization of the subject and the presentation of the latest facts on AI as applied to creative projects. Instead, it seeks to initiate a slower debate and establish a longer-term view of the creative concerns of artists, musicians, designers and coders making creative AI.

In preparing this book, we did not make any kind of distinction between the different technologies used by the artists and authors who contributed to the chapters. We believe that dividing the vast panorama of creative AI into drawers such as "art made with symbolic artificial intelligence", "made with neural networks", "GAN art", would be a taxonomic academization that does not help to understand the vast aesthetic panorama of the languages that constitute the universe of creative AI.

What Is in the Book

The book is split into three parts: (1) *Aesthetics and Context,* (2) *Structures and Frameworks* and (3) *Practices.* The first part introduces the basic context of creative AI by introducing these core topics: purpose, challenges, creativity, ethics and language. In the second part, creative AI researchers, practitioner–researchers and artists outline the structures and frameworks they have built through deep engagement with the theories, philosophies and practices underpinning their work. These are not specific to individual works, or practitioner–researchers, but offer a set of guiding principles with which technologists and artists can construct their own creative AI. The final part brings together a range of practitioner–researchers who have had a long-standing relationship with creative AI. Through discussions of their works, they explain the processes and imperatives that they use to guide their creative AI constructions.

Aesthetics and Context

Harold Cohen's chapter *On Purpose* is a reprint of a formative essay by Harold Cohen first published in 1974.[1] It is reprinted here with kind permission by the Harold Cohen foundation. By reprinting it in a book in 2022, we highlight that the central concern of Harold's original—that of the "ultimate purpose" of an artist—is still as relevant and of critical importance to creativity with AI as it was in the 1970s. In fact, this essay serves as a guiding light by which this book was edited, and, as you will read, something that binds all the authors and their approaches with creative AI.

The second chapter in this part by Fabrizio Poltronieri reflects on notions such as language, intelligence, abduction, synechism, creativity and the role of aesthetics in a future where our creative relationship with intelligent artificial systems will become one of increasing symbiosis. Using perspectives from semiotics, he argues that it is possible to delegate tasks to increasingly intelligent autonomous systems, including playful and aesthetic ones, the ones that speak to our hearts as humans.

The third chapter by Lucia Santaella discusses the creative modes of artists incorporating AI in their work. She argues that placing AI as a current resource in the historical context of the development of the technological arts is a well-founded way of entering the debates that the topic may arouse. By doing so, she is not seeking a point of arrival that minimizes the importance of discussions around the topic, but rather raising arguments that can bring the discussion to the specific field of the new challenges that are presented to human creativity.

AI, Creativity and Art by Ernest Edmonds reviews a personal history of research and art practice in which AI systems played a significant part. His key observation was that interacting with such systems provided support for creative thinking. This history is told in relation to a series of meetings held on Heron Island in Australia over 30 years in which computational creativity was explored.

The Ethics of Creative AI by Catherine Flick and Kyle Worrall argues that creative AI has had and will continue to have immense impact on creative communities and society more broadly. They outline how these techniques come with significant ethical responsibilities in their set-up, use and the output works themselves. Their chapter sets out the key ethical issues relating to creative AI: copyright, replacement of authors/artists, bias in datasets, artistic essence, dangerous creations, deepfakes and physical safety and looks towards a future where responsible use of creative AI can help to promote human flourishing within the technosocial landscape.

Structures and Frameworks

The first chapter in Part Two is *Ecosystemic Thinking: Beyond Human Narcissism in AI* by Cesar and Lois (A. K. A. Lucy H. G. Solomon (California) and Cesar Baio (São

[1] This article was first published in Studio International, Vol 187, No. 962, January 1974, pages 11–16.

Paulo)). This chapter introduces the reader to critically oriented artworks by Cesar and Lois that propose that AI move away from anthropocentric modes of processing information and towards more ecologically oriented decision-making. In this chapter, the artists argue that the layering of ecological and biological logical inputs within a relational system (ecosystem) has potential as an environmentally aware model for artificial intelligences. In doing so, they frame microbiological systems as intelligent networks. By extension, they question what an AI built on knowledge that predates human beings would answer to and the form that its logic would take.

The second chapter *Embodied AI and Musicking Robotics* by Craig Vear discusses a hypothesis for embodied AI and its development with music robots. It proposes a solution to the challenge: *if we want AI/robots to join us inside the creative acts of music then how do we design and develop systems that prioritise the relationships that bind musicians inside the flow of musicking?* This requires significant thinking around some core questions such as "what does AI need to do in order to stimulate an embodied interaction in music?", "what sort of intelligent agent does the AI need to be?" and "what does the machine need to learn; what is to be modelled?" This chapter argues for a definition of embodied AI and outlines foundational theories such as embodiment, embodied cognition, flow, musicking, meaning-making and embodied interaction to argue for such a concept in music.

Latent spaces: a creative approach by Matthew Yee-King explores the creative possibilities offered by latent spaces. Latent spaces are machine-learnt maps representing large media datasets such as images and sound. With a latent space, an artist can rapidly search for interesting places in the dataset and then generate new artefacts around and between places. These unique artefacts were not in the original dataset, but they relate to it. The chapter presents a detailed explanation of what latent spaces are and how they fit into a series of developments that have taken place in digital media processing techniques. Furthermore, he outlines four examples of machine learning systems that provide latent spaces suitable for creative work.

In the fourth chapter in this part, Carlos Castellanos examines artists' experimentations with linkages between intelligent computational systems and non-human living organisms. He outlines the theoretical structures created for these unusual hybrid systems and showcase models for how we can bridge heterogenous lifeworlds. In doing so, they offer new perspectives on AI non-human alterities. Additionally, they serve to question the anthropocentric divisions between humans, human technology and the human world, while also pointing towards a model of art-making where encounters between living organisms and intelligent machines can serve not only as vectors of novelty and unexpected variety, but also as a step towards developing a system of ideas focused on showcasing alternative possibilities of human–machine–non-human relations.

The final chapter in this part by Silvia Laurentiz outlines a philosophical perspective on the role of logical–symbolic reasoning in creative AI systems. She presents a novel representational model influenced by information processing that will be able to point out questions about computational algorithms. Using this, she illustrates her model with examples of art and creative procedures so as to highlight the relevancy of such a perspective.

Practices

This part of the book presents a series of case studies from prominent creative AI artist–technologists. Each chapter discusses the aesthetics and structures, concerns and processes of working on a specific project. The discussion is not about what was made, but about how the underlying language of creative AI affects the ways of working and relationships between creators, their technology and the work.

Creative AI, Embodiment, and Performance by Rob Saunders and Petra Gemeinboeck explores the relationship between creative AI, embodiment and performance with reference to their artistic practice. Through this, they have shifted their focus from the development of computational systems as models of creative agents towards the realisation of skilful performers able to facilitate the emergence of creative agency between humans and machines.

In *Musebots and I: Collaborating with Creative Systems* Arne Eigenfeldt describes *musebots* and their specific use as collaborators within systems as designed by an artist. He views collaboration as an equal partner in the creative process and describes the unique relationship between composer and artificial agents in the creation of artworks that exist as artworks, rather than examples of computational creativity.

Eduardo Reck Miranda's chapter *Composition with Computer Models of the Brain: An Alternative Approach to Music with Artificial Intelligence* proposes a novel angle to harness the Neurosciences for composition. Rather than building musical ANN to learn how to compose music, he introduces his forays into harnessing the behaviour of a type of neuronal model referred to as *spiking neuronal networks* to compose music. The discussion revolves around a piece for orchestra, choir and a solo mezzo-soprano entitled *raster plot*.

In *Tuning Topological Morphologies: Creative Processes of Natural and Artificial Cognitive Systems* by Johnny DiBlasi, the AI model used in creative practice serves as a model or framework for thinking about the aesthetics and structures of creative processes. Through a discussion in relation to his artwork *432 Hz*, he highlights a framework for a foundational language of creative AI which affects the creative process and the relationships between creators, technologies and the resulting works.

The final chapter of this part, and the book, is by the artist Sougwen. In her work as an artist and researcher, she has developed AI systems and robotics that explore phenomenological constructions, interrelations and alternative configurations of human and machine. The work, vestiges of presence in the form of artefacts, research and performance, is an ongoing process of inquiry and invention. The operations investigate the computable and uncomputable, interrogating the promises and pitfalls of meaning-making metaphors for understanding complex systems. Her practice focusses on the contours of where AI ends and the 'I', the individual human subject begins, using technology of the present day to ask questions of authorship on a space of a canvas and a performance over time. For Sougwen, perhaps even every author in this book, this question of agency within systems is a microcosm of the wider implications of technological governance and its entangled relationship to the human subject.

Introduction

Simply put, creative AI is the application of artificial intelligence (AI) in creative activities. This definition encompasses a broad range of pursuits, approaches and perspectives, but for us the overall goal seems to be to enhance human creativity. For example, creative AI can encompass artistic practices that use AI to suggest or collaborate with the human in the shared making of an artwork; or installations that use AI to connect viewers movements to generative sound and abstract image; or music creation that uses a neural network modelled on the work of a composer such as Mozart; or a dancing robot that improvisers in real time by itself or with another partner. The list goes on as each artist or creative technologist finds new ways of expressing themselves through or exploring the creative potential in, the application of creative AI. But a core tenet endures:

> it is creative and intelligent *within its context*.

Throughout this book, many of the authors offer their definition and scope of creative AI. These range from creativity support tools, building autonomous models of creativity, agents for creative collaboration, environments to evoke emergent creativity, symbolic and strategic systems to help guide creativity into new directions, co-creative artistic partners, interactive creativity and systems to support a creative experience. Whatever the application understanding, the context of how the creative AI is to be considered *creative* and *intelligent* is of the upmost importance. For example, it is very easy to dismiss a creative project that employs behaviours of ants protecting a colony to train a neural network that generates artworks because it does not correspond to a definition of intelligence based on the workings of the human-mind. Or to dismiss a music composition that co-creates in real time with a human musician using short-term memory recall and fuzzy logic as "not resembling any reasonable definition of AI as it does not learn". Or, if a petri-dish of slime mould is connected to a sketch-plotter and through its daily life, a simple AI translates its state into small marks and squiggles on a piece of paper as "not creative" because it "does not look like an artwork hung in a gallery" or that it could not possibly be considered creative because it was "generated by mould and AI".

What these examples show—and these are real responses that ourselves and others have received—is that without *context* prejudice and bias on the part of the observer is allowed to get in the way of the reception of these works. And by extension, can bias and prejudice creativity too. This is mostly because the two core terms of *creative* and *AI* are infused with human bias and prejudice, especially when used together in the creative and cultural acts of art, music, sculpture, interactive museum design, fashion, etc.

Critically, both these terms "are 'suitcase' words with several meanings that can be confused, and must be unpacked" (Sturm 2022).[2] But, as illustrated throughout this book, this 'unpacking' needs to be contextualised within each of the works. Sturm continues:

> In the context of Ai,[3] "intelligence" is a quite brittle thing that bears very little resemblance to human intelligence. And "learning" is merely the prosaic computation of numbers in an algorithm that represent probabilistic relationships in data. Falling into this trap can give power to the Ai that it does not actually possess, leading to dystopian fantasies. (Sturm 2022)

Sturm is introducing into his argument, here, another key issue when presenting creative AI to others: that of fear and misconception of the power and future of AI in our society. And, this is another crucial part of understanding the context of creative AI. This is because, art is somehow seen to be a pure artefact achieved only by the result of human endeavour. Involving a machine, or even a thinking machine, in this process can challenge or disrupt this sense of purity. And yet, when we speak with artists and creative technologists, the tangible artifact is not the priority, but it is the intangible experience of the process and activity, or the feel of the thing, that is of greater concern.

This is one of the main reasons why this volume focuses on the language of creative AI. Moving the focus from technical discussions to the field of language allows us to deepen the discussion on creativity and AI. This is because language, in the semiotic conception that served as a guide for some of the initial discussions around this book, goes beyond merely limiting discussions to the representation of things. More importantly, language allows artists to project new things. From this perspective, the history of art, music, etc., can be understood as the narrative of how these projections and the world become enmeshed as one thing. Language is a powerful transformative technology.

Dealing with creative artificial intelligence based on language also allows us to establish a common denominator, since we understand that the mechanisms that build thought, language, art and technology depend on symbols, which are the essential elements of the production of ideas in art and science. To deal with symbols is to deal with language at its most primordial level.

---

[2] Sturm, B (2022) https://tunesfromtheaifrontiers.wordpress.com/about-me/ (accessed 24 May 2022).

[3] Sturm prefers to use a small i in 'Ai' to 'highlight the fact that its "intelligence" is questionable'. Sturm, B (2022).

Technology and creative tools have had a long history of expanding and enhancing human creativity and practice, precisely because, ultimately, they expand the human symbolic universe. Although originally these were passive in the sense that the piano needed a human operator, the brush needed guiding by a hand and eye, the introduction of computation, and lately AI, has meant that our creative technologies can now be endowed with varying degrees of *creative agency* (Bown and McCormack, 2009)[4] and *autonomy* (Boden, 2010).[5] Creative agency joins in with us as we produce, invent, generate or consume cultural objects. For example, they can be collaborative (i.e. a useful partner), combative (i.e. challenging decisions), concurrent (i.e. sharing a core aesthetic but independent) or co-creative (i.e. expanding human creativity and potential).

Either way, it is important to note that in all these modalities, technological systems work creatively in partnership with humans. We, particularly, do not believe in a dystopian future where AI will replace artists, poets and musicians. Our belief and our research point to a symbiotic, more challenging future where intelligent systems will play in real time with humans in creative contexts, challenging the limits of creative languages and being challenged by us.

This scenario is inspiring and can lead us to new areas of human culture, new understandings of the human condition and new ways of using our brains. Creative AI can present new kinds of creativity and can engage in new forms of human–machine collaboration. But it can also change the nature of relationships, hierarchy, authenticity, quality, the artists 'essence' and credibility of culture.

Machine learning and datasets can be constructed with bias that can reflect on societies prejudices. The universe of art, due to its disruptive history and breakthroughs, can offer new paths and solutions to these and other problems encountered in the current stage of development of AI. In a world where AI systems become ubiquitous, it is important to try to ensure that all voices and colours and shades are contemplated by them. We believe that art and creativity are powerful tools in the path towards greater representation and inclusion in the growing world of AI.

This book has brought together a critical mass of artists and researchers coming from distinct backgrounds from around the world in order to expand the context with which creative AI has been applied. Furthermore, it is through this explanation that their meaning of the terms *creative* and *AI* are to be understood and also the *value* their work brings to society, art, politics, ecology, culture and other areas of human activity.

Craig Vear
Fabrizio Poltronieri

[4] Bown, O. and McCormack, J. (2009). Creative agency: A clearer goal for artificial life in the arts. In Kampis, G., Karsai, I. and Szathma´ry, E., editors, *ECAL (2)*, volume 5778 of *Lecture Notes in Computer Science*, pages 254–261. Springer.

[5] Boden, M. A. (2010). *Creativity and Art: Three Roads to Surprise.* Oxford University Press.

Contents

Practices

List of Figures

Embodied AI and Musicking Robotics

Latent Spaces: A Creative Approach

Intersections of Living and Machine Agencies: Possibilities for Creative AI

Conformed Thoughts, Representational Systems, and Creative Procedures

Creative AI, Embodiment, and Performance

Musebots and I: Collaborating with Creative Systems

Composition with Computer Models of the Brain: An Alternative Approach to Music with Artificial Intelligence

List of Tables

Aesthetics and Contexts

On Purpose: An Enquiry into the Possible Roles of the Computer in Art

Harold Cohen

Abstract This is a reprint of a formative essay by Harold Cohen first published in 1974. It is reprinted here with kind permission by the Harold Cohen Trust. By reprinting it in a book in 2022, we highlight that the central concern of Harold's original—that of the 'ultimate purpose' of an artist—is still as relevant and of critical importance to creativity with AI as it was in the 1970s. In fact, this essay serves as a guiding light by which this book was edited, and, as you will read, something that binds all the authors and their approaches with creative AI.

Keywords Creativity · AI · Purpose · Computation · Art

1 This is not Another Article About 'Computer Art'

The development of the computer has brought with it a cultural revolution of massive proportions, a revolution no less massive for being almost silent. We are living now in its early stages, and it would be difficult to predict—certainly well outside the scope of this article—what changes will be effected within the next two or three decades. I think it is clear, however, that well within that period, subject to such issues as public education, the computer will have come to be regarded as a fundamental tool by almost every conceivable profession.[1] The artists may be among them. That will be the case, obviously, only if it shows itself to have something of a non-trivial nature to offer to the artist, if it can forward his purposes in some significant way.

[1] I wish I had more space here to develop and justify what may seem to be extravagant views. Readers wishing to pursue the issue for themselves will find these views to be almost timid compared to the current rate of growth and technological development within the industry. There is extravagance indeed! Of an estimated 80,000 computers now operating in the US alone, 13,000 were installed in 1972 by a single manufacturer. Spending on *small* computers is projected by a leading magazine to rise to $600,000,000 a year in the US by 1975.

This article was first published in Studio International, Vol 187, No 962, January 1974, pages 11–16.

H. Cohen (Deceased) (✉)
IOCT, De Montfort University, Leicester, UK
e-mail: cvear@dmu.ac.uk

C. Vear and F. Poltronieri (eds.), *The Language of Creative AI*, Springer Series on Cultural Computing, https://doi.org/10.1007/978-3-031-10960-7_1

There is little in ‘computer art’ to justify such an assumption. On the other hand, I have come to believe, through my own work with the machine, that there may be more fundamental notions of purpose, and a more fundamental view of what the machine can accomplish, than we have seen so far; and this article is intended as a speculative enquiry into that proposition.

Speculation is cheap, of course, as the popular media have shown. If you fantasize any given set of capabilities for the computer, without regard to whether the real machine actually possesses them, then you can have it achieving world domination or painting pictures, falling in love or becoming paranoid; anything you wish. I would hope to offer something a little more rigorous, if rather less romantic. Thus, I propose to proceed by describing the machine’s basic structure and functions, and by giving a simple account of programmes of instructions which it can handle with those functions. It should not prove necessary to make any speculation which cannot be stated in terms of these.

All the same, the undertaking is not without its difficulties. There is no doubt that the machine can forward artists’ purposes. It has forwarded a reasonable range of specific purposes already—some have been trivial, some have not—and there is no reason why that range should not be extended. But the significance of the question would seem to point to the notion of Purpose rather than purposes, implying, if not a hierarchical structure with Ultimate Purpose sitting on top as its informing principle, certainly a structure of some sort which *relates* all of an artist’s individual purposes.

The chain of interrogation: Why did you paint this picture blue? Why did you paint this picture? Why do you paint? Is thus a good deal less innocent than it might seem at first glance. I suspect that the notion of Ultimate Purpose enjoys little currency today: but then it must follow that Purpose is not to be arrived at by backtracking up a hierarchical structure from the things that an artist does, much less from the objects he makes. The problem is rather to propose a structure which can be seen, as a whole, to account for the things the artist does. The notion of Purpose might then reasonably be thought to characterize that *structure*, as a whole (Fig. 1).

In what terms, then, would it be possible to maintain that the use of the computer might ‘advance the artist’s Purpose’? Any claim based upon the evidence that ‘art’ has been produced would need to be examined with some care, and in the absence of any firm agreement as to what is acceptable as art we would probably want to see, at least, that the ‘art’ had some very fundamental characteristics in common with what we ordinarily view as art. This could not be done only on the basis of its physical characteristics: merely looking like an existing art object would not do. We would rather want to see it demonstrated that the machine behaviour which resulted in the ‘art’ had fundamental characteristics in common with what we know of art-making behaviour.

This is already coming close to a more speculative position: that the use of the machine might be considered to advance the artist’s Purpose if, following the earlier argument, it could be seen that this use might itself generate, or at least update, an appropriate notion of structure.

In either of these cases, it must be clear that my definitions have much in common with the curious way in which we ordinarily make our definitions of art. We would

Fig. 1 The Hewlett Packard 2100. A computer is a small, fast, general-purpose machine characteristic of the 'minis' now on the market

probably agree, simply on the evidence that we see around us today, that the artist considers one of his functions to be the redefinition of the notion of art.[2] Or we might say that the artist uses art in some way to redefine, i.e. modify himself. But since he is the agency which is responsible for the art process which effects the modification, we could restate this: the artist who uses art to modify the artist who uses art to modify.

These are recursive[3] structures. I think it will become evident in due course that my definition of Purpose is recursive also; and the balance of this article may suggest that it has, in fact, been generated by my use of the machine. For the moment, though, I propose to adopt the earlier position, and to argue that the machine's behaviour shares some very fundamental characteristics with what we normally regard as art-making

[2] But not necessarily for other times and other cultures.

[3] Recursion is a powerful mathematical concept which is difficult to describe in non-mathematical terms: indeed, the above examples are as good as any I have been able to find. If you think of a mathematical function as being a structure which operates upon something provided to it, then a recursive function is one which provides *itself* with the 'something' by its previous operation. Since the 'something' will be different for each operation, this is not to be confused with circular structures: e.g. art is something produced by an artist, an artist is someone who makes art. Also, the idea of the boy holding the bag of popcorn on which there is a picture of the boy holding the bag of popcorn on which ... actually represents a hierarchical structure rather than a recursive one.

behaviour. Let us now look at the computer itself and then examine what some of these characteristics may be.

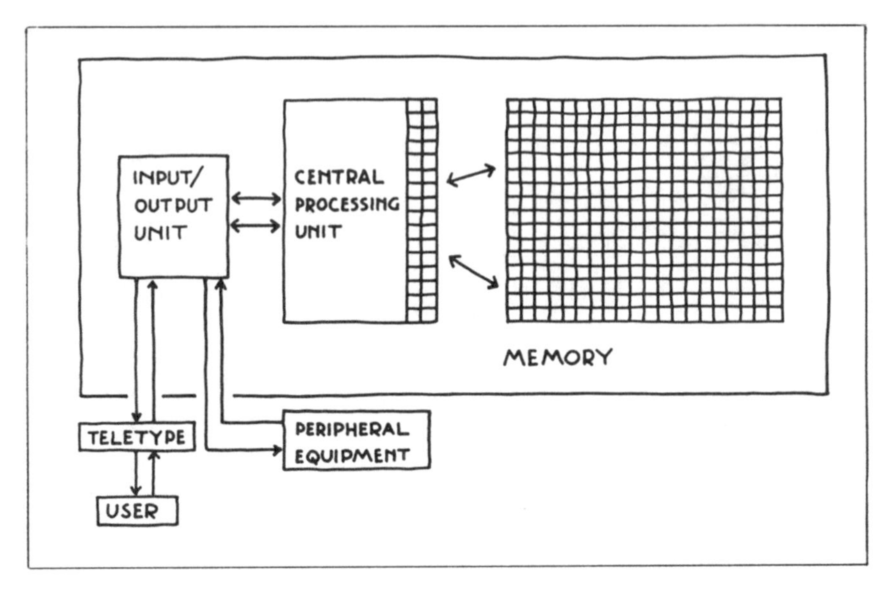

Fig. 2

There is an increasing diversity in computer design today. At one end of the spectrum, machines are getting smaller, at the other end they are getting much, much bigger; at both ends, they are becoming much faster. Yet it remains reasonable to talk about 'the machine' because, big or small, fast or slow, all computers do much the same things and consist, diagrammatically at least, of the same parts. Part of it, usually called the Input/Output Unit, takes care of its communication with the world outside itself. Part, as you probably know, is used for storage—it is the computer's 'memory'. The 'operations room' of the whole machine, appropriately enough called the Central Processing Unit (CPU), is concerned with the processing of the stuff the machine handles, and for shifting this stuff around inside the machine. If you think of 'memory' as a very long string of numbered boxes, or cells, then the CPU looks after the business of storing things in the cells, labelling the cells, keeping up an index of where all the labels in use are to be found, retrieving the contents of cells with particular labels and so on.

What are these 'things', this 'stuff' the machine handles? There are different ways of answering this question, and their relationship demonstrates one of the most significant features of the computer. Physically speaking, what the machine handles is pulses of electrical current, which are triggered by switches, and in turn trigger other switches. But the configurations into which the switches are set actually represent numbers, and the numbers represent… well, just about anything that can be represented numerically: quantities, dimensions and values obviously, but also anything which can be given a numerical code, like alphabetic characters, colours or instructions. The computer is a general-purpose symbol-manipulating machine, and it is capable of dealing with any problem which can be given a symbolic representation. If its accelerating use in our society rests upon its remarkable versatility, then its versatility rests in part upon the fact that a very large number of problems—much larger than you might suspect—do indeed lend themselves to symbolic, even numerical, representation.

The on–off switch might not seem too promising as a device for counting, since it can only record 'zero'—off, or 'one'—on. But a race of creatures with two-hundred and seventy-nine fingers might consider our own ten-position-switch system pretty limiting also. We still need to add a second switch to get up to 99, a third to get up to 999 and so on. Whatever 'base' you use for counting, how high you can count depends upon how many switches—each with the 'base' number of positions you put together. When the 'base' is two, you will need a large number of switches to get very far, but each of them need only have two positions—on or off: obviously an ideal situation for counting electrically (Fig. 3).

If you were to take a somewhat less metaphorical look at those little cells in the computer's memory, you would see that each one was in fact a string of switches. Most small modern computers have adopted sixteen as a standard, though not all: and you can figure out that this sixteen-switch cell—or 'sixteen-bit word', to use the jargon—will be able to hold any number up to 216! In a very rough sense, the size of the machine is measured by how many of these words it has in its memory, and its speed by how long it takes to retrieve one. There would probably be between four and thirty-two thousand sixteen-bit words in a small machine: up to a quarter of a million sixty-four-bit words in a big one (Fig. 4).

The Central Processing Unit is responsible for moving these words around and for performing certain operations upon them. Ingeniously, it knows from the words themselves what it is to do, since several bits of each word are actually reserved for instruction codes. Thus, part 'A' of a word might tell the CPU, 'put the number shown in part "B" into memory'; or 'get the number which is in the cell in memory specified by the number in part "B"'; or 'add the number in part "B" to the number you are now holding, and put the result back in memory'. A machine might recognize and act upon as many as fifty or sixty such instructions, but in fact most of them will be concatenations of simpler instructions, like 'add', 'subtract', 'multiply', 'divide', 'compare', 'move this into memory' and 'move this out of memory'.

The user sees nothing of all this going on. Sitting in the outside world, the set of instructions he composes for the machine will almost certainly be written in a 'higher level' language, like Fortran or Algol, and before the machine can execute that programme of instruction it must first run a programme of its own to translate it into its own numerical code. A single line of code—a 'statement'—in any higher-level language will normally break down into a large number of machine instructions, and these are executed electronically, literally by switching electrical currents, with consequent speeds measured in millionths of a second per instruction. Yet the computer's phenomenal speed is probably less significant in accounting for its versatility than the fact that it can break down any user's programme into the same instruction set. While the machine is running a user's programme, it cannot do

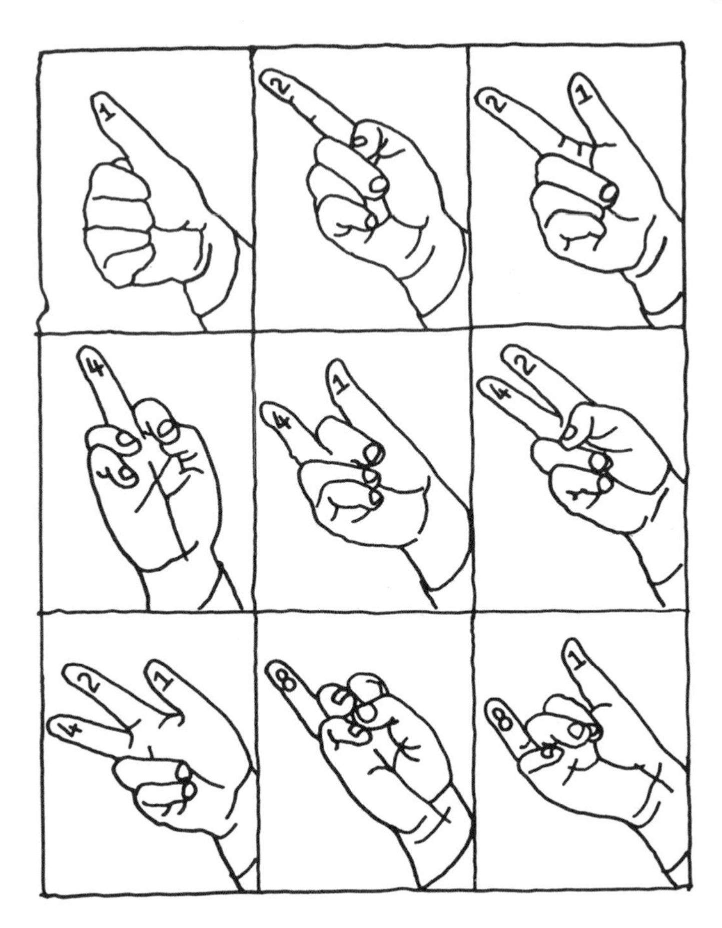

Fig. 3 ‘Binary’ counting is illustrated here by hand, using each successive finger in its ‘on’ or ‘off’ positions to count successive powers of two. The total is given in each case by adding the ‘on’ fingers together

anything else, so that you might say the machine is identified by the programme. But it can take on a new identity in the time it takes to clear one programme from memory and load a new one, and in a single day, a moderately sized computer installation may run a thousand different programmes. A thousand different tasks, a thousand ‘different’ machines.

The man–machine relationship I am describing here is a very curious one, and not quite like any other I can think of. Nor is it possible to deal meaningfully with questions relating to what the machine can do except in terms of that relationship. It is true that the machine can do nothing not determined by the user’s programme; that the programme literally gives the machine its identity. But it is true also that once it has been given that identity, it functions as independently and as autonomously as if it had been built to perform that task and no other. Whatever is being done, it is being done by the machine.

Fig. 4 This memory module taken from the Hewlett Packard 2100. A computer illustrates the development of miniaturization in recent technology. The module holds 8000 sixteen-bit words—128,000 switches in all. The switches are minute doughnut-shaped ferrite 'cores' strung on wires. Courtesy Hewlett Packard

When we talk of the computer doing something, it is implied that it is doing it, or controlling the doing of it, in the outside world. For the computer, this outside world consists of any or all of a large number of special purpose devices to which it may be connected through its Input/Output Unit, varying widely in their functions from typing or punching cards, to monitoring heart beats or controlling flow valves. Some of these 'peripheral' devices serve the computer in the very direct sense that they provide communication channels to the user, allowing him both to get his programme into the machine and receive its response to it. The ubiquitous teletype and its many more sophisticated modern equivalents serve both needs: combinations of punched-card reader and line-printer, or paper-tape reader and punch, do the same. Several peripherals function as extra memory for the machine, but then memory simply means storage, and a deck of punched cards, or a punched paper tape, is as much a storage medium as is magnetic tape or the more recently developed magnetic disc. Once a programme has been entered via the teletype or the card reader, the computer can permanently record it in any of these media and reload it from them when required to do so. Obviously, these media can be used also for storing large quantities of information.

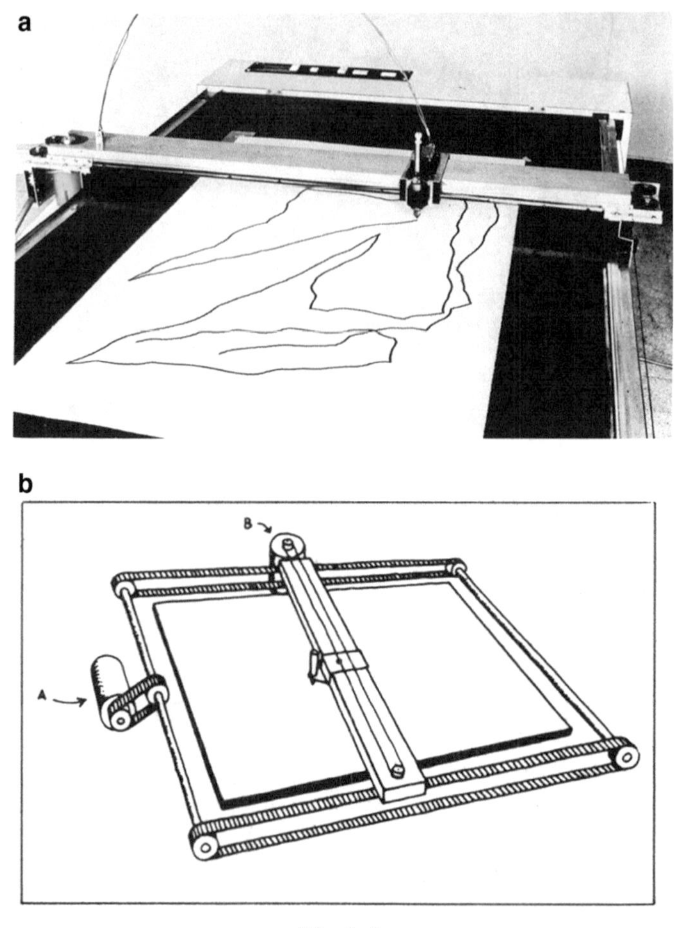

Fig. 5a, b

In general, you might say that the computer may receive messages from any device which is capable of putting an electrical voltage on a line and may control any device which can be switched by a change in voltage generated by the computer. The user today has a host of peripherals at his disposal, covering a wide range of sophisticated abilities: perhaps for that very reason, it is important to recognize that the use of more sophisticated peripherals does not necessarily imply more sophisticated use of the computer. If you wanted to make an animated sequence, say of a cube revolving in space, then a television-like device which could display individual frames at the rate of thirty per second would have much to commend it over a mechanical device like a plotter, whose pen only moves at five or six inches per second as it draws the frames one by one. As far as the computer is concerned, however, the task is to generate a series of views of a cube rotating in space, and it will use literally the same programme to do so regardless of what device it is addressing.

The point would seem obvious enough not to need underlining, were it not that many writers appear to hold the view that the failure of 'computer-art' to achieve images of notable stature can be ascribed to the lack of peripherals appropriate to the artist's needs! Incongruously, the kind of peripheral upon the basis of which some of these writers project rosy futures for 'computer-art' do not relate to new needs, but to old ones. All will be well when the artist can communicate to the computer with a paint brush.[4]

Failure to produce significant images arises from lack of understanding, not from lack of machines. The truth is that it has been, and remains, extremely difficult for any artist to find out what he would need to know, either to use the computer, or even to overcome his certainty that he could not possibly do it for himself. He will almost inevitably find himself confronted by professionals who are more than anxious to help him, but that might be a large part of his problem. 'What will the machine do?' he asks. 'Well', he told, 'it will do A, B, C, or D. You just choose which you want and we will programme it for you!' The specialist is well intentioned, and it seems unreasonable to blame him if he is less than well informed about what the artist wants. Surprisingly, he will probably assume the artist to be incapable of learning to programme, or at least unwilling to do so. Less surprisingly, he will probably hold the notion that art is principally involved with the production of 'exciting' images and that he will best serve the artist's needs if he can enable him to produce a large number of widely differing images, all 'exciting'.

How would it be to try to write poetry by employing a specialist in rhyme-forms? Each time you get to the end of a line you call him up to ask what word *he* thinks would best convey what *you* have in mind. The process sounds rather more promising than trying to produce art by getting a specialist to write computer programmes on your behalf. If we are to get past 'computer-art', as I am sure we shall, to art made with the help of computers, it will need to be on the basis of a massive change of mental set on the part of the artist.[5]

Suppose, now, that I have a computer whose abilities are like those I have described. Suppose also that it is connected to a teletype and to a drawing machine (Fig. 5a, b). Assume that the computer has already been loaded with the programme by means of which it will be able to interpret my own instructions. (My instructions here will not be phrased in any existing 'higher-level' language but in a fictitious one designed to make clear what is being done. In fact, I will describe programmes diagrammatically, by means of what are known as 'flow-charts', rather than in the line-by-line form required by every language.) Let us see if the machine works:

---

[4] I am not making this up. See 'Idols of Computer Art' by Robert E. Meuller, *Art in America*, May, 1972: and my own reply in 'Commentary' in the following issue.

[5] Under grant number A72-I-288 from the National Endowment for the Arts, Washington, DC, I am currently investigating the feasibility of setting up a Centre for Advanced Computing in the Arts. One might speculate that, among other things, such a centre might enable artists to use the machine for their own purposes, rather than presenting them with a cookery book of possibilities.

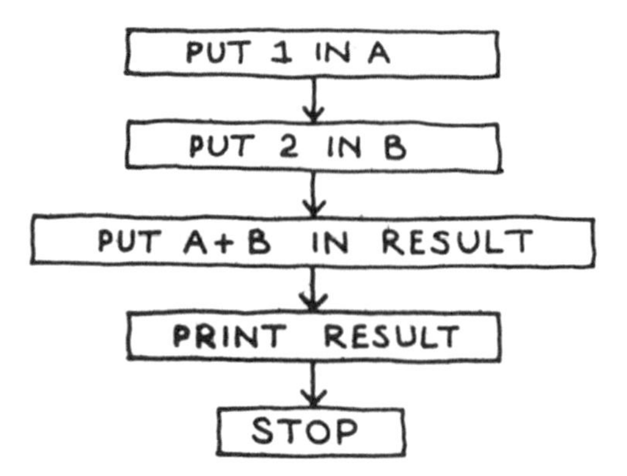

Now that this programme has been loaded, I type 'RUN' on the teletype, and the machine responds … 3. The programme has taken around I/50,000 of a second to run—the teletype, being mechanical, takes much longer to operate, of course—and we know that the machine can figure out that $1 + 2 = 3$. Let us try something a bit more complicated:

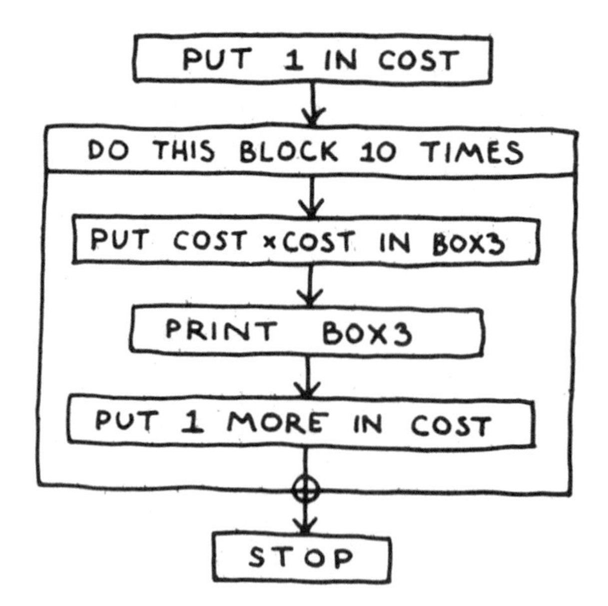

This time, when I have loaded the programme and type 'RUN', the machine will get the 1 it has just put in the cell labelled COST, square it, store the result in BOX3 and then print out that result. But then, instead of stopping, it will add 1 to the 1 already in COST, and go through the whole cycle again, printing out 4 this time, and then 9, 16, 25 and so on until it has completed the ten re-iterations called for.

This is pretty simplistic, of course, involving a lot of unnecessary PUTting and GETting into and out of memory. If the machine's language were a little more sophisticated, we could have written the programme:

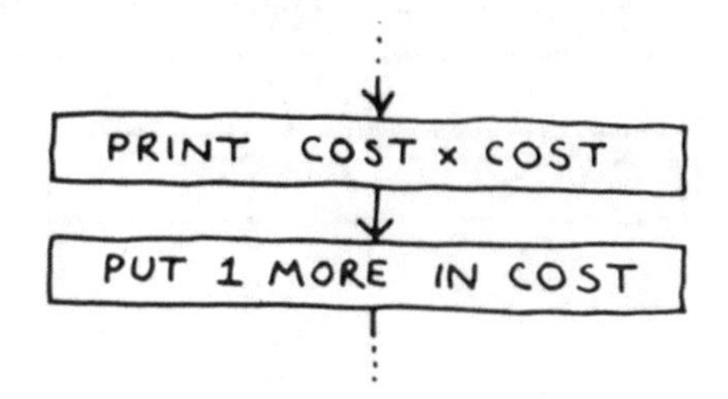

With exactly the same result. Note how powerful a device it is that instead of saying first 'print the square of 1', then 'print the square of 2', then 'print the square of 3', we need only say, 'print the square of whatever is in the cell labelled COST', repeating the same instruction every time. All that changes is the contents of the cell COST. This notion of referring to a number by the name on its cell is fundamental to programming, and in fact, it is something we do all the time ourselves. Saying that a carpet is ten feet long and seven feet wide is essentially like saying:

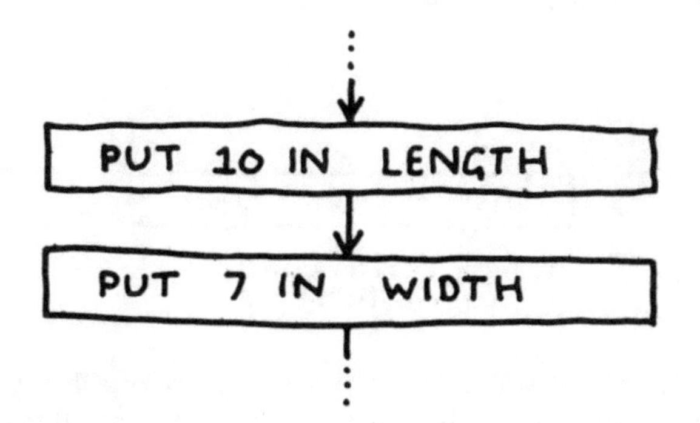

And we could obviously build this into a programme for finding the area of the carpet by adding.

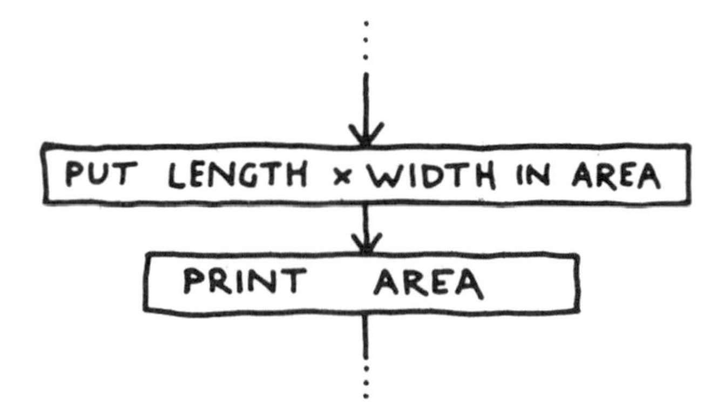

The important thing here is the level of generality, since the programme will now work for whatever values we put in the cells labelled LENGTH and WIDTH.

We should be able to get the drawing machine to draw something now. You will probably remember the idea that you can describe the position of any point on a sheet of paper by two distances, or coordinates: how far the point is horizontally from the left-hand edge and how far it is vertically from the bottom. Suppose we were to reserve two cells labelled HOZ and VERT for storing the two coordinates for any point to which we wanted the pen to go. If the pen is sitting in the bottom left-hand corner, and our programme says:

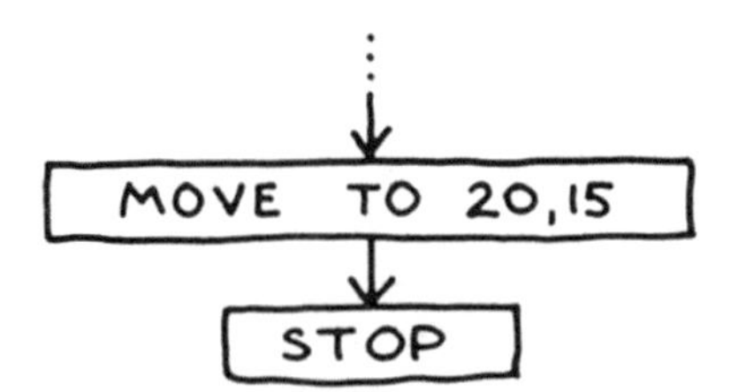

The computer will recognize from the command MOVE that it must send its instructions to the drawing machine, not to the teletype, and will thus send out the commands required to make the pen move to the centre of the bed. The only problem with this programme is that it did not specify whether the pen was to be down or up. The programme should probably have read:

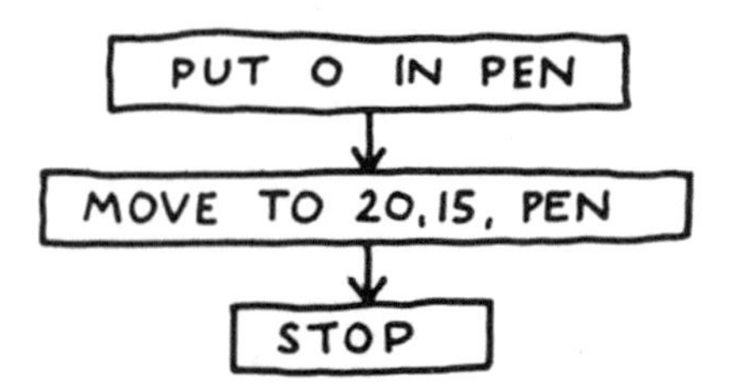

Where the cell PEN will hold r as a code for 'pen down' and o as a code for 'pen up'. We might also have generalized a step further and said:

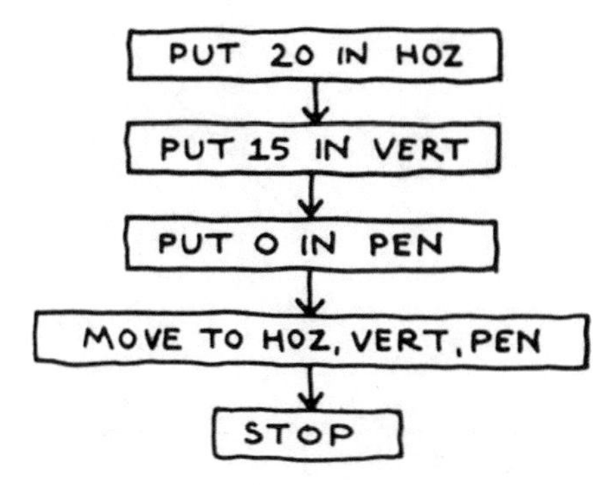

Because now we might want to write the sort of reiterative programme we looked at earlier, to draw a whole series of points. In writing such a programme, we will now use a shorter notation for PUT, so that instead of writing PUT 5 in HOZ, we would write HOZ < –5.

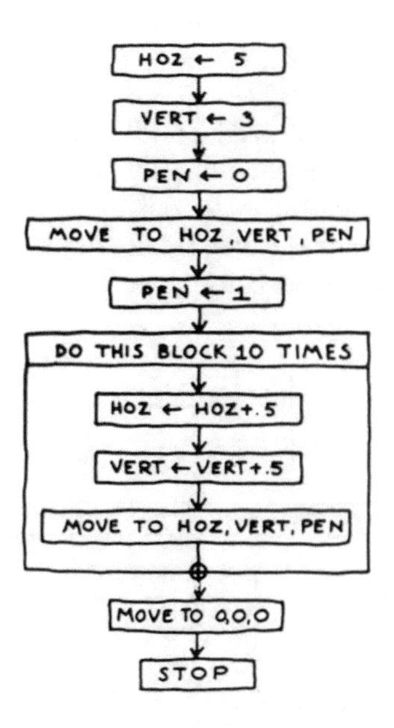

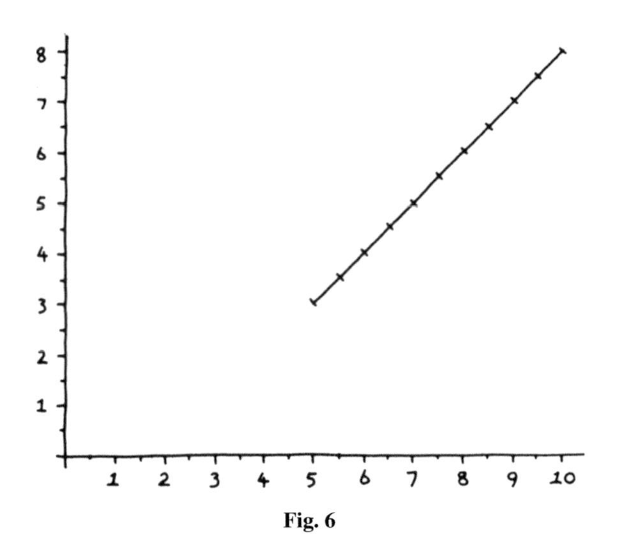

Fig. 6

Not a very exciting drawing, but it does illustrate a lot of principles. You might be surprised by the statement.

$$\mathrm{HOZ} < -\mathrm{HOZ} + 0.5$$

But of course this is not algebra, and it is not an equation. It means, simply, 'take what was in the cell labelled HOZ, add 0.5 to it and put it back in the same cell'. The pen has drawn a series of ten short line segments which in this case make up a straight line and has then lifted and gone back to the bottom left-hand corner. The same general form will draw lines which are not straight, if we can simply think of a way of generating the appropriate pairs of coordinates. For example:

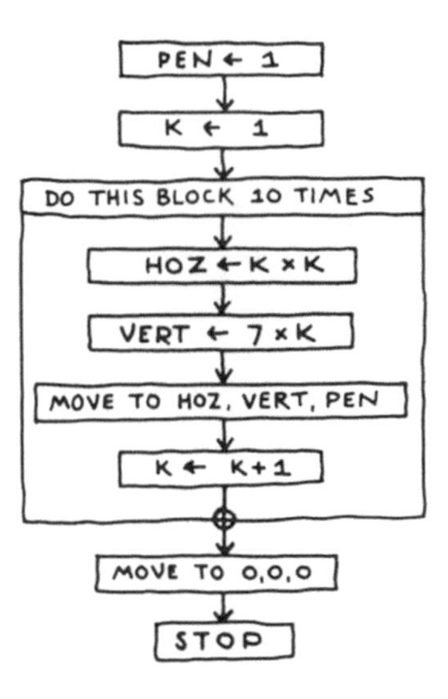

Will produce this curve.

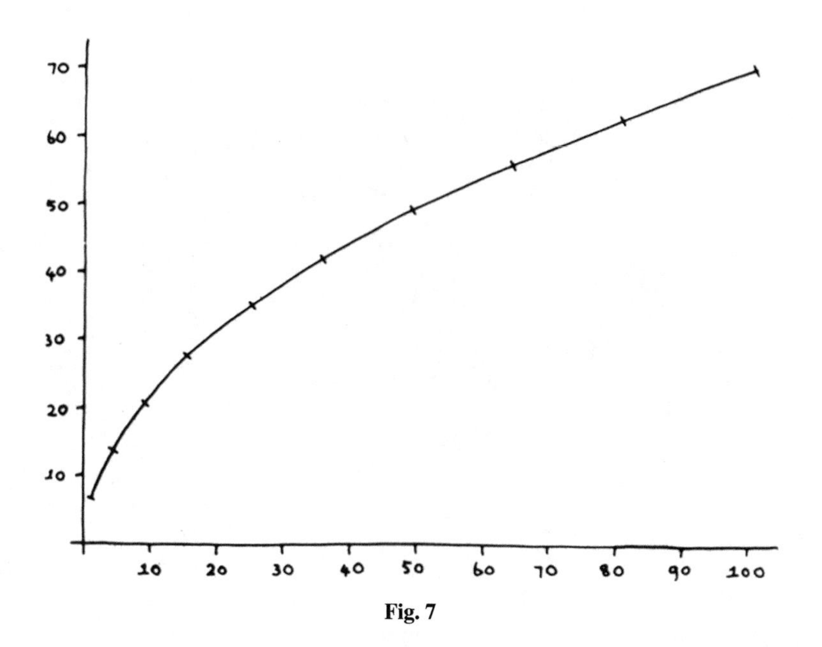

Fig. 7

The thing is that any pair of statements which relate the horizontal coordinate to the vertical in a coherent way will produce some sort of curve, and it is quite easy at this point to start popping in all kinds of trigonometrical functions and stand back to see what happens. This one (in Fig. 8) was written by a passing computer-science student—I hesitate to say 'invented', since it is almost entirely a matter of chance whether it will produce anything pretty, which I think it does.

No doubt the introduction of this sort of technique for drawing curves into 'computer art' owes much to the mathematics-oriented programmer, who would tend to view a curve essentially as the graph of a mathematical function. But not all curves can be handled in this somewhat simplistic way, and artists wishing to handle more complex curves have been obliged mostly to use an entirely different approach, if anything even more simplistic. Since it is possible to describe any point by its HOZ-VERT pair, it follows that any drawing can be approximated by a set of points, each of which can be treated in the same way, so that the whole drawing can be described in purely numerical terms. Imagine then that you have already done a drawing, that you have reduced it to a string of points, and that you have typed the HOZ and VERT values of each point together with its PEN code, on a series of punched cards (or, of course, any other storage medium for which the computer has the appropriate peripheral). A programme like this

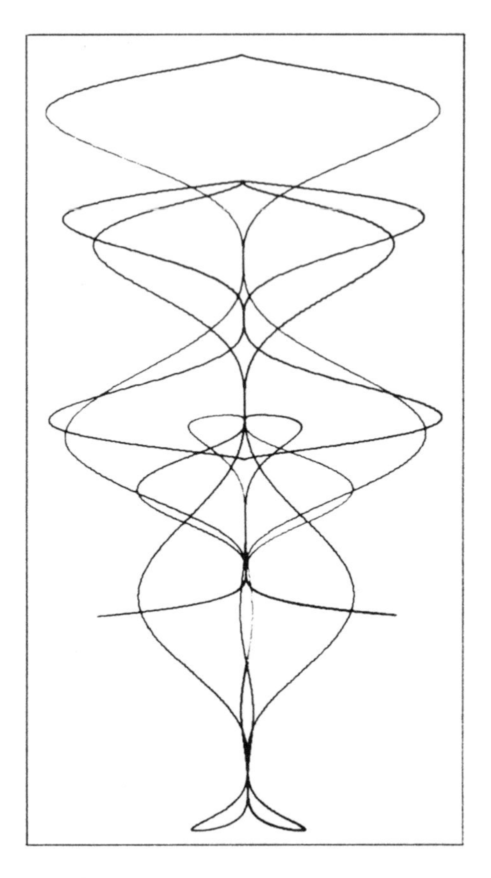

Fig. 8 Computer science student

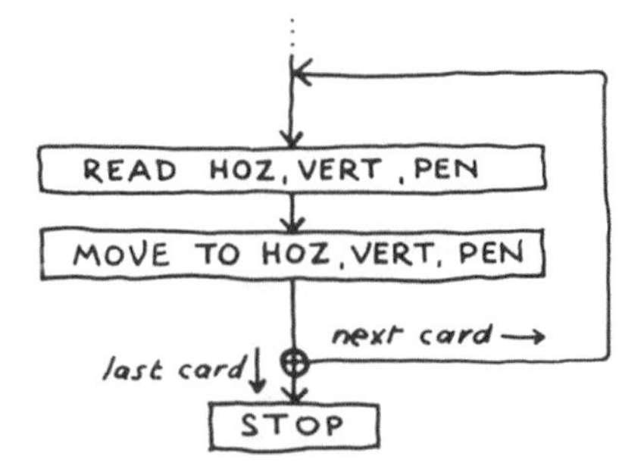

would then simply read the first card to find the first point in the drawing (PEN would presumably be o until it gets here), move the pen to that point, read the next card for the next point and so on until it has done all the points. The machine has duplicated your drawing from your numerical description.

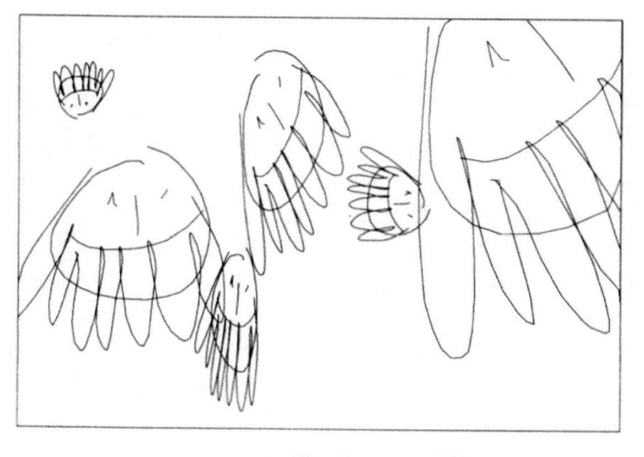

Fig. 9

It may not be clear why anyone would want to use such elaborate means to reproduce a drawing he has already made. The answer is that quite a lot of things can be done to the drawing by suitable programmes. Not only can it be reduced, enlarged, shifted, rotated, squashed up, pulled out (Fig. 9), but it can also be transformed as if it were drawn on a sheet of rubber which was then stretched in various irregular ways. None of these operations, or transformations as they are called, is difficult to programme, and since they can be applied to any set of points whether generated from mathematical equations or read in from cards, they have tended to become the stock-in-trade of 'computer art'. Indeed, it would be difficult to see how any computer animation involving drawn images could proceed without such transformations.

For our purposes, however, the question to be asked is whether the notion of a picture processor, operating upon some previously generated image, corresponds in any useful way to what we know of human art-making behaviour. I think the answer has to be that it does not. To achieve that correspondence, the machine would need to generate the image, not merely to process it.

Intuitively, it seems obvious that the human process involves characteristics which are quite absent from these procedures, and in particular, I think we associate with it an elaborate feedback system between the work and the artist; and dependent upon this system are equally elaborate decision-making procedures for determining subsequent 'moves' in the work. Our enquiry might reasonably proceed by examining whether the machine is capable of simulating these characteristics.

Before going on, I must explain that the computer possesses one significant ability which was implied by the earlier examples but never explicitly stated. It is able to compare two things, and on the basis of whether some particular relationship holds between them or not, to proceed to one of two different parts of the programme. In practice, this primitive decision-making device can be built into logical structures of great complexity, with the alternative paths involving large blocks of programme, each containing many such conditional statements, or 'branches'.

It would be quite difficult to demonstrate a complex example here in any detail. The drawing on the cover of this issue of Studio International was generated by a programme of about 500 statements, of which over 50 were concatenated from these simple conditionals, equivalent to about 85 branches. We might look at one part of that programme, however, about 50 statements in all, which generates the individual 'freehand' lines in the drawing. Obviously, the flow-chart is a much-simplified representation.

The argument behind the sub-programme runs like this: in any 'sub-phase' of a line's growth, it will be swinging to the left or to the right of its main direction ('straight on' if given by swing = o). This swing may be constant, accelerating or decelerating, and both the rate of swing and the rate of *change* of swing may be either slow or rapid. Overall, the line must not swing beyond a certain pre-set angle from its main direction. A single full phase will consist of two sub-phases normally swinging in opposite directions. Both of these, and the phase itself, may vary in length, and normally, the starting direction for each full phase does not depend on that of the previous one.

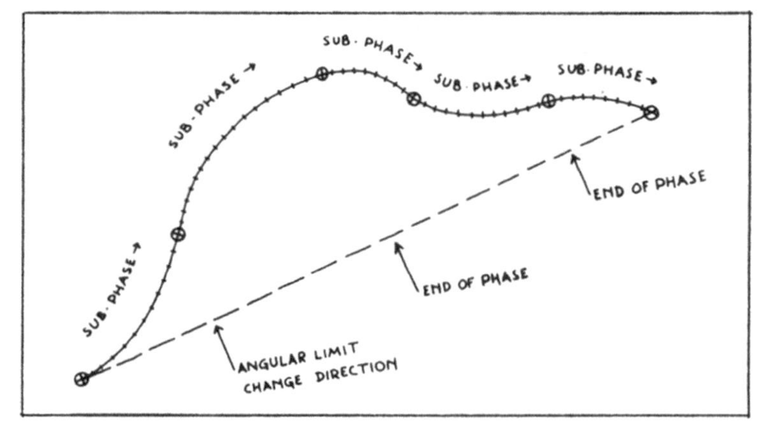

Fig. 10

If the line swings beyond its angular limit, however, all the factors controlling the current phase are immediately reset and a new full phase is initiated, starting off in the opposite direction. It should be noted also that the line has some definite destination and corrects continuously in order to get to it. The programme would look something like this:

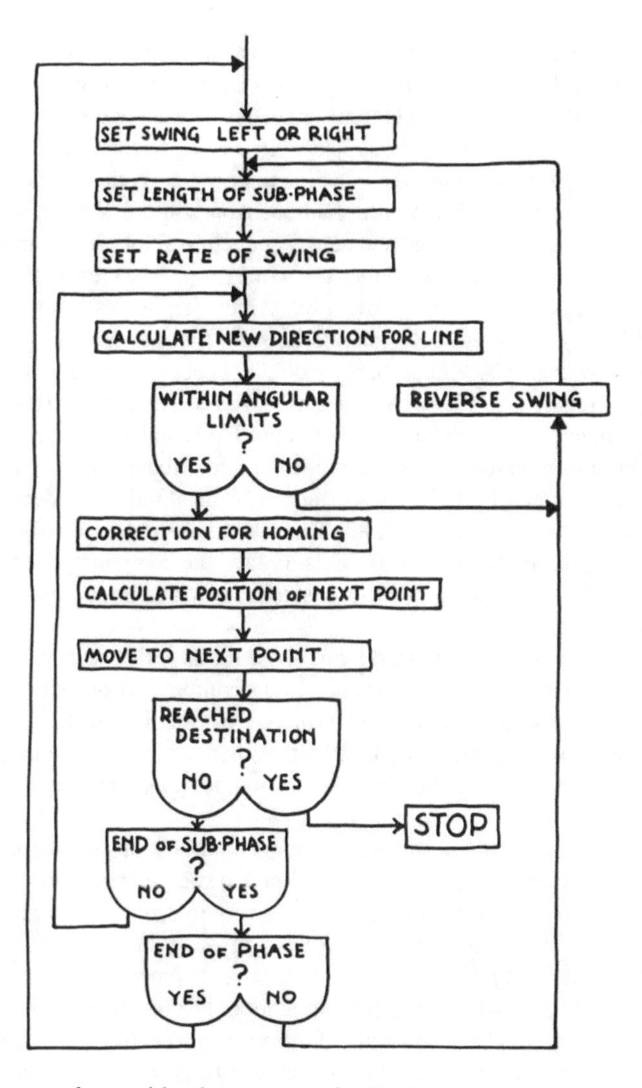

This structure does evidently possess a feedback system not unlike the kind we employ in driving a car. There is an overall plan—to reach a destination—which breaks down into a succession of sub-plans, which are in turn responsible for generating a series of single movements. But if an 'emergency' is signalled, the current sub-plan is abandoned, and a new one set up.

The *quality* of the line is directly related to the way in which the factors for each new sub-phase are reset: if the length of each sub-phase varies enormously, or if the rate of change of swing varies greatly from one to the next, the line will tend to be quite erratic. If the angular limits are set quite small—by the overall plan—then the line as a whole will be more 'controlled'. How does the programme 'decide' on new factors for each new sub-phase? The ranges permissible for each factor are precisely determined in relation to what the range was last time, indicating another level of feedback. Within that range, the machine makes a random choice.

There seems to be so much popular misunderstanding about the nature of randomness that a word might be said on the subject before going further. Contrary to popular belief, there is no way of asking the machine to draw 'at random', and if you try to specify what you mean by drawing 'at random' you will quickly see that what you have in mind is a highly organized and consistent behavioural pattern, in which *some* decisions are unimportant *provided* they are within a specified range of possibilities. This is characteristic of directed human behaviour: if you plan to rent a car, you will probably be concerned that it should be safe, that its size and power will be appropriate to your needs. You probably would not care too much what colour it is, and in being prepared to take whatever comes you are making a 'random choice' of colour: although you probably know it is not likely to be iridescent pink, matte black, or chromium plated. The same might be said—though with much narrower limits—of the painter who tells his assistant to 'paint it red'; or indeed the painter who uses dirty brushes to mix his paint. They are all examples of making a random choice within specified (or assumed) limits. In fact, the computer generates random numbers between zero and one, which must then be scaled up to limits specified by the user's programme.

You might consider that, in human terms, these limits will be narrow where precise definition is required, wide where it is not. For the computer, the existence of limiting ranges rather than specified values will result in the possibility of an infinite number of family-related images being produced rather than a single image made over and over again. There might be some difficulty in demonstrating the case to be otherwise for the artist.

While it would seem obvious that any complex purposeful behaviour must make use of feedback systems, there is no suggestion that such systems alone can account adequately for the behaviour. Moreover, the ability to satisfy some given purpose, as the 'freehand' line generator does in homing on its destination, accounts for only slightly more. The *formulation* of the purpose is something else and we would expect to find in human art-making behaviour not only a whole spectrum of purpose-fulfilling activities, but also a spectrum of purpose-formulating activities. If I am to pursue my enquiry, I must now try to demonstrate the possibility of such a structure occurring in machine behaviour, although the strategies employed within the structure may or may not correspond to the strategies the artist might employ. Certainly, no such claim will be made for the programme I am about to describe.

This programme is one of a series in which the principal strategy is devised in relation to an 'environment' which the programme sets up for itself. An example would be one in which the programme first designs, and then runs, a maze: the resultant drawing being simply the path generated by the machine in performing the second part. In the present programme, the environment is a rectangular grid of small cells, into which are distributed sets of digits (Fig. 11). The strategy adopted in the second part involves starting at a '1' from there seeking to draw a line to a '2', then to a '3' and so on. The digits are considered as a continuous set, '10' being followed by '1', so that but for three things the programme would continue indefinitely. The first is that no digit may be used as a destination more than once, and since a digit is also cancelled if a line goes through its cell, the number of destinations steadily reduces, and the programme terminates. The second is that a destination will not be selected if getting to it involves crossing an existing line, so that finding a destination becomes more difficult as the drawing proceeds. And the third is that there are certain 'preferences' operating in choosing between those destinations recognized by the machine as viable. As a consequence of these constraints, the machine will eventually find itself unable to continue to the next digit, and it will then back up to the previous digit on its part and attempt to go on again from there. The drawing will be complete when the back-up procedure has taken it all the way back to the original '1'.

Now, it is possible, by manipulating the factors controlling the machine's 'preferences'—I will say more about those in a moment—and by appropriately setting various other factors, to produce a very wide range of characteristics in the drawings produced.

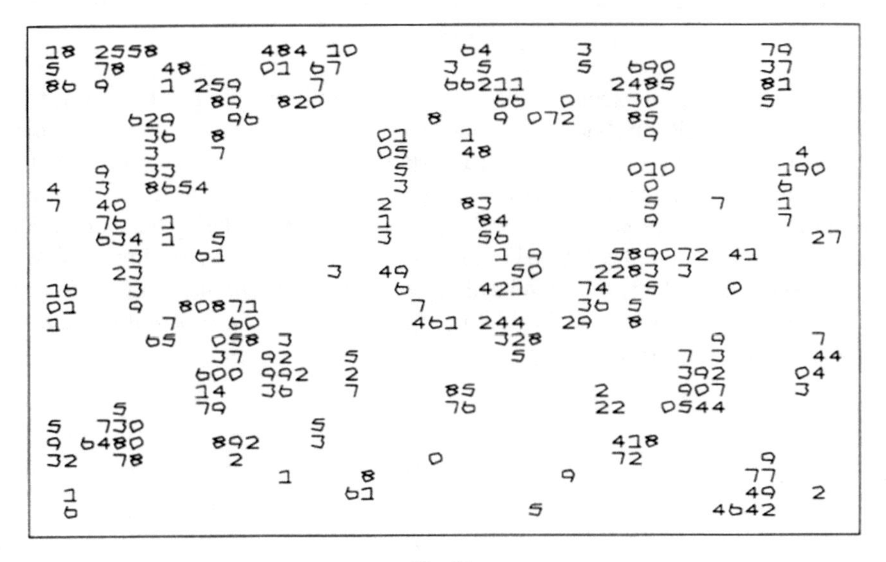

Fig. 11

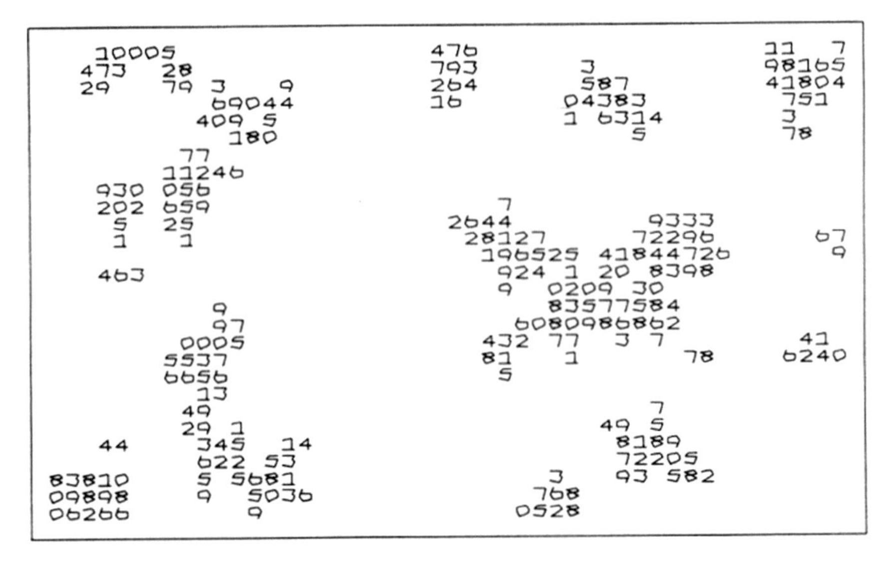

Fig. 12

For example, the number of cells in the grid and the number and type of distributions of the digits are both critical factors in determining the complexity of the drawing. Under studio conditions, I have varied these factors myself, but for a recent exhibition I wrote an executive programme which took over that task and under its control the machine produced almost three hundred drawings during the four weeks it was in the museum. These varied from a few squiggly lines to quite complex drawings, from a single large image to anything up to twelve small ones on a page; and they required no human participation beyond changing the paper and refilling and changing pans (Fig. 13.).[6]

[6] 'Machine Generated Images', La Jolla Museum, California. October–November 1972. The drawings reproduced here are taken from the show. The machine was able to make drawings in several colours, but the museum staff had some difficulty in following its instructions for mounting the appropriate pens. In the event, it was limited to asking for the correct size pen to be mounted.

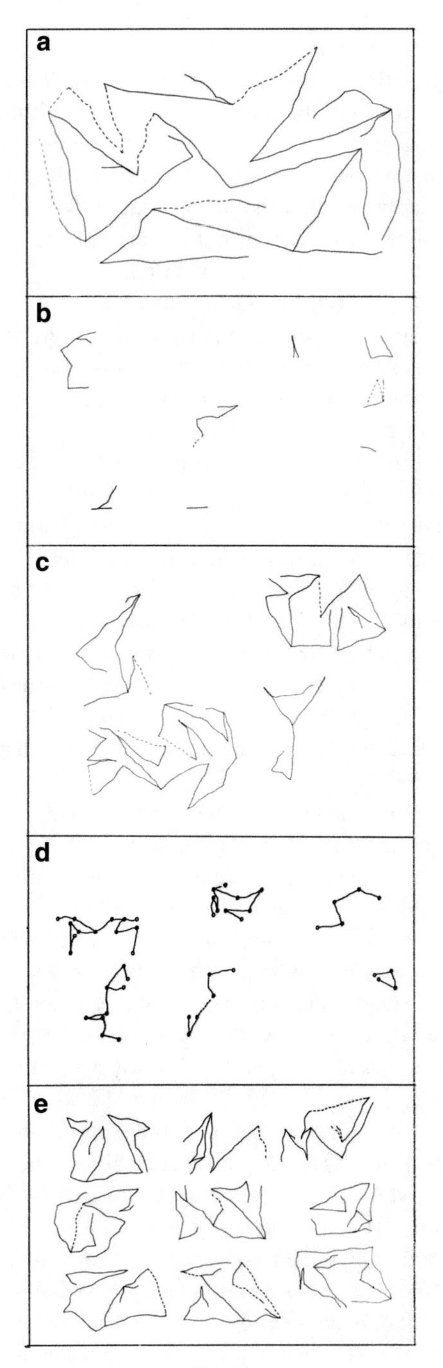

Fig. 13

What I have described as being controlled by the executive is, in a very general sense, the purpose-formulating mechanism for the 'freehand' line generator, the structure that determines where the lines are to be drawn. You might say that *I* am the purpose-formulating mechanism for the programme as a whole, but the executive programme makes my own part in the process rather more remote, if no less significant. In fact, I doubt whether the main programme will be changed much at this point, since what is at stake for me is not what it does, but what determines what it does. I am referring to the 'preferences' mentioned earlier.

As it reaches each destination, the machine has to choose between anything up to twenty-five next destinations, depending upon the state of the drawing. In the present state of the programme, its preference is for destinations within certain distance limits, but it is easy to see how it might 'prefer' long lines to short ones, a destination near the centre of the picture, or in highly active parts of the picture: or it might 'prefer' the one involving minimum change of direction; or the reverse of any of these. Obviously, the character of the resultant drawings would vary enormously as the machine exercised one 'preference' rather than another, but in fact I am suggesting something more complex than simply switching 'preferences'. Suppose, rather, that the machine exercises its whole range of 'preferences' by scoring each possible destination for its ability to satisfy each preference and taking the destination with the highest total score as its choice. It might then choose the destination which was relatively far away, did not involve too much deviation from the current direction and was in an area of high activity quite close to the centre of the drawing. I think this would be a much closer simulation of the way in which human preference-structures are exercised.

Let us go one step further, and suppose the machine to be capable of weighting its scores for its different 'preferences', and of modifying these weightings itself. This possibility is by no means speculative: readers familiar with the development of the field of Artificial Intelligence will recognize its similarity to Samuel's now classic programme for a checkers-playing machine (1959). They will recall also that the programme enabled the machine to learn to play, by having it play against itself, one part always adopting the best strategy found to date, the other varying the weightings of the 'preferences' which determined that strategy until it found a better one, and so on. In a short time, it was able to win consistently against any human player.

We might recognize a significant difference between applying a learning programme of this sort to successful game playing and doing so to successful art making. Of course, the difficulty is ours, not the machine's, since we ourselves would be in some doubt as to the nature of the criteria towards the satisfaction of which the machine might aim. Art is not a deterministic game like checkers, to be won or lost by the 'player'; and though we acknowledge, empirically, that some artists are 'better' than others, that some artists do improve, the problem of formulating general criteria for improvement may be no different in relation to the machine than it is for the teacher in relation to the art student. It is probably reasonable to assume that there do exist criteria at levels even more remote from the work than any I mentioned: in which case we should be able to formulate them and the machine should be able to satisfy them. But there remains the suspicion that satisfactory performance in art is not to be measured solely by the satisfaction of explicit criteria and would still not be so no matter how far back one pushed.

As to those explicit criteria: there would seem little reason to deny that the machine behaves purposefully at every level described. Yet no level defines its own purpose. The learning level would—but for the difficulties mentioned above—advise the preference-structure as to the best way of defining the manner in which the executive commands the main programme to select the points between which the 'freehand' line generator is to draw lines. One might say even that the purpose of each level is to formulate the purpose for the next level. It is true, of course, that the machine's organization is that way because it has been set up that way, but in considering the nature of explicit criteria in human art-making behaviour we might reasonably adopt the machine's organization as a model and say that these criteria relate to the formulation of new levels of purpose in *satisfaction* of prior purposes. This can be maintained without any suggestion that the machine can move higher and higher up the ladder until it is finally in possession of the artist's Purpose. On the contrary, it seems to me that pushing back along the chain of command—either for the machine or for oneself—is less like climbing a ladder than it is like trying to find the largest number between zero and one: there is always another midway between the present position and the 'destination'.

It should be evident, then, that I do not consider 'serving the artist's Purpose' to be equivalent to 'talking over the artist's Purpose', or identify the machine with the artist. I identify the artist with the whole Purpose-structure, the machine with the processes which are defined by the structure and in turn help to redefine it. Since under other circumstances these processes too would be played out by the artist, I am also identifying playing-out with the computer with playing-out without the computer. For the machine to serve his Purpose, the artist will need to use it as he uses himself. There is no reason to anticipate that the use will be more or less trivial than the use he makes of himself, but every reason to suppose that the structure will change in ways which are presently undefinable.

The step-by-step account of the computer's functions and its programmes was intended, of course, to try to demonstrate that the machine *can* be used in this way. The original question—whether the machine can serve the artist's Purpose—is more redundant than unanswerable and is in any case not to be confused with asking whether artists might see a need to use it. It is characteristic of our culture both that we search out things to satisfy current needs and also that we restate our needs in terms of the new things we have found. Nor is it necessarily immediately clear what wide cultural needs those things might eventually serve. The notion of universal literacy did not follow immediately upon the development of moveable type, but it did follow that development, not demand it. Up to this point, the computer has existed for the artist only as a somewhat frightening, but essentially trivial toy. When it becomes clear to him that the computer is, in fact, an abstract machine of great power, a general-purpose tool capable of delimiting his mind as other machines delimit him physically, then its use will be inevitable.

(Photographs for this article by Becky Cohen.)

Towards a Symbiotic Future: Art and Creative AI

Fabrizio Poltronieri

Abstract The emergence of new AI systems brings not only technical challenges but also new questions to the realm of art. Even though the use of AI is not a novelty in the art world—Harold Cohen (1928–2016), perhaps the most famous and relevant example, worked on his AARON program for more than four decades—the new scenario, bursting with supervised, unsupervised and reinforced learning techniques make it possible for algorithms to have sufficient autonomous decision-making power to choose paths that hitherto were only available to programmers and artists. It can modify how we creatively use computers. Cohen and AARON actively collaborated with one another. The artist devoted a substantial amount of time employing Symbolic AI techniques to teach AARON to draw and colour in, hard-coding the decisions the system could make. In the later years, the program provided lines and basic abstract compositions for the artist to colour in the spaces, taking the program's input as a starting point. Currently, a growing set of frameworks provide neural networks out of the box, providing the logic for systems capable of learning tasks semi-autonomously by analysing a large amount of data. Thus, we can delegate tasks to increasingly intelligent autonomous systems, including playful and aesthetic ones, the ones that speak to our hearts as humans. Given these circumstances, this chapter reflects on notions such as language, intelligence, abduction, synechism, creativity and the role of aesthetics in a future where our creative relationship with intelligent artificial systems will become one of increasing symbiosis.

Keywords Creative AI · Semiotics · Abduction · Synechism · Creativity

1 Introduction

Over the decades, culture has helped consolidate an imaginary about the role of artificial intelligence that is dark. In the dystopian societies presented by films such as

F. Poltronieri (✉)
De Montfort University, Leicester, UK
e-mail: fabrizio@fabriziopoltronieri.com

C. Vear and F. Poltronieri (eds.), *The Language of Creative AI*, Springer Series on Cultural Computing, https://doi.org/10.1007/978-3-031-10960-7_2

Metropolis (1927) and *2001: A* Space Odyssey (1968), the role of artificial intelligence systems is to take over and replace humans. More recently, the term singularity has re-awakened fear with the idea that ordinary humans will someday be overtaken by artificially intelligent machines, cognitively enhanced biological intelligence, or both (Shanahan 2015).

While such narratives are compelling as a form of entertainment, my vision as an artist and researcher working with Creative AI for over two decades differs substantially from fiction. I believe in an increasingly symbiotic future, where artificial intelligence and humans play together in real time, sharing transparent processes and decisions, having collaboration as a critical concept. The main objective is to challenge and enhance human creativity and the creativity in computers. I do not think computers will replace human creativity, but they will collaborate even more to enhance our perceptions of, and through creative acts.

Although it still sounds strange to talk about creativity in computers, this text aims to present the theoretical basis of my vision, showing how computers can be types of creative minds. For that, I will use the semiotic and philosophical theories of Charles Sanders Peirce (1839–1914) about the concepts of intelligence and abduction, which is the kind of thinking that is the foundation of creative processes.

The history of the creative use of artificial intelligence brings us significant examples of this symbiotic vision. The British artist Harold Cohen (1928–2016) is a good example, having pioneered the use of Symbolic AI for creative purposes. It is important to note that the differentiation between Symbolic AI—also called Strong AI or GOFAI (Good Old-Fashioned Artificial Intelligence) (Haugeland 1993)—and neural networks is not essential for my argument. In my artistic practice and research, I intersperse the use of both paradigms, as I will demonstrate later in this chapter.

Cohen started exploring Creative AI very early on, reinventing his already well-established career as a painter and aiming to develop an autonomous system that would eventually be equivalent to his creative processes. In the 1950s and 60s, he delved into abstract art, intending to make a painting system that could paint itself. When he encountered his first computer in 1968, the artist could bring this idea to fruition by coding software that somehow emulated human cognition. His aim was that the software could learn what an artist's everyday practice involves, starting from the concrete world of objects to the very abstract levels of computer-generated images.

Cohen started to set up his own Creative AI generator, trying to teach the computer how to make compelling images according to his artistic standards. The artist was not interested in the precision that the computer could offer but in the possibility of designing routines that would enable the machine to make artistic decisions, working together with the human creative process. In its most recent versions, the software—called AARON—began to provide the artist with only basic sets of visual information, such as lines and abstract shapes. Using these basic elements, Cohen would play with the computer, filling the image with additional colours and shapes.[1]

[1] The aim here is not to detail Harold Cohen's vast oeuvre, a topic covered by an extensive bibliography, but rather to point to the fact that the relationship between Cohen and AARON was already

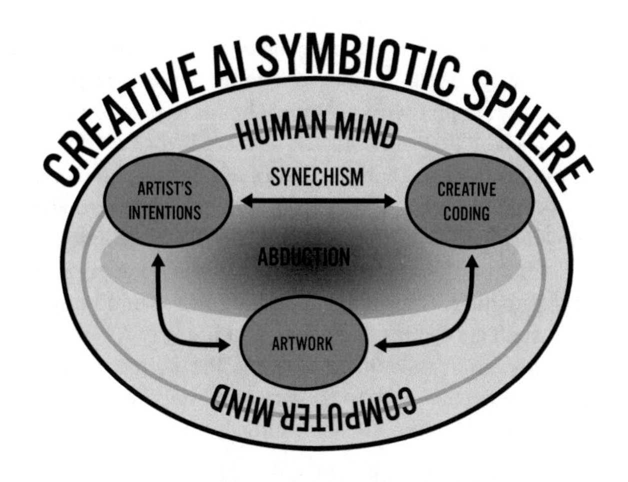

Fig. 1 Creative AI symbiotic sphere

Figure 1 illustrates what I call "Creative AI Symbiotic Sphere", where the artist's intentions (human) are challenged and changed by the creative coding (computer), producing an artwork which challenges and changes both the artist's intentions and the creative coding, in an infinite playful loop, mediated by processes of abduction and synechism. Therefore, if, in practical terms, we observe that Creative AI can play symbiotically with human creativity, what would explain such a possibility? To answer that, it is necessary to understand the concepts of creativity and synechism.

2 Abduction: A Semiotic Approach to Creativity

One aspect that characterizes creativity in Peirce's theory of semiotics is that it can only exist where there is room for chance. For Peirce, creativity has its basis in a type of reasoning that he calls abductive, which is a kind of inference that introduces new elements to already existing arguments. According to Peirce's semiotics, creativity is governed by laws of logical order, in which abductive reasoning reigns.

Creativity can be conceived as a syntactic logic that creates unique recombinations using existing elements, giving rise to new semantics in a process that expands our previously structured set of beliefs. For Peirce (CP 1.383), there is an inner compulsion leading the mind to unite disparate ideas to achieve greater intelligibility of reality through the connections of ideas made by the mind. The creative abductive process is set in motion by perceiving anomalies, surprises and questions regarding what is already known, triggering new mental hypotheses that may solve the questions posed by the newly observed phenomena.

symbiotic. For a more thorough introduction to Cohen's career and work, see Poltronieri and Hänska (2019).

Peirce says that when something we believe in—which he calls "belief"—is embraced as accurate, this something turns into a habit, a source of reliability determined by its predictive nature. The mind is a dynamic system whose main activity is the production of habits. Beliefs are firmly consolidated habits: "For belief, while it lasts, is a strong habit, and as such, forces the man to believe until some surprise breaks up thc habit" (CP 5.524). When a behavioural habit starts to show insecurities, showing alterations in its known pattern, an opening for creative opportunities arises. The feeling of surprise generated by the perception of an anomaly is the first step of the abductive reasoning, stimulating the mind to initiate an investigation process until such anomalies disappear, making way for new beliefs.

A dynamic movement towards the adjustment and expansion of pre-existing concepts is necessary to acquire and establish a new set of beliefs. This movement articulates the three logical inferences described by Peirce:

1. the already familiar abduction,
2. deduction and
3. induction.

Although I am focussing solely on abduction in this chapter, it is important to mention that these three modes of reasoning enable the mind to think in a structurally logical and formal manner. Abduction generates hypotheses which must be justified and tested during the development of the two other modalities of reasoning.

For Peirce, the composition of reasoning's cognitive structure is not static but instead formed by layers of processes that gradually gather, establishing a network that relates the inferences of abductive reasoning to empirical conditions. The act of experimenting creatively, that is, abductively, forms the logical basis of any rational process since whenever one acts rationally, one acts according to a conviction guaranteed by an experimental phenomenon (CP 7.337). The creation of new convictions and knowledge starts with abductive reasoning, which triggers experimental processes that test new conditions that may or may not become a reality. Among the three types of logical inference, the abductive one is the most original, as it is the only one capable of generating new hypotheses. As it represents something new, abductive inference cannot guarantee its validity as a general behaviour law. Peirce states that abduction's characteristics are distinct from the other two types of inference in that it is not based on prior knowledge but rather on an experimental process.

Abduction is, therefore, the form that rational thought takes when, for example, the study of new forms of musical composition that have not been previously approached. Abduction involves studying facts and devising a theory or practice to explain them. Peirce presents abductive reasoning as the sole logical operation capable of introducing new ideas, explaining that the mind's creative capacity springs neither from nothing nor an innate ability but rather from its cognitive structure (CP 5.171). It is important to emphasize that creativity, this mental faculty based on abductive reasoning, is linked to the creation, change and expansion of a set of beliefs that form habits. The creative process is triggered when a form of a creative mind is confronted with a problem, causing surprises and uncertainties that initiate the abductive process, which will generate hypotheses to solve the problems in question creatively.

If creativity is something logical, therefore mental, how do we explain the existence of creative computer systems or Creative AI? To address this, we would have to justify that computers are also forms of mind. This is where synechism comes in.

3 Synechism: The Correlation Between the Human Mind and the Computational Mind

In logical terms, if I claim that there is a symbiotic connection, or creatively collaborative essence, between the human mind and AI systems, this means claiming that there is a continuous connection between the human mind and the mind that exists in computational apparatuses. It is necessary to clarify this topic and make it clear why I also attribute a mental form to computers.

To accomplish this task, I will use the branch of Peircean metaphysics called synechism, which deals with continuity, with the connection between mind and matter. The philosophical conception in the foundation of synechism is generous for not separating mind–spirit–body, thereby breaking with the dualist Cartesian philosophical tradition, which seeks to affirm the supremacy of human intelligence over the natural intelligence observed in the cosmos.

In the Peircean conception, the human mind differs from other mental forms because it is the most plastic reality of the known universe, presenting the highest level of malleability. When I use the terms "human mind" and "computational mind", I presuppose that these two mental forms with different levels of malleability (man and computer) can establish effective communication. For Peirce, there is no separation or division, but differences of degree, between nature and culture, between the organic and the inorganic, the physical and the psychic, the natural and the artificial.

The difficulty in understanding and accepting this conception comes from the fact that the semiotic field of signs, from computers to living systems, has often been analysed in terms of dualisms: "tools versus instruments", "instruments versus machines" and above all "machines versus living beings". Rather than confirming such dualisms, Peirce describes this field as a continuum of signs, from the simplest to the most complex semiotic systems. Among the least complex systems are those mediated by instruments or technical devices such as a thermometer, a sundial, a thermostat or an automatic traffic signalling system. The most complex semiotic systems take place in living beings.

Synechism is a crucial concept for understanding Peircean metaphysics, based on the pragmatic idea of continuity and evolution. Therefore, it should be understood as a continuity between mind and matter, the latter also being a form of mind, but more exhausted, especially if placed next to, for comparison purposes, the human mind. This connatural, correlational aspect enables a creative communicational flow between the various mental types we observe in the universe. There is a kind of mind in the work of art, which allows the contemplation of art to be a dialogue, a two-way street and not just a discourse to be observed. Such an approach explains why the observer contributes to the construction of the work.

The universe and everything that makes it up is a mental form because it has its physical laws derived from psychic models. The great law of the universe can therefore be considered as the law of the mind. For Peirce, all reality is governed by the law of mind, that is, the law to acquire habits, from the purely physical world to the human mind, with the difference that the human mind does not submit to the law in the same rigid way that matter does. Therefore, matter is a mind so closed in habits, so regular that it stops exhibiting the same spontaneous behaviour which is so abundant in the human mind.

Peirce notes that all that we can know, in any way, is purely mental since intelligence can only act upon what is intelligible. Synechism opens the door to an unfettered understanding of the universe since the connaturality between representation and real objects eliminates the nominalist[2] barrier between subject and object, between consciousness and world. This free boundary is already present in Peirce's phenomenology since all phenomena pass undifferentiated through interiority (the phenomena as they really are) and exteriority (appearance). Thus, to understand and use the doctrine of synechism, one must admit that the material universe is provided with habits of conduct in the form of natural laws. In other words, it is a form of mind.

This is, in fact, the central argument of the doctrine that Peirce calls objective idealism. Such doctrine postulates that matter is also a form of mind, but exhausted, transformed into crystallized habits. However, even among minds with a more significant number of crystallized habits, we can observe differences concerning the degree of crystallization presented. In the territory of material artefacts that expand the universe of languages, computers are the most unrestrained mental forms, enabling the fertile emergence of new syntaxes and semantics.

Access to these minds takes place in its most sophisticated degree through metalanguages, such as programming languages, which allow the complete dissolution of one mental form into another, representing a superhighway for the circulation of signs.

4 Exploring the Creative AI Symbiotic Sphere: A Case Study

I have been doing research and artworks with artificial intelligence for over 20 years, almost the same time I had the first contact with Peirce's ideas. The Creative AI

[2] Briefly, nominalism is a philosophical doctrine that holds that things called by the same name have nothing in common except that. This doctrine is usually associated with the thesis that all that exists are particular individuals, and therefore, there are no universals. What all chairs have in common is that they are called "chairs". In opposition to nominalism, Peirce adopts a strong realist position, where nature is self-denominating, independent of human will or languages.

Symbiotic Sphere (see Fig. 1) is a methodology I created and have been improving over these decades. It is not a guide or a model to be strictly followed but a theoretical platform for experimentation, for exploring my restlessness as a researcher and artist, activities that are interconnected for me. I do not see my activity as a researcher as disconnected from my artistic practice. On the contrary, one informs the other.

My first contact with computers happened very early when I was eight years old, and my father returned at home with a mysterious box containing a Brazilian clone of the classic ZX SPECTRUM. Since then, programming has been an activity I do almost every day. During my childhood and adolescence, computers and programming languages were my playgrounds, spaces where I explored creative possibilities and expanded the universe of my languages. I never perceived computers as a substitute for my intelligence or creativity. On the contrary, they were always devices that challenged me, making me a more creative person. I can say that the conception of a symbiotic, mutually beneficial existence with computers and their languages has always been present in my life.

When I started to have contact with Peirce's philosophy more systematically and discovered the concepts of abduction and synechism, it was as if what I had been doing all my life intuitively gained new colours, enhanced by the discovery of a theoretical set that responded perfectly to my practical activity. I had always believed that programming a computer meant, ultimately, accessing a kind of mind plastic enough to allow itself to be reconfigured, reprogrammed. At the same time, I felt contact with the computer-triggered parts of my brain that remained dormant while performing other activities. Later, I discovered that the same thing happened to other people when using computers in a creative context. In a conversation in his studio in Encinitas, California, Harold Cohen told me that when he started using computers, he felt as if parts of his brain that had previously been switched off were being activated.

Computers are ideal apparatuses for the execution of a series of rules. What seems, at first glance, as something limiting is in fact a creative enhancer because nothing says that these rules cannot be subverted or used for the generation of new languages. Furthermore, this is the basis of my symbiotic collaboration model with artificial intelligence:

- I start by delimiting a problem, usually some philosophical uneasiness that I wish to explore aesthetically, in the process of abduction.
- Quickly I start to think algorithmically about what would be the minimal rules for the solution of such a problem; already accessing the computational mind through programming codes.
- This step triggers elements of synechism since the real-time output I get from the computer informs my creative process.
- Next, I am rapidly refining through prototypes the aesthetic language obtained through the communicational game between my mind and the computer's mind, reprogramming it until I reach the result I deem satisfactory.[3]

[3] For a more in-depth description of my working methods see Poltronieri (2021).

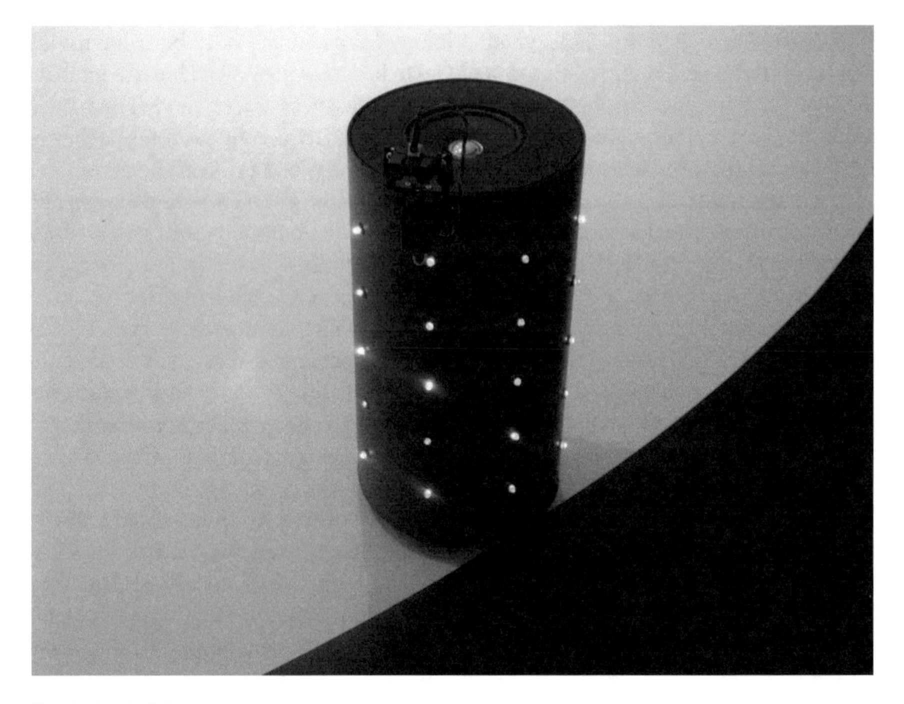

Fig. 2 Amigóide

Many times, I delegate this aesthetic judgement to the programming code, letting the computational mind decide if the obtained result is satisfactory or not.

4.1 Amigóide

An example of this process is my automaton *Amigóide*[4] (Fig. 2). It is an apparatus which performs a range of functions in order to search for humans to engage in a friendship. It is an award-winning project initially commissioned in 2010 by the Itaú Rumos Prize, one of the most important art prizes in Latin America.

Two versions (1.0 and 2.0) have been developed, using different AI approaches. The first one (2010–2011) uses GOFAI, whereas version 2.0 (2019) was built utilizing a mix of GOFAI and deep learning, taking advantage of modern machine learning frameworks, which allowed the use of computer vision and image recognition techniques in real time. This case study describes the process behind the automation's design and developments, including the benefits encountered in the transition from GOFAI to a mixed AI approach.

[4] The made-up word *Amigóide* is derived from the Portuguese word "Amigo", which means "friend".

The automaton interacts with people through movements, LEDs, a pre-recorded synthetized voice and demonstrations of digital feelings, which in this case are very simple, with *Amigóide* reacting through its lights when the infrared sensor readings (version 1.0) or computer vision analysis of the camera video stream (version 2.0) indicate whether a human interactor gets close to it or not after a round of interaction.

Once the automaton finds a human, it starts to follow this potential friend, intending to start an endless, flawless friendship. It gets closer to the eminent friend and poses the question: "Fabian, do you want to be my friend?". Fabian is an imaginary friend, a super one, programmed into the automaton's mind. After the initial contact, *Amigóide* tries to conquer the friend through a series of phrases such as "Fabian, what do you search in a friendship?"

Amigóide can be described as a rational agent, i.e. an agent "that acts so as to achieve the best outcome or, when there is uncertainty, the best expected outcome" (Russel and Norvig 2016: 4). Hence, *Amigóide* is a rational agent whose best expected outcome is to establish contact with humans and conquer their friendship. The concept of rational agent (Russel and Norvig 2016) is pivotal to this project because it leaves room for uncertainty, opening a space for abduction to act.

Although deemed successful in the public-facing installation, *Amigóide's* first version had some limitations related to the technological choices employed in its construction. We need to bear in mind that this initial version was designed from scratch in 2010, before the actual wave of machine and deep learning algorithms was widely disseminated and available. These technological constrains had a negative impact on the way version 1.0 behaved and the level of interaction with humans. It was a reflex agent, acting on the basis of its current percept, which informed the series of condition-action rules that handle its behaviour. The major limitations of this approach, however, started to show when I tried to improve many of *Amigóide's* basic capabilities—such as its environmental sensing—and to implement new features, like the ability to perform in any environment, and not only in a special room.

These problems were hard to be defined by a list of formal hard-coded rules. It can be said that the process of synechism was there, as always, but not so efficient due to the fact that the technologies used were not as malleable. We need to remember that the human mind is the most flexible one, and it is part of the symbiotic process to be challenged by the limitations found in less plastic kinds of mind. A proof of this is the fact that these questions proved to be not only technical but rather philosophical ones, intimately related to the approach used to design the automaton.

My attempt to use high-level symbolic, human-readable representations of problems was not flexible enough to provide the level of versatility I needed, at least not at a reasonable cost–benefit level: GOFAI was holding me back, and it took me a long time to realize that. For this reason, and because the history of GOFAI proves that it has been providing a useful experimental framework for the creative use of computers, it is important to have an in-depth discussion about this topic.

According to Haugeland (1993: 112), "GOFAI, as a branch of cognitive science, rests on a particular theory of intelligence and thought – essentially Hobbes' idea that ratiocination is computation". For Hobbes, the reasoning was like numerical computation, in the sense that we add and subtract silently in our heads. GOFAI systems attempt to implement theoretical basis that emulates human behaviour through the construction of software and hardware, such as *Amigóide* 1.0. Historically, it aims to provide cognitive results at least comparable, and preferably superior, to the expected behaviour of intelligent human beings in similar circumstances, and in all the human cognitive domains, including the abilities to write, read, draw, talk and interact. Thus, a major flaw in GOFAI's philosophical architecture is the very idea of overcoming human intelligence rather than promoting the symbiotic idea of working together to achieve a common goal. According to Floridi (1999), GOFAI tries to achieve its goal by seeking to find a balance between two distinct approaches:

1. A Cartesian one, based on a rationalist dualism, where the presence of intelligence is completely detached and independent of the presence of a body, constituting a complete disembodied cognitive system. It would, at least in principle, be implementable by other disembodied and stand-alone kinds of cognitive systems. However, these notions of intelligence are also completely mind-dependent—mind here meaning the human mind—making it impossible for a machine or even an animal to achieve sophisticated intelligence in this manner.
2. A materialist monistic attitude, which considers that intelligence is solely a complex property of a physical body. In this vision, mental processes play a secondary role, being caused by physical brain processes. This view draws support from the philosophical theories of epiphenomenalism and mechanistic materialism. According to this view, mental events are caused by physical events in the brain but have no effects at all upon any physical events, i.e. mental states or events are by-products of states or events in the brain, necessarily caused by them but exercising no causality themselves. Thus, a thought, belief, desire, intention, or sensation is produced by a specific brain state or event but in no way affecting the brain or the body to which the brain is connected. Hence, this approach would, theoretically, be implementable by other kinds of embodied cognitive systems, including computers and robots.

Facing such paradoxes, where the initial propositions in both philosophical views are favourable, but the outcomes are not achievable, GOFAI implements a form of computational materialism, described in the first half of the materialist monistic approach (item 2 above). Therefore, GOFAI should be possible to be achieved by any brainless, mindless and lifeless general-purpose, symbolic system, deprived of psychological or embodied experience, which is not the case, as Amigóide demonstrates.

In order to implement consistent computational materialism, a GOFAI-based system has to adopt an extreme reductionist and abstract version of computational materialism, assuming the following presupposition:

$$\text{Intelligence} \rightarrow \text{Ratiocination} \rightarrow \text{Symbolic Processing} \rightarrow \text{Computation}$$

Although GOFAI has suffered recurrent criticisms, I am not claiming that it should be abandoned or saying it is no longer relevant. I use GOFAI techniques in combination with neural networks in my experiments and artworks. This mixed approach is supported by several AI front-runners, as highlighted by Ford (2018), and fits very well in my Creative AI Symbiotic Sphere.

4.2 *Amigóide 2.0*

After years with the 1.0 project shelved, and having worked with deep learning in different artistic contexts in the last few years, I decided to give *Amigóide* another shot, rewriting all the code base with my Creative AI Symbiotic Sphere in mind, meaning to do not fight with technology, but to use it "in the flow", as a continuous extension of my own mind.

Amigóide 2.0 takes advantage of deep learning, which brings a different mindset, a distinct philosophical approach that is beneficial for the field of Creative AI, notably when mixed with GOFAI techniques. In contrast to GOFAI, machine learning—of which deep learning is a subfield—seeks to allow computers to "learn" from experience and come to some understanding about the world in terms of a hierarchy of concepts, where each concept is defined in terms of its relation to other simple concepts (Goodfellow et al. 2016).

The term "learning" can be misleading, as, at the actual stage of development, machines do not really learn. Machine learnt algorithms typically apply mathematical formulas to a collection of inputs—the training data in supervised learning, the most used type of machine learning nowadays—in order to produce the desired outputs. A machine learning model is the result of "brute force" curve-fitting. When the training is successful, the model can map new inputs to the patterns found in the training data. That is the reason why it is expected that these mathematical formulas also generate the correct outputs, or predictions, for most other inputs distinct from the training data, respecting the condition that those inputs are from the same or a very similar statistical distribution as the ones found in training data. It is not proper leaning, in the human sense, because even a slight distortion or change in the inputs is likely to produce completely wrong outputs, what can be interesting for creative and artistic purposes.

The first learning algorithms were intended to be computational models of biological learning, trying to mimic the way learning possibly happens in the brain. According to Goodfellow et al. (2016), deep learning is motivated by two main ideas:

1. That the brain provides a proof by example that intelligent behaviour is possible. From this perspective, a possible path to building real intelligence in apparatuses would be to reverse engineer the computational principles behind the brain, duplicating its functionality.

2. Deep learning models can help us to understand the brain and the principles that underlie human intelligence. These models can shed light on fundamental scientific questions about how the brain works, besides their ability to solve engineering questions and help in artistic and creative ones.

Deep learn as a philosophical concept is much more aligned to synechism than GOFAI. Despite its limitations, by gathering and analysing knowledge from experience, machine learning algorithms do not require human operators to formally specify all the knowledge that the computer needs, as opposed to the GOFAI descriptive paradigm.

As an aesthetic decision, the plastic black cylinder was kept as the main body, making both versions look practically the same. In terms of its inner-logic, the new outputs originated from the deep learning algorithms fed into a simplified GOFAI structure which is similar to the one deployed in version 1.0. The perceived results of implementing this approach in *Amigóide* 2.0 were palpable. Although the core behaviour of *Amigóide* 2.0 was not significantly different from *Amigóide* 1.0, the perceived relationships with it did shift. This, I believe, is because of its continuity of thought through the machine learning models, and the symbiotic connections it engendered in me. More experimentation is to be done here, but as a solution for the issues revealed in *Amigóide* 1.0, the machine learning approach is the correct way forward.

5 Conclusion

While the human mind remains the most plastic mind we find in the universe, the conception that there is a "mind in matter" found in synechism opens new frontiers for exploring creative collaborations between humans and artificial intelligence. Synechism, the doctrine and metaphysical theory that all that exists is continuous, is the view that the universe exists as a continuous whole of all of its parts, with no part being entirely separate, determined or determinate, and continues to increase in complexity and connectedness through semiosis and the operation of an irreducible and ubiquitous power of relational generality to mediate and unify substrates (A Walk to Meryton 2022).

Together with the Peircean theory of abduction, the theoretical basis for the explanation of creative phenomena, we have a robust theoretical and practical arsenal with which to break the Cartesian barrier that prevents the modern world from seeing the multiple possibilities of collaboration existing in a symbiotic universe, where the simple idea that something always replaces another—e.g. artificial intelligence replacing human creativity—gives way to a more complex conception of the universe, where everything is continuously related.

References

A Walk to Meryton (2022). https://aeigenfeldt.wordpress.com/a-walk-to-meryton/
2001: A Space Odyssey (1968) Directed by Stanley Kubrick [Film]. MGM, USA
Floridi L (1999) Philosophy and computing: an introduction. Routledge
Ford M (2018) Architects of intelligence. The truth about AI from the people building it. Packt Publishing
Goodfellow I, Bengio Y, Courville A (2016) Deep learning. MIT Press, Cambridge, MA
Haugeland J (1993) Artificial intelligence: the very idea. MIT Press, Cambridge, MA
Metropolis (1927) Directed by Fritz Lang [Film]. UFA, Germany
Poltronieri F, Hänska M (2019) Technical images and visual art in the era of artificial intelligence: from GOFAI to GANs in ARTECH 2019: Proceedings of the 9th international conference on digital and interactive arts. Article No.: 38, pp 1–8
Poltronieri F (2021) Dreaming of utopian cities: art, technology, creative AI, and new knowledge in the Routledge international handbook of practice-based research In: Vear C (ed) Routledge, London
Russel S, Norvig P (2016) Artificial intelligence: a modern approach. Pearson, Essex
Shanahan M (2015) The technological singularity. MIT Press, Cambridge, MA

Artificial Intelligence and Creativity Under Interrogation

Lucia Santaella

Abstract Given the increasing ubiquity of artificial intelligence (AI) in all human activities, it is not surprising that artists, designers, and other creators in the field of culture and creative economy are making use of AI capabilities in their productions. Avoiding the controversial debates about the benefits on the one hand and the negative externalities of AI on the other, this chapter will focus on the specific modes of incorporation of AI by artists. The chapter argues that placing AI as a current resource in the historical context of the development of the technological arts is a well-founded way of entering the debates that the topic may arouse. Thus, the framework of photography, followed by electronic and computational resources, and the plethora of possibilities opened to the artist by the digital universe constitute assumptions of continuity for the understanding of the technological partnerships that artists have sought over at least two centuries and that today culminate in artificial intelligence. This does not mean seeking a point of arrival that minimizes the importance of discussions around the topic, but rather raising arguments that can bring the discussion to the specific field of the new challenges that are presented to human creativity.

Keywords Creative AI · Semiotics · Creative process · Radical art

1 Introduction

Researchers working in the Artificial Intelligence (AI) development centers, especially in the United States and China (the two countries that are taking the lead in this newest race of capitalism), are unanimous in assuring that we are only at the dawn of AI: weak AI as it is called. This already indicates that it is at the beginning of its development. Nevertheless, AI is already acting, almost always invisibly, in nearly all fields of human activity. When the topic starts to appear on websites, newspapers, and magazines for the general public, it means that it has already found a home

L. Santaella (✉)
São Paulo Catholic University, São Paulo, Brazil
e-mail: lbraga@pucsp.br

C. Vear and F. Poltronieri (eds.), *The Language of Creative AI*, Springer Series on Cultural Computing, https://doi.org/10.1007/978-3-031-10960-7_3

in the most capillary tissues of human society. In fact, art and questions related to creativity could not be absent from this capillarity. Thus, it is the questions related to the specificity of the modes of incorporation of AI by artists that this chapter aims to put on the agenda.

1.1 The State of the Art of AI

AI studies began in the 1950s when John McCarthy quoted the term at a seminar at Dartmouth University in the United States. However, the English mathematician Alan Turing came before that. He gave a lecture on it in 1947, and he is also taken to be the first to decide that AI was best researched by programming computers rather than by building machines. In 1950, Turing published the study "Computing Machinery and Intelligence" in which he presented the *Imitation Game* also known as the *Turing Test*: a set of questions in which it is possible to discriminate whether the respondent is human or machine. By that time, the seeds in the field of AI had already emerged as being strongly associated with the area of genetics in biological sciences.

For a few decades, research on AI in the context of cognitive science has gone through ups and downs until it found its promising path a few years ago. This is explained by the convergence of several factors: the exponential increase in computer processing capacity and the gigantic growth in the speed, volume, and variety of data gathered in the networks, which, together with the functional increment of neural networks, led to the explosion of AI, an explosion which is being transformed into an implosion of previous human productive and cognitive configurations.

To get started in the field of AI, especially where it is today, the first step is to find a definition of intelligence that is reliable. There is some consensus among experts that AI means the simulation by computer systems of human intelligence processes. It is a branch of computer science aimed at creating intelligent machines. This implies the machinic development of skills such as learning, knowledge, acquiring information including the rules for using it, reasoning used to reach definite or approximate conclusions, self-correction, problem-solving, perception, linguistic recognition and processing, planning, and the ability to manipulate and move objects. To accomplish these purposes, the computer needs access to objects, categories, properties, and relationships. With this in mind, AI is today an umbrella for an ever-increasing multiplicity of applications.

Undoubtedly, AI's resources today spread across a variety of human activities. Intelligent personal assistants organize routines, document "automatizers" assist with a variety of tasks, software analyzes online behavior, algorithms are able to predict the success of audio-visual narratives, advanced software is aimed at perceptual recognition, and deep learning is employed for medical diagnosis and machine learning for health treatments; there is software for autonomous aerial systems and also robots with human faces, who talk sympathetically. The advances do not stop there. However, the aim of my chapter does not go in that direction. My point of departure coincides with Broeckmann's (2020):

> AI is not a unified phenomenon, a *something* to be handled, understood, addressed, but rather a conceptual construct, a discursive tool that both facilitates communication about the technoscientific phenomenon, and over-simplifies it. The current urge *to get to grips with AI* is understandable, given the radicality with which the related technologies challenge an established understanding of technics that presumes tool-like passivity, rather than active techno-logical agencies, which co-determine what humans can do in the world. But such skewed terminologies, which claim monolithic notions of "intelligence" or "learning" and pitch "human" against "machine", affirm mythical conceptions of technology and the related schemata of human subjectivity, rather than open them up to new and alternative narratives.

Thus, my discussion does not go in the direction of the disturbing or rather shocking occurrence which Bogost (2019) called the "AI-art gold rush" when the "New York auction house Christie's sold *Portrait of Edmond de Belamy*, an algorithm-generated print in the style of nineteenth-century European portraiture, for $432,500." It is not this narrative and others concerning the commercial aspects of AI and art that matter to this chapter. Rather, it is of my interest to show that, in the field of arts, we are testifying the emergence of a new mode of creative and artistic production that has been incorporated by artists and has aroused the interest of theorists and critics of culture and the arts. My aim is again in tune with Broeckmann (2020) when he says that

> the art world participates in this discourse through a flurry of exhibitions and public debates, with a noticeable emphasis on the technical and the social, rather than the particular aesthetic and artistic aspects, placing an awkward, and at times, playful or dilettante-like focus on the technical medium. Art criticism perpetuates this tendency when it highlights the societal concerns instead of engaging with the artworks and their aesthetic affordances.

For him, the emergence of AI and art "deserve, and require, critical scrutiny not only as reflections on a technical paradigm, but as artworks in their own right", since "they develop their own scenarios, projecting their own rules and raising their own, hard questions." In fact, these are thorny questions, which this chapter aims to face, without any desire to exhaust the subject, but only to bring some contribution to the debate.

2 AI as an Adjunct to the Creative Process

Far from being a bizarre and astonishing phenomenon, the relationship between art and AI brings a type of creation perfectly tied to the already secular history of technologically inseminated arts, trends that were progressively accentuated after the digital revolution, in a multiplicity of developments such as net art, web art, digital art, computational art, algorithmic art, interactive art, robotic art, and so on. The art that is produced today through the artist's creative sharing with AI resources, in particular machine learning (ML) and deep learning (DL), is not an isolated phenomenon.

If we abandon the addiction to what may be called "presentism," as if recent trends and explosions in both the sciences and the humanities had fallen by parachute—there from some unknown height—directly onto today's culture, it will be possible to see that culture is continuous and that new scientific, technical, artistic, and cultural phenomena gradually open up their space until their clear emergence in the present occurs. According to Liu (2018):

> Technological revolutions have brought crucial influences in the history of mankind. Our understanding of technology and our relationship with machines are also changing. McLuhan famously defines technology as media, and media as the extensions of our senses and bodies. Our relationship with machines in this light is not simply instrumental, since 'we shape our tools, and thereafter our tools shape us' (McLuhan 1964). In an age of machine intelligence, the ubiquitous machine-learning-based products and services are augmenting various aspects of our lives. Our relationship with intelligent machines is evolving, given that machines are increasingly moving away from the role of passive objects into the position of active subjects.

This is what is happening with AI especially in its creative production face. In fact, since 1968–69, when the first computational art exhibitions took place, at the Howard Wise gallery in New York and at the large-scale exhibition Cybernetic Serendipity in London, more than 50 years have gone by in the development of this kind of art. The beginnings can be found in the 1950s in the work of some artists and designers who were using mechanical devices or analog computers to carry out their work.

Coincidentally or not, in the mid-1950s, the cognitive sciences began to blossom, having as one of its scopes, among others, the development of AI. Such development never failed to attract the attention of artists working in partnership with computational algorithms, so much so that, while research was faltering, artists were already incorporating what they had at their disposal, genetic algorithms. There are remarkable examples of works that used these resources in the 1990s as the one that became famous for its ingenuity, *A-Volve*, 1993–94, by C. Sommerer and L. Mignnoneau. Therefore, since then, artworks that incorporated AI resources have followed *pari passu* the technical development of this field. In the last ten years, research in AI has exploded and at the same time, of course, artistic works have started to project themselves more and more in this field.

It is not my intention to explain technical issues of AI, nor will I limit myself to the presentation of works and artists who are producing works that use AI algorithms. What I intend to discuss are the aesthetic questions that this type of art is raising.

2.1 The Aesthetic Debate

Artificial intelligence methods open up new possibilities both in art and in the creative economy and even in entertainment, allowing for rich and deeply interactive experiences (Santaella 2021). As AI opens up new fields of artistic expression, AI-based art itself becomes a fundamental creative and research agenda, raising and answering aesthetic questions that would not have arisen without the advent of AI.

Creativity is a fundamental feature of human intelligence and functions as a challenge to AI. According to Boden (1998), AI techniques can be used to develop new creative proposals in three ways:

- producing new combinations from existing productions
- exploring the potential of conceptual spaces
- making transformations that allow the generation of forms, images, and ideas that would be otherwise impossible.

There is no lack of questions and concerns about the nature and fate of AI. In discussions far from consensus, the most frequent questions are as follows: *But what is intelligence? Is there a solid definition of intelligence that doesn't depend on relating it to human intelligence? Does AI aim at human-level intelligence? Which aspects of AI cannot compete human capacities?* and so on.

When uncertainties about the conditions and fate of AI are transferred to the field of art, aesthetic questions of all kinds are intensified, especially issues concerning creativity and the status of art, when the artist finds in AI a partner for his/her creative abilities. What questions does this bring to our traditional conceptions of aesthetic creation? The most common among them are as follows: *but will a true artificial artist ever exist?* Can we foresee that one day aesthetics will be generated entirely by the machine, without any design commanded by a human agent? Will AI ever have an aesthetic of its own? (Nakazawa 2018).

Bogost states that "Given the general fears about robots taking human jobs, it's understandable that some viewers would see an artificial intelligence taking over for visual artists, of all people, as a sacrificial canary" (Bogost 2019). This is how the advance of ML is recreating old debates about how to define not only what art is, but what creativity is as well. Similar debates took place during the rise of the YouTube generation, when anyone could suddenly be a creator. Now, new generations of AI are raising the question: What is the role of the artist? If a machine can make visual art, edit a movie, write a script (Sunspring) or compose a song (Daddy's Car), what is the artist's value? What about the creative ability that has always characterized the artist?

Creativity is a skill that we generally consider uniquely human. Throughout history, we have been the most creative beings on planet Earth. Birds can make their nests, ants can make their hills, but no other species on the biosphere comes close to the level of creativity that we humans demonstrate. In recent decades, and in the wake of an exploratory tradition that belongs to art, we have acquired the ability to do amazing things with computers and their substitutes such as robots. With the AI boom in the 2010s, computers can now recognize faces, translate between all languages, receive calls for you and beat players in the world's trickiest board game, to name a few. Suddenly, we must face the possibility that our ability to be creative is not unrivaled in the universe. So, creativity is no longer an exclusively human trait? Can AI be creative? (Kulpaki 2018).

It is not necessary to return to the post-Duchamp debates about what is art and what is not art, because, if we take as a reference the growing pluralism and heterogeneity of the arts throughout the twentieth century, such debates become tedious. Given this, the argument I defend is that the historical development of the technological arts suggests a good path to understand the current aesthetic phenomenon of creativity in AI. After all, the issue of creativity in connection with the computer has been transforming the traditional conceptions of aesthetics for some decades, so that there should be no surprise in relation to the creativity that today finds a valid supporting role in AI.

2.2 Creative Amalgamation Between Humans and Machines

Since the advent of images that Flusser (2019) calls *technical*,[1] that is, images that, since the photographic camera, are produced by the mediation of a machine, art has started to develop creative amalgamations between the human and the machine. Without going that far, when the computer became our ally in a multiplicity of tasks, the possibilities of creation via technologies multiplied, causing a continuous growth of heterogeneity. Examples of this can be found in the production of images in 3D modeling, in cyber installations, in works in telepresence and telerobotics, in network art, and in the creation with databases, in addition to works in virtual reality and those produced with artificial life algorithms, besides genetic art and transgenic art in their use of genetic engineering techniques linked to gene transfer (natural or synthetic).

It is a fact that heterogeneity and multiplicity came to command the vectors of artistic production to the point where being technological or non-technological, being digital or non-digital, art is no longer a matter of order. The trend toward heterogeneity, boosted by hybridity, has increased due to the convergence of technologies, cultures, knowledge, and forms. It is a convergence of such an order that it blurs the boundaries between techno, games, movies, the spoken word, dance, literature, music and sound design, theater, visual arts, the sciences of perception, architecture, physics, psychology, sociology, biology, religion, and medicine. Virtually, all fields of knowledge are contributing to this convergence that energizes human creativity.

Creative multiplicity appears in innumerable types of production. A few examples are enough to prove this statement: alternative reality games, augmented reality, mixed reality, virtual reality, gestural interfaces, live cinema, interactive books, connected immersion, video mapping, and generative art. It is in this context and preserving its continuity that creativity in AI is installed today.

Many examples can be listed, such as the robot works by Ken Rinaldo, or Eduardo Kac's genetic art, to prove that it is not—and has never been—the role of art to slip into conservative and frightened tendencies. Art is risk, exploration of yet unknown territories, and adventure along the paths of estrangement for the transfiguration of human sensibility. While the camera extends the human capacity to see, stimulating the eye to perceive what escapes distracted attention, while productions with mixed, augmented, and virtual realities multiply sensory experiences with possible worlds,

[1] I prefer to call these *technological*, since the camera is already a technological and not only a technical artifact.

artificial intelligence introduces a new alliance that increases the human cognitive potency. When starting from a multifaceted view of cognition, it is not difficult to see the crucial cognitive role played by imagination in human creativity. What consequences can the imaginative partnership with AI bring to the creative capacity of artists and what are its implications for the re-accommodation of our being in the world?

After all, mechanical, electronic, and digital arts have been transforming traditional concepts of aesthetics for over a century. Although art may seem disruptive, it in fact continues a tradition of ruptures that the arts have always provoked. We find ammunition for this proposal in the initiative of the International Society for Arts, Sciences and Technology, *Leonardo*, which, in collaboration with the Open University of Catalonia, dedicated, in 2020, an entire issue of the *Node Journal* to more than a dozen chapters focused on the theme of creation in AI in the context of culture.

2.3 The Voices of Experts

The number and ingenuity of the chapters written by artists and specialists are impressive and very illustrative of the state of the art in AI. The axis of the discussions is aimed precisely at critical questions about ML that, far from being based on theoretical abstractions, are based on the teachings provided by concrete examples. For the organizers of the issue (Burbano and West 2020), there is no way to ignore the exponential growth of ML applications in all areas of the arts (visual, sound, performance, spatial, transmedia, audio-visual, and narratological). Activities in this field are growing so fast that publications cannot keep up. Seeking to meet this growth, the authors of the chapters question the crucial problems involved in authorship and ethics, autonomy, and automation, by exploring not only AI's contributions to art, but also the reverse, to what extent art contributes for AI.

Also included in the discussions are algorithmic biases, control structures, machine intelligence in public art, new formalizations of aesthetics, the production of culture, sociotechnical dimensions, relations with games, and the democratization of creative tools based on machines. Despite the diversity of issues, they end up revolving around the backbone of the volume's proposal: Machinic creativity in arts and design represents an evolution of artistic intelligence or is it either a metamorphosis of creative practice that generates forms and distinct modes of authorship? For the editors, the complexity of this situation is not a symptom that the world is changing, but that it has already changed (Burbano and West 2020).

Fundamental to thinking about the democratization of tools for creation in collaboration with ML is the chapter by Mazzone and Elgammal (the latter is the director of the Artificial Intelligence and Art Laboratory, in New Jersey). In the chapter, the design for an easy-to-use Web-based system is presented, similar to digital imaging applications. What is sought is to allow ML to be used as easily as filters or the digital composite for 3D imaging. Interviews with various artists, using the system while in beta, provide information on ways to work with the design, called *Playform*, while discussing unresolved issues inherent in the recent emergence of ML in its nature as a creative content generator in the visual arts, texts/narratives, and musical composition. The question remains: Is ML a *medium*, a tool, or a creative partner?

Caldas Viana (2020), in turn, mapped some important issues of neural networks within the framework of the generative art tradition, emphasizing the emergency of a paradigm shift in creative procedures. Idárraga (2020) questions the neutrality of databases, delegitimizing the automatism of algorithms and even criticizing the assumptions embodied in their functioning. Faced with this, he proposes that, in the field of AI, art builds places where one can think and create other realities. Galanter (2020) discusses the relationship between ethics and AI in art and culture. For Forbes (2020), creative AI consists of a range of activities at the intersection of new media arts, human–computer interaction, and AI. Technics of ML introduce a new contribution to bring meaning to the world from the judicious choice of examples and definition of mappings that enable applications for new forms of creative expression.

Forbes is the director of the Creative Coding Lab at the University of California, Santa Cruz. This laboratory incorporates an interdisciplinary team of researchers and artists affiliated with the Computer Media department. The focus of the works is on applied research in interaction and visualization, and on the exploration of experimental and creative works based on current techniques of human–computer interaction, scientific and information visualization, graphics, computer vision, immersive environments, and ML. A core philosophy of the laboratory is that, by incorporating research methodologies from media arts, design, and computer science, new solutions to interdisciplinary problems can be developed. Furthermore, it is believed that creative results generated at the intersections of artistic and empirical research can significantly elucidate issues in science and technology relevant to contemporary culture. The lesson that remains from this and other advanced research and creation laboratories is that ML and DL procedures did not suddenly fall by a miracle from the skies, but are intruding and incorporating themselves into a tradition of capable art, science, and technology innovations able to illuminate essential cultural issues.

In his chapter entitled "Creative AI," Forbes (2020) clarifies that creative projects in the laboratory he directs are developed by imitating existing data, mapping resources found in one database to another, or mapping inputs to outputs in unusual ways, visualizing or otherwise probing the inner workings of the algorithm and analyzing or speculating on the social impact of ML systems. These activities can enable new types of generative works of art that replicate or incorporate existing works of art, or they can create entirely new artistic productions. In doing so, other ways of analyzing and experiencing cultural artifacts and data are introduced. Finally, the ML algorithm—its computational architecture, the input it requires and the resulting output, and the analysis structure of which it is a part—can be thought of as a cultural artifact in itself, enabling new forms of critical investigation.

Discussions about various aspects of AI inside and outside academies have been frequent in Brazil. For example, in the field of AI and art, Venâncio Júnior (2019) published a chapter on "Art and artificial intelligences: implications for creativity." This chapter proposes a reflection on works of art endowed with AI, stressing issues such as autonomy and creativity. First, some works are offered as examples, to discuss the problem of creative autonomy under which these initiatives are commonly interpreted. References from evolutionary algorithms and cybernetics culminate in a particular model for analyzing works in terms of syntax, semantics, and pragmatics. Such a model offers possible segmentation of the spectra of human creativity, while clarifying some challenges for the development of creative machines. Finally, an artistic proposal is presented that uses AI resources to generate drawings, bringing a situation in which the machine influences, interferes, and redefines a creative process that dilutes the artist's intentions.

At UNESCO's invitation, I, (Santaella 2021) published a policy chapter on "Artificial Intelligence and Culture. Opportunities and challenges for the Global South," in which I argued that the impact that AI provokes on culture, art, and the creative economy is of great proportions. The first impact comes in the form of the challenge facing the hegemony exerted by gigantic data companies or big techs over the functioning of culture on a global scale, with strong repercussions in Latin America and the Caribbean. As a necessary counterpoint to this impact, in the Global North there is a growing mobilization of AI in alternative creative and value chains, which brings to the culture of Latin America and the Caribbean a second challenge that is added to the first: the risk of deepening a digital divide between the North and the Global South. With this in mind, my chapter discussed the implications of the identified challenges and presents recommendations regarding possible strategies to face them.

Undoubtedly, the development of renewed critical thinking is one of the demands that the advent of AI is bringing to the agenda of discussions. A critical thinking that is capable, above all, of freeing us from the dysphoric and gloomy litanies about AI that have taken over the debate and which do nothing to contribute to the multifaceted understanding of the nefarious forces and counterforces that are at play. As always and as expected, it is the artistic creations that are at the forefront of the counterforces.

3 The Variety of Artistic Productions with AI Mediations

While theorists, historians, and critics debate, artists do what is their role to do: They create. There is a wide variety of artistic productions that make use of AI. Apparently, artists seeking AI collaboration are not imbued with competitiveness. It is, above all, an exploratory search for the development of a new expansive form of human creativity. The variety of productions begins with works that fall within the tradition of pictorial arts and are distributed in the following types.

3.1 Style Transfer

Experiments can be as simple as teaching machines to understand and replicate human-made art. This technique is called style transfer. It uses deep neural networks to replicate, recreate, and mix art styles. One of the examples, which went viral, was performed by Chris Rodley in a work that, thanks to algorithms, managed to mix dinosaurs and flowers in a single syntax (Sukis 2018). Another example, a little more complex, makes use of existing paintings that are mixed in an unprecedented fusion. This type of transfer can also be applied to videos and music, when musical genres are mixed with more mathematical compositions such as those by Bach, whose work has a very consistent structure of patterns, which facilitates replication by AI (Sukis 2018). Also, according to this author, there is still a form of imitation similar to style transfer, but in this case the algorithm convincingly changes the appearance of a photo or video, allowing users to edit the context of the image according to the time of the day, the season, or the weather.

3.2 From Transfer to Collaboration

The next step in complexity goes from mere transfer to collaboration. In this case, AI enters as a partner in the ideation of the work and the process, when an algorithm, which is constructed to generate an artistic result, becomes an art form in itself. AI does not just come in as a collaborator by processing images and sounds through mathematical equations. It can equally "inform and inspire artists who want to come up with new insights, connections, or patterns through a huge set of data points" (Sukis 2018).

3.3 From Collaboration to Creation

The works that are developed at the Art and Artificial Intelligence Lab in Rutgers, New Jersey, give us an idea of the meaning that can be extracted from AI as a creator on its own. In this laboratory, researchers created an AI system for art generation that does not involve a human artist in the creative process, but rather involves human creative products in the machine learning process. The director of the laboratory, Ahmed Elgammal (undated), when interviewed about the work being carried out there, argues:

> We're trying to show the world two things: first, what the machine can create by itself. Second, that these are creative partners for artists in the future. I think this is analogous to the creation of photography in the 19th century, because when it was invented the definition of art back then was depicting the world on canvas, but then you have this device that can capture the world for you with the click of a button. So, what's your job as an artist? The definition of art changed as it was influenced by photography. Art focused more on the

> conceptualization and abstraction of the world rather than just depicting it. We now have a tool that can create things for you. It won't take the jobs of artists away. It can explore a space of possibilities for you as an artist. You're framing it in terms of what details to feed to the machine, what you want to do with the data. Your job as an artist is the same — to control the process — but now you have a partner.

One of the works carried out at the laboratory, for example, consisted in presenting to a group of people, on the one hand a mixed set of images created in the laboratory with a communicative and inspiring visual structure, and on the other, images created by artists. The hypothesis was that human subjects would rank art created by human artists on higher scales. To great surprise, the results showed that the images generated by AI received higher ratings (Sukis 2018). On the face of it, that AI will be able to create original artwork seems quite possible. However, at the point of its current development, it must be considered that AI production is completely guided by what humans consider art. This is because, to produce images considered artistic, the machines are powered by a profusion of examples of works of art made by humans. This does not quite minimize the fact that the advance of machine learning is recreating, under a new tone, old debates about how to define art.

However, AI brings a new complication to the issue, such as when intelligent algorithms work like parasites, using source materials from a millennium of human creativity to find patterns and samples that are remixed and blended into something contemporary. Given this, some argue that this process is similar to what artists already do when taking advantage of past collections. In fact, in his laboratory, for artistic AI creation, Elgammal uses the WikiArt database, and the results generated turned out to be extremely similar to those performed by humans (Sinclair 2018a, b).

3.4 More Complex Projects

Current artistic achievements using AI resources are not limited to the imitation or transfer of pictorial and imagistic works of art. According to Sinclair, artists Tara Shi and Sam Kronick, for example, hope that the art they produce can help explain the mysterious workings of artificial neural networks. Likewise, the Splinter art group is incorporating neural networks into their work in order to help the public better understand this technology that is increasingly part of our lives and making decisions for and about us and the world around us.

Artist Memo Akten developed the project *Learning to See: Hello World!*, a series of works that use ML algorithms as a means for us to reflect on ourselves and how we make sense of the world. For the artist, the image we see in our conscious mind is not a mirror image of the outside world, but a reconstruction based on our own prior expectations and beliefs. *Learning to See* is an interactive installation that uses live cameras to demonstrate how ML works. Using a deep neural network, the machine quickly compares patterns and produces an image of its own. But beyond the technique, the work aims to show that this artificial neural network, weakly inspired by our visual cortex, looks through cameras and tries to make sense of what it sees. Of course, she can only see what she already knows, just like with us (Atken 2017).

A similar example was created by Shi and Kronick in which the artists fed the AI program with 3D scans of natural matter (e.g., rocks). The program maps the contour of rocks, learns to recognize this type of matter, and generates an artistic image of nature. By taking AI to produce art from nature, the intention of the artists was also to discover the limits of computational creativity, in the current state when the work was produced. To do this, they used a neural network, that is, a computer program loosely modeled on biological neural systems like the human brain. A given neural network needs to be trained on data; in this case, data could be the shape of many rocks, a huge collection of Google images, or hundreds of thousands of search terms, depending on how the neural network will be used. So basically, the machine thinks in layers, with each layer working on a different aspect of what the network is analyzing, in this case identifying rocks. Thus, an algorithm can try to find the texture of a rock, other different colors on its surface, and so on, layer by layer until it arrives at a convincing result (Chiel 2016).

Another type of AI project is the *New Dimension in Testimony (NDiT/New Dimension of Testimony)*, aimed at storytelling. The project made use of an advanced natural language algorithm that allowed an audience to verbally interact with a 3D image of a holocaust survivor. Powered by a complex algorithm, the hologram responded to the audience's questions in real time, giving the impression of a realistic conversation. This form of AI can be used in many ways, leading us to imagine conversations with holograms of family photos, including our transport to related immersive environments (Sinclair 2018a, b).

More and more, artists are embracing the new challenges AI is bringing to artistic creation. Reflections, analysis, and evaluations are also beginning to appear on the way to the constitution of a theory of art in AI.

There are several Brazilian artists who are experimenting with ML techniques. Among them, it is worth mentioning the work called *Sentimentos da virada*, by Marilia Pasculli and André Gola (2021). Animations of the character 'Suadinho' by the artist André Gola are used to represent the emotions of the inhabitants of São Paulo. Emotions are inferred through the use of an AI that analyzed Instagram photos with geolocation in São Paulo. The "selfies" and corresponding animations were projected on a large scale on urban buildings in three city districts during the cultural turn. The main idea of the installation was to reflect on how we demonstrate our emotions through social networks and how this flow of data that we generate when posting it also ends up influencing the way we see the world and ourselves.

To make this work possible, a bank of images of faces was used, separated according to the seven categories of emotions proposed by Paul Ekman (anger, disgust, fear, joy, sadness, surprise, and neutral). This bank was used to train a neural network (NN) capable of inferring the emotion associated with the image of a face. Once the NN was trained, it analyzed approximately 10,000 georeferenced selfies posted on Instagram in the city of São Paulo in order to obtain a kind of measure of the distribution of feelings in the set of posted images, in addition to projecting the analyzed faces themselves along Suadinho's animations, which interacted with the selfies.

Due to their complexity, Cesar Baio's works in collaboration with Lucy HG Solomon stand out, as their proposal was explained in the chapter "An Argument for an Ecosystemic AI: Articulating Connections across Prehuman and Posthuman Intelligences." (Baio and Solomon 2000) As an art collective Cesar & Lois develop projects that examine sociotechnical systems, attempting to challenge anthropocentric technological pathways while linking to intelligences sourced in biological circuitry. In their role as artists, they imagine new configurations for what we are led to understand as (social, economic, technological) networks and intelligences. With this ecosystemic approach, they consider the possibility of an AI that supports well-being in a broad sense, accommodating relationships across different layers of living worlds and involving local and global communities of all kinds. The artists grounded their thinking on interdisciplinary researches, including communication and media theory, microbiology, anthropology, decolonial studies, social ecology, sociology, and environmental psychology. "At a time when human beings and their ecosystems face grave threats due to climate change and a global pandemic, we are rethinking the basis for our AIs, and for the resulting decision making on behalf of societies and ecosystems." Hence, their work provides alternative conceptual models for thinking across networks, reframing the artists' and potentially viewers' understanding of what motivates and shapes societies.

4 Conclusion

In the end, as we stand today, the issue to remember is that AI is just taking its first steps. Prognostics say its destiny is to grow in complexity. This means that artists will have many new horizons ahead of them that open up to be explored with the artists' own insignia: their sensitive wisdom.

References

Atken M (2017) Learning to see. http://www.memo.tv/works/learning-to-see-hello-world/. Access 10 Oct 2020

Baio C, Solomon LHG (2000) An argument for an ecosystemic AI: articulating connections across Prehuman and Posthuman intelligences. Int J Community Well-Being. 2000. https://doi.org/10.1007/s42413-020-00092-5. Access 1 Oct 2021

Boden M (1998) Criativity and artificial intelligence. Artif Intell 103:347–356

Bogost I (2019) The AI-art gold rush is here. An artificial-intelligence "artist" got a solo show at a Chelsea gallery. Will it reinvent art, or destroy it? https://medium.com/abovethefold/the-ai-art-gold-rush-is-here-a3dda143563. Access 20 June 2020

Broeckmann A (2020) Inordinate images. On the machine aesthetics of AI-based art. ESPACE art actuel No. 124, Montreal, pp 16–23

Burbano A, West R (2020) Introduction. In: Burbano A, West R (coord.) AI, arts & design: questioning learning machines. Artnodes 26:1–8. UOC
Caldas Viana B (2020) Generative art: between the nodes of neuron networks. In: Burbano A, West R (coord.) AI, arts & design: questioning learning machines. Artnodes 26:1–8. UOC
Chiel E (2016) Meet the artists who have embraced artificial intelligence. https://splinternews.com/meet-the-artists-who-have-embraced-artificial-intellige. Access 10 April 2020
Elgammal A (2020) Elgammal on using artificial intelligence, Undated. https://aiartists.org/ahmed-elgammal. Access 10 April 2020
Elgammal A, Mazzone M (2020) Artists, artificial intelligence and machine-based creativity in Playform. In: Burbano A, West R (coord.), AI, arts & design: questioning learning machines. Artnodes 26:1–8. UOC
Flusser V (2019) Elogio da superficialidade. O universo das imagens técnicas. Monografias, vol XVI. É Realizações Editora, São Paulo
Forbes AG (2020) Creative AI: from expressive mimicry to critical inquiry. In: Burbano A, West R (coord.) AI, arts & design: questioning learning machines. Artnodes, 26:1–10. UOC, Leonardo/ISAST
Galanter P (2020) Towards ethical relationships with machines that make art. Artnodes 26:1–9. UOC
Idárraga HF (2020) Identificación, clasificación y control: estrechos vínculos analizados desde las prácticas artísticas en el corazón de la inteligencia artificial. In: Burbano A, West R (coord.). IA, arte y diseño: Cuestionando el aprendizaje automático. Artnodes, 26:1–6. UOC
Kulpaki S (2018) Can AI be creative? A comprehensive look at the state of computers and creativity. https://towardsdatascience.com/can-ai-be-creative-2f84c5c73dca. Access 15/09/2021
Liu Y (2018) Seeking a new human-machine relationship
Mateas M (2002) Interactive drama. Art and Artificial Intelligence School of Computer Science. Ph.D., Computer Science Department, Carnegie Mellon University, Pittsburgh
McLuhan M (1964) Understanding media: the extensions of man. McGraw Hill, New York
Nakazawa H (2018) O que falta para a inteligência artificial produzir obras de arte? Folha de S. Paulo. 20 April 2018. https://www1.folha.uol.com.br/ilustrissima/2018/04/o-que-falta-para-a-inteligencia-artificial-produzir-obras-de-arte.shtml. Access 10 Sept 2021
Pasculli M, Gola A (2021) Sentimentos da virada. https://drive.google.com/drive/u/1/folders/1JLcvN2fB4AL9vq8WeRqE0duTzC7bs_BG. Access 10 March 2021
Santaella L (2021) Inteligência artificial e cultura. Oportunidades e desafios para o Sul Global. UNESCO 2021. https://commons.wikimedia.org/w/index.php?curid=45488198
Sinclair K (2018a) Categories of emerging media. Em: https://medium.com/vantage/categories-of-emerging-media-9c8d3c96004a. Access 05 March 2019
Sinclair K (2018b) The zeitgeist of emerging media. https://medium.com/vantage/the-zeitgeist-of-emerging-media. Access 15 March 2019
Sukis J (2018) The relationship between art and IA. https://medium.com/design-ibm/the-role-of-art-in-ai-31033ad7c54e. Access 05 March 2019
Turing AM (1950) Computing machinery and intelligence. Mind LIX(236)
Venâncio Júnior SJ (2019) Arte e inteligências artificiais: implicações para a criatividade. ARS ano 17(35):183–201

Lucia Santaella is a 1A researcher at CNPq, a full professor in the graduate program in Communication and Semiotics and director of the graduate program in Technologies of Intelligence and Digital Design (PUCSP). Ph.D. in Literary Theory from PUCSP and Ph.D. in Communication Sciences from USP. She was a guest professor and researcher at several European and Latin American universities. She published 53 books and organized 27, in addition to publishing almost 500 chapters in Brazil and abroad. She received the Jabuti awards (2002, 2009, 2011, 2014), the Sergio Motta award (2005) and the Luiz Beltrão award (2010). She is currently professor of the Oscar Sala Chair at São Paulo University Institute of Advanced Studies (2021–2022).

AI, Creativity, and Art

Ernest Edmonds

Abstract The chapter reviews a personal history of research and art practice in which AI systems played a significant part. The key observation was that interacting with such systems provided support for creative thinking. This history is told in relation to a series of meetings held on Heron Island in Australia in which computational creativity was explored. Meetings were held over a 30-year period, the last one in 2019, where an earlier version of this chapter was presented. The chapter reports on those meetings and on the advances that have taken place since 1989.

Keywords Creative thinking · AI · Interactivity · Human-centred · Art

1 Introduction

The chapter presents a personal history of work on computation and creativity, placing it in a broader context. It reviews a set of contributions to meetings about computation and creativity held on the Great Barrier Reef Island, Heron, starting in 1989 and concluding in 2019, where an earlier version of this chapter was presented. The chapter briefly states the key arguments that I made at the earlier Heron Island meetings and reviews the progress of that thinking, including a discussion of how the ideas developed in relation and the debates held on the island. A central concern has been to take a human-centered perspective on the value of computational and cognitive models. This goes back to a research agenda first announced in a joint chapter at a 1970 Computer Graphics conference. The title was "the creative process where the artist is amplified or superseded by the computer", where the two alternatives provided defined the end points of the field to be investigated. There has been considerable progress in creative artificial intelligence applications and the cognitive modelling of creativity since 2005 and the chapter looks at these earlier arguments in the context of that progress, re-assessing them in light of our understandings of 2019.

E. Edmonds (✉)
De Montfort University, Leicester, UK
e-mail: ernest@ernestedmonds.com

C. Vear and F. Poltronieri (eds.), *The Language of Creative AI*, Springer Series on Cultural Computing, https://doi.org/10.1007/978-3-031-10960-7_4

2 The Earlier Heron Arguments

In this section, I briefly restate the key arguments that I made in my con*t*ribution*s* to the earlier Heron Island meetings. I will do this chronologically without reflecting on those arguments, which will be done later in this chapter.

My chapter in the first Heron Island meeting (Edmonds 1993) was titled "Knowledge-Based Systems for Creativity". The key claim was that knowledge-based systems offered new mechanisms for the support of human creativity. The term "knowledge-based systems" refer to AI and at that time was generally used to refer to symbolic AI systems as against the connectionist systems that are so popular today. This *argument* was supported by two very different case studies. One was from my own computer-based art practice, and the other was of a speech scientist investigating speech recognition. The first example was concerned with making time-based generative artworks. That is, artworks that change over time as defined by an internal set of rules specified by the artist. I will return to this domain slightly later in this section. Gerhard Fischer and Robert Coyne, with Eswaran Subaran, also advocated the study and development of creativity support systems in the 1989 meeting (Fischer 1993; Coyne and Subrahmanian 1993).

In the second meeting I presented a paper jointly written with Basel Soufi (Edmonds and Soufi 1992) and inspired by Bill Mitchell's contribution to the first conference (Mitchell 1993). Mitchell had pointed out that the recognition of emergent shapes in drawings was often a key factor in the creative discovery of what might be. The paper investigated the computational implications for support systems and showed how they might be implemented. An important conclusion from this *investigation* was that models of creativity need to be open systems, not closed ones that are essentially Turing machines.

At the 1995 Heron conference I presented another paper by Basel Soufi and myself in which we went into more detail on the cognitive issues of emergence and their implications for our computational models (Soufi and Edmonds 1995). It was shown that, in order to support the creative practitioner's interaction with a support system, different mechanisms need to work in parallel so that emergence can be fully handled.

Kelvin Clibbon and I also produced a paper for the 1995 meeting that looked at another important aspect of creative design. The paper was titled "Strategic Knowledge in Computational Models of Creative Design" (Clibbon and Edmonds 1995). The key problem addressed was to deal with the fact that creative design cannot be seen as problem solving within a well-defined problem space. that is normally known as "routine" or "variant design". We took an approach that used formal logic as the basis of the computational modeling and showed how the use and modification of strategies could be incorporated by extending first order logic to include meta-rules representing strategic knowledge. We argued that an explicit multi-layered representation of design knowledge within the computational support system was important in giving the designer the freedom that they need for creative exploration.

In 1998, I co-wrote a paper with Linda Candy (Edmonds and Candy 1999). We returned to the concerns of my 1989 Heron paper, the influence on human cognition of interacting with a computational system. In this study, we observed an artist working collaboratively with a VR specialist in order to build virtual sculptures. The key observation was that the interaction with the technology had a significant influence on the direction of the artist's thinking. Following the computer-based work, he went on to make paintings (by hand) that he only conceived of as a result of the computational experience. A new direction for his painting emerged during the interactions.

In 2001, I contributed to a paper presented by Linda Candy, "Model-ling Creative Collaboration: Studies in Digital Art" [an expanded version was later published (Candy and Edmonds 2002)]. In this paper we added to the regular Heron topics by considering collaborative creativity, which we were studying in the context of digital art and had found to be very significant. An initial report was given of findings about the various flavors of collaboration that we observed, together with the identification of success factors for good practice.

It was also in 2001 that I presented a paper, written with John Dixon, that returned to computational support for making generative art (Edmonds and Dixon 2001). As I had argued on Heron Island in 1992, a valid and useful model of the creative process has to be an open system. In my personal exploration of the implications of the computer for art practice, first announce in 1970 (Cornock and Edmonds 1970) and discussed more fully below, I was very interested in computer-based artworks that were themselves open systems. In this paper we showed how the logic of a closed generative artwork could be extended by adding inter-relationships between the rules and external events. In the examples discussed, those events were movements by observers detected and analyzed through video capture. By making such artworks, we postulated that "the "significant trigger for creative thought" in the artist, reported in the 1989 Heron Island meeting, might here be provided to everyone".

It seemed natural to move on to consider the nature of the experiences that the artist, designer and everyone else have when interacting with these various computational creativity models and support systems. So, in the 2005 Heron meeting, I presented a paper, written jointly with Lizzie Muller, that investigated human creative engagement in such contexts (Edmonds and Muller 2005). The motivation was largely to come to a better understanding of the relevant human experiences and engagement in order to better inform the design of these systems, whether they were design support environments, artworks, scientific aids, or anything else. Based on field studies conducted in the Powerhouse Museum, Sydney, we proposed an initial model of creative engagement, and an approach to further study, that began to provide the answers required.

3 Reflections on Personal Heron Contributions

None of the papers that I presented at Heron provided definitive answers or completed studies. That was never appropriate. They each defined a step in the development of the thinking that my colleagues and I were gripped by. Most importantly, the meetings between 1989 and 2005 provided a forum for debate, for the refinement of concepts and for the birth of new ideas. Thus, in each case, there are later papers published elsewhere that elaborate on the ideas, but these Heron papers typically represent my first public testing of them. In this section, I will provide an overview of the trajectory that provides a basis for the next part of this paper, where I look at what has happened since 2005. First, though, it is helpful to step back and describe certain earlier publications that form an important background to this work.

3.1 Context for Heron

At the 1970 computer graphics conference, CG70, held at Brunel University, UK, Stroud Cornock and I presented a paper with the title, "the creative process where the artist is amplified or superseded by the computer" [later published in Leonardo (Cornock and Edmonds 1973)]. As the title implies, a key concern was to consider whether the computer would become the creative practitioner, the artist, or whether the computer would amplify the artist's capabilities. One prediction that we made was that a major development would be computer-based interactive art and we identified several models of different interactive scenarios that could apply. We also showed an example interactive artwork. It took another decade before we had personal computers and all the opportunities that they brought with them, but the basics were clear enough then. In brief, the conclusion was that the artist, and by implication the designer as well as other creative practitioners, would be amplified. This meant, however, that much research was needed into how to build effective human–computer (or man–machine as they were termed then) systems.

I immediately started working on human–computer interaction and chose design, and architecture in particular, as the domain within which to research. The very early papers, following Cornock and Edmonds (1973), made a number of points that framed the work that I have discussed on Heron. In a 1972 paper with Christine Daniels and Martin Humphrey (Edmonds et al. 1972), we presented a simple demonstration support system for designers that concentrated on helping them define the problem as well as solve it. This was important because, from my experience within art practice and in talking with designers, I saw problem specification as a key part of the creative process and that would have to be supported by any helpful system.

I then worked with the architect John Lee, developing ideas about how computer systems might successfully support creative design. For example, in 1974 we presented a paper at the EUROCOMP online computing systems conference with the title "an appraisal of some problems of achieving fluid man–machine interaction" (Edmonds and Lee 1974). This was a study made in the context of computer aided architectural design. We argued that it was important to match the structure of the computer system to the human processes that it was to support. We observed the consequential need for flexibility, the provision of backtracking the opportunity to change one's mind. We showed an approach to dealing with this problem and elaborated on the software and human–computer interaction issues the following year (Edmonds and Lee 1975).

When I came to the first Heron Island meeting, therefore, I took it that a significant contribution that computational models of creativity could make was within interactive creativity support systems. I also saw that such interactive systems should handle poorly specified problems—in fact help to formulate the problem—and should handle the process flexibly, allowing the creative user to develop their thinking and understanding as time progressed. With those starting points, I can now summarize the trajectory of my Heron Island contributions from 1989 to 2005.

3.2 The Heron Trajectory

I began by showing how interacting with a knowledge-based system, AI, could be a significant trigger for creative thought. I showed how, given the right computational system and the right user interface, the creative *practitioner's endeavors could be amplified*. To achieve this, it was clearly necessary to understand the human side of human–computer interaction as well as the computational one.

Then, considering such interactive systems it was clear that the human ability to discover *emergent shapes and properties* was critical in creative thinking, so it needed computational support. Doing this requires relatively deep technical decisions to be appropriately taken. As part of this investigation, it also became clear that any adequate model of creativity had to be *interactive*, an open system.

Later, the need to also model and interact with knowledge about design strategies became clear and so I presented ideas about *strategic knowledge*, showing how multi-layered logics could be used for this purpose.

Then, we reported on studies that we had been undertaking of *collaborative creativity*. In practice, it seemed that much design, and art in the digital domain, was conducted in some form of collaboration or other and the characteristics of these collaborations needed to be investigated.

Returning to the technical aspects of building interactive creative systems, I next demonstrated how a closed generative art system, built using logic, could be made open and so interactive. In the example used, image analysis of video captured from the artwork's environment was used to modify parameters in the generative rules being followed. Thus, I demonstrated formal systems that were used to make *interactive generative art*. See Fig. 1.

The 2005 meeting changed the name from "Computational Models of Creative Design" to "Computational and Cognitive Models of Creative Design", so it was particularly appropriate that the concern I always had for the human, cognitive, side of the problems being addressed became the prime focus of the paper presented then. It reported empirical work that was leading toward a *cognitive model of creative engagement* with interactive, computationally driven, art systems.

Fig. 1 Interactive video construct at Kettles Yard, Cambridge, 2001

Taking all of the above contributions together, we can see a core argument concerning computational and cognitive models of design. I take a very broad view of the scope of *design*, in relation to this work. In particular, I take it to include *art*, while recognizing that the focus of design might normally be different to that of design. In the last 100 years, since the founding of the Bauhaus, it has been particularly common to see these two fields as at least firmly overlapping.

The core argument, then, is as follows:

> An adequate computational model of creative design needs to be an open system. It is important to consider both the internal characteristics of the computation and the characteristics of what that system interacts with. An interesting application of such models is in creativity support systems, and hence, it makes sense to look at computational and cognitive models of creative design together. Emergence and strategic knowledge both need to be addressed as part of this work, and finally, if their application is to be a practical value, collaborative creativity must also be better understood.

4 What Progress Has Been Made?

4.1 The Computation and Creativity Community

Just after the second Heron Island meeting Linda Candy and I started the *Creativity and Cognition* conference series. This was inspired by the first Heron meeting and intended to complement it by having a focus on the human dimension of the computational support for creativity, including creative practice. Because of the interest in practice, an art exhibition has been a normal feature. Later in that decade the series was adopted by the ACM Special Interest Group on Computer–Human Interaction, who continue to run it, most recently in Venice this year (C&C 2022).

The Heron meetings would seem to have also inspired other initiatives, more focused on the computational modelling aspect. For example, in 2009, a Dagstuhl Seminar was held on "Computational Creativity: An Interdisciplinary Approach" (Dagstuhl 2009) and the next year a series of conferences on the subject began and a society was formed, the Association for Computational Creativity, which runs regular conferences, most recently in 2021 in Mexico City (ICCC 2021).

There is no doubt that the Heron Island innovation has led to at least two ongoing and overlapping community initiatives.

4.2 Developments of My Heron Argument

It is encouraging to see that, since 1989 (indeed since 2005) there has been a notable growth in the literature about the Heron Island topics. A full survey is beyond the scope of this chapter, so I will just refer to a few notable books and the latest conferences in the two series mentioned above.

As a background point, it is important to note the undeniable resurgence of artificial intelligence in recent years, both in terms of public interest and success in applications. It is also important to notice that this has been largely driven by progress in connectionist AI, rather than in the symbolic AI that was more commonly discussed at the earlier Heron meetings. What does this mean? Does it imply the need for a re-think? Richard Coyne, Sidney Newton and Fay Sudweeks actually discussed these questions at the first Heron Island conference (Coyne et al. 1993). Their conclusion was that objections to connectionist modeling of creativity did not stand up and that it had significant potential. I will comment on this issue before a more specific look at what has happened in relation to the points in my argument. The issues, then, are:

- Connectionist models
- Interaction
- Creativity support
- Emergence
- Strategic knowledge
- Collaborative creativity
- Experience.

4.2.1 Connectionist Models of Creativity

Of the objections to connectionist models that Coyne et al. covered, the one that is now most widely discussed and seen as problematic is the obscurity of such models:

our inability to question them as to why they have generated a particular result. This is often seen as particularly concerning when we suspect an inappropriate bias in the data set that the system has been exposed to (the *training set* as it is often termed). Examples of serious problems are frequently reported, for example in facial recognition applications (USA Today 2015; Wired 2019).

Whichever approach one takes to our subject, building autonomous models of creativity or building creativity support tools, and the inability to question the model is a problem. If the model is autonomous and connectionist, it may be able to generate a creative outcome, but what else can it tell us? It will not be a very valuable model, in the sense of a theory about that creative act, because we cannot inspect the decision-making that led to it. If it is a connectionist support tool, then if it is just throwing up proposals for the user to evaluate, it may well be helpful. If it is being used for pattern recognition, then it will be as useful as the training set is unbiased. However, if the user needs to interact with the knowledge in the system, as for example was done in my 1989 paper (Edmonds 1993), then it is as useless today as it was then. I would argue that the recent great advances in connectionist AI do not offer a revolution or even a notable opportunity in our area of the computational and cognitive modeling of creative design.

4.2.2 Interactive Systems

The discussion about creative models and open systems has continued. The most significant contributions have probably been philosophical. In her book "Creativity and Art" (Boden 2010), Margaret Boden considered autonomy, integrity, and authenticity in relation to computer art as well as posing the question "is metabolism necessary"?. In each of these cases she implicitly contributed to our open/closed system debate. Her arguments are too long and complex to summarize here, but I will cover the important context in which she makes them. Boden lists three different ways in which we can be surprised. They are, briefly:

1. Combinational: the generation of unfamiliar combinations of familiar ideas.
2. Exploratory: finding new ideas in a known conceptual space.
3. Transformational: changing or extending a known conceptual space.

It is the third category that we would normally consider creative, rather than just innovative, although this is a matter of definition rather than fact so not everybody will agree. It is also the case that Boden is not taking much interest in surprises that have no relation to anything that we know already, so she dismisses "…the undisciplined outpourings of a schizophrenic's "word-salad"—which, despite being unpredictable and occasionally suggestive… is not in itself an exercise in creative thinking".

By definition, if a computer program is a closed system, it can only reach states within the conceptual space defined by that system. Only by interacting with some other system(s) can the space be changed. Hence, from this perspective, transformational creativity can only be exhibited by an open system. The question of whether a computer program, rather than a human being, could be said to be creative is something else that Boden discusses, but this is beyond my scope. In the Heron context we would at most argue that such a program modeled human creativity.

4.2.3 Creativity Support Systems

Just before the 2005 Heron meeting, in June of that year, a workshop, sponsored by the U.S. National Science Foundation, met in Washington to discuss creativity support tools. They generated a white paper that set out a research agenda as well as various carefully debated propositions about how such support systems should be developed, evaluated and deployed (Shneiderman et al. 2006). Partly as a result of that initiative there has been a widespread growth of work in the area, much too large to survey here. Suffice to say that we now have an active research community working on computational systems that support human creativity. Much of that work addresses the support of "everyday" creativity rather than the work of expert artists, designers, scientists, etc., but a significant proportion is addressing the kind of problems that we debated on Heron. For a recent review, see for example Frich et al. (2019).

4.2.4 Emergence

Emergence was discussed in a range of Heron papers and is often considered in creativity research. One notable example is the book on Emergence in Interactive Art by Seevinck (2017). Although the examples described in this book are artworks, the principles investigated are more general. The research can be seen in the context of creativity support where the interest is in "…facilitating emergence in creative practice as well as characterizing it in people's interaction …".

Seevinck's results lead her to list certain mechanisms as particularly important in the building of this kind of support system:

1. Structural transparency: giving the user access to the system's underlying mechanisms
2. Priming: exposing the user to examples prior to engaging them in the interactions that might lead to emergence
3. Combined directly responsive and influencing interactions: combining action-response with (possibly delayed) interaction over long periods (Edmonds 2007a).

She also points out that, at least in her work, it was important to make empirical studies of the interactions so as to get a handle on the thinking going on in the heads of her users.

4.2.5 Strategic Knowledge

I have worked on strategic knowledge myself, making significant use of the concepts in a range of interactive artworks. The theoretical foundation was described in a paper already mentioned (Edmonds 2007a). In that paper a systems view of interaction is taken and contrasted with an action/response model. A refined view of such interactions is proposed in which artwork and audience are said to influence one another. All interaction involves an exchange but need not necessarily leads to a significant change in behavior: at least not at that time.

In the case of my application of the idea to art works, I created the *Shaping Form* series of interactive pieces (Edmonds 2007b, 2017). In the *Shaping Form* works, images are generated using rules that determine the colors, the patterns and the timing. Figure 2 shows a moment in the interaction with an installation version, *Shaping Space*.

The strategic knowledge employed in these works is codified in meta-rules that can modify the basic rules that run the system. Interaction aside, a rule-based computational system produces a never-ending stream of graphics, with an equally changing timing pattern. However, these are interactive generative works that evolve by the influence of the environment around them, as detected by image analysis of data from a Webcam. Movement in front of each work is detected and leads to continual changes to the rules being used. The data produced from the image analysis are used

Fig. 2 Shaping Space at the Site Gallery, Sheffield, 2012

by meta-rules to modify the generative system as well as to cause an occasional direct response. People can readily detect these immediate responses to movement but the changes over time are only apparent when there is more prolonged, although not necessarily continuous, contact. A first viewing followed by one several months later will reveal noticeable developments in the colors, the timings, and the patterns. The *Shaping Form* works use meta logic to implement, or one might say model, the systems-based "influence" form of interaction mentioned above, contracting with an "action-response" one as is commonly used, for example, in computer games.

4.2.6 Collaborative Creativity

The work that Linda Candy and I introduced in 2001 (Candy and Edmonds 2002) has been expanded and reported in various ways, see for example Mamykina et al. (2002) and, in particular, in the book "Explorations in Art and Technology", now in its second edition (Candy et al. 2018). In that book, as well as reporting on the empirical research on collaboration in creative computational systems, a range of practitioners report on their experiences of making computer-based art systems in collaborative teams. The results of the studies reported in the first edition (and briefly announced on Heron) identified various success factors:

- Shared language: the need to evolve shared terminology
- Common understanding: of intentions and visions
- Open discussion: facilitating free "what-if" conversations
- Establish relationships: giving time for the team to develop, for example feeling free to make mistakes and recover.

These were found to largely stand but the successful work of described by collaborative teams, such as Christa Sommerer and Laurent Minonneau (Candy et al. 2018: 363–370) Andrew Johnston and Andrew Bluff with Stalker (Candy et al. 2018: 341–352), Anthony Rowe and his Squidsoup team (Candy et al. 2018: 333–340), etc., gave a much stronger argument for the importance of collaboration in much creative work in the area.

4.2.7 Experience

The study of creative experience has demanded the development of appropriate research methods. Most notably, these have to be methods that can be applied in the field, in real rather than laboratory situations. Our interest is not in what experiences people have in artificial situations, we need to know what happens in real contexts. One approach to this problem was to conduct investigations in a dedicated space within a museum. The space was known as Beta_Space and was part of the permanent exhibition area of the Power House Museum, Sydney. Much of the work has been described in detail in a book (Candy and Edmonds 2011). As a result of conducting many studies of creative engagement with interactive art systems in Beta_Space,

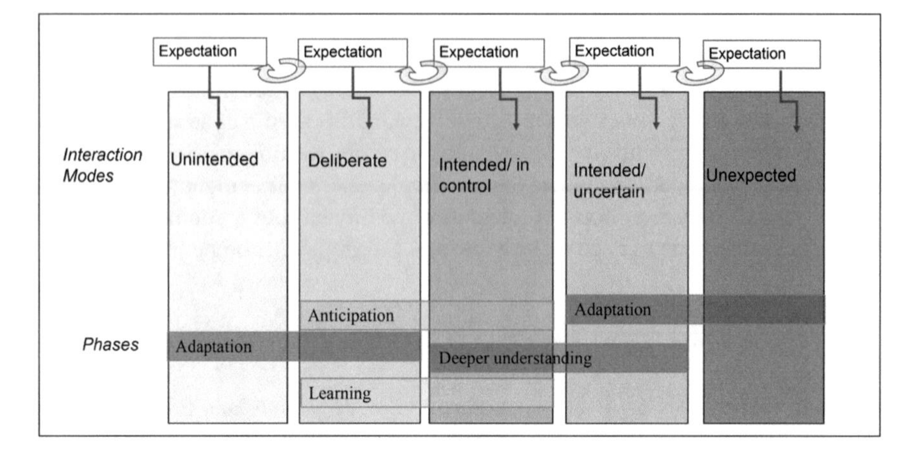

Fig. 3 Bilda's model of creative engagement

Zafer Bilda developed a model of engagement that showed, in particular, how the nature of that engagement changed over time (Candy and Edmonds 2011:163–181) See Fig. 3. The important lesson is to see that experience changes the form of engagement and, consequently, the design criteria that work for instant engagement will be different to those that work for long term, even life long, engagement.

4.3 Where Are We Now?

Looking at the most recent relevant conference series mentioned above (C&C 2022; Heron 2019; ICCC 2021) we see that the issues covered in my above "Heron argument" are still live in these communities. Collaboration is quite a significant topic, often in terms of "co-creation" and the nature of sharing. Interaction (using an open system) is everywhere with explicit studies of computer support systems and a nice clear observation, drawing upon Lubart (2005), that what might at times be seen as a "failed AI system" is, in fact, a successful human–computer interactive system. Having the human in the loop is not such a bad thing, a point that goes back to my first contribution on Heron (Edmonds 1993).

As to my own current position, apart from the points made above, I have argued strongly that rather than always starting with the science and observing the creative arts—and design—there is much to be gained by starting with the increasingly impressive volume of research within those arts to see what science can learn. I have argued this in a recent book, by taking a selection of research projects in interactive art and showing how they offer human–computer interaction a great deal (Edmonds 2018). The lesson here is equally valid if other creative arts are considered in light of what can be learnt by other scientific disciplines, such as computational creativity.

In a more extensive study, I have worked with Margaret Boden on a book that covers the history, philosophy and practice of generative art: art which focuses on the writing of code as a key medium (Boden and Edmonds 2019). Following on from the point made in the previous paragraph, a non-trivial contribution of the book is a set of interviews with artists who use code, artists for whom computational models in effect are central to their work. It is impossible to summarize the book in this chapter, but I will mention two points that seem to me to be significant in relation to our research into computation and creativity or, as I prefer to think about it, computation for creativity.

Margaret Boden and I spend some time placing computational art in its broader context, both philosophically and historically. We show how the computer, software and computational modelling are being used to support and enable creativity, in many forms, in ways that have a strong relationship with what was done before the computer existed. Of course, computation offers something new and specially but understanding just what that is can only be done by seeing it in its true perspective. The fact, for example, that very many computer-based artists only show their work in computer art contexts, rather than in mainstream exhibitions, and in so doing risk marginalizing their work. More significantly, the risk is that the evaluation criteria and the critical apparatus that we use in considering art and the art making process might be by-passed in relation to the computational arts.

The second point from the book is the change in attitude and perception observable in different age groups. The interviewees range from pioneers who started using computers in the 1960s to artists who regularly programmed computers well before they discovered art. This second category is "digital natives", and it was clear that they see code and computation quite differently to their elder colleagues. For example, Alex McLean, a live coding artist, said "code is the most suitable way of thinking about music". For many current creative practitioners, code and computational models are an integral part of their work and something that they can hardly imagine not using. We need to understand the creative thinking, the cognitive processes, of digital natives and that means making new studies, not based on the work of the established pioneers. The best route for pushing this research forward is, I suggest, to interrogate that thinking from the inside. The best approach that we have available for this has to be based on the creative reflective practice (Candy 2019), where the creative practitioner themselves articulates the processes of their thinking.

5 Conclusion

The Heron Island meeting, from 1989 to 2005, provided an exceptional forum for proposing, debating, and innovating ideas about computational and cognitive modeling in creative design, including the arts. In itself, the conception of the series was creative and, as I have briefly indicated, it has led to many new paths of research and application. The developments in technology and in research since 2005 have only made that earlier work more relevant. In my discussion above, as well as reviewing my own "Heron" thinking, I have pointed to some examples of where we have moved on and where we are going. Looking overall at this work, it seems that the human element, the cognitive, and the computational/cognitive interactions have proven most interesting and most promising. A start has been made, but as I see it, the list of interesting questions has only grown.

References

Boden MA (2010) Creativity & art. Oxford University Press, Oxford

Boden MA, Edmonds EA (2019) From fingers to digits: an artificial aesthetic. MIT Press, Cambridge, MA

Candy L (2019) The creative reflective practitioner. Routledge, London

Candy L, Edmonds EA (2002) Modelling co-creativity in art and technology. In: Proceedings of creativity & cognition 2002, ACM Press, NY, pp 134–141

Candy L, Edmonds EA (eds) (2011) Interacting: art, research and the creative practitioner. Libri Press, Oxford

Candy L, Edmonds EA, Poltronieri F (2018) Explorations in art and technology, 2nd edn. Springer, London

Clibbon K, Edmonds EA (1995) Strategic knowledge in computational models of creative design. In: Gero JS, Maher ML, Sudweeks F (eds) Preprints of third international round-table conference on computational models of creative design. University of Sydney, Australia, pp 319–342

Coyne RF, Subrahmanian E (1993) Computer supported creative design: a pragmatic approach. In: Gero JS, Maher ML (eds) Modelling creativity and knowledge-based creative design. Lawrence Erlbaum, Hillsdale, NJ, pp 295–327

Coyne RD, Newton S, Sudweeks F (1993) A connectionist view of creative design reasoning. In: Gero JS, Maher ML (eds) Modelling creativity and knowledge-based creative design. Lawrence Erlbaum, Hillsdale, NJ, pp 177–209

Cornock S, Edmonds EA (1970) The creative process where the artist is amplified or superseded by the computer. In: Computer Graphics'70. Brunel University, UK

Cornock S, Edmonds EA (1973) The creative process where the artist is amplified or superseded by the computer. Leonardo 16:11–16

C&C (2022) https://cc.acm.org/2022/. Accessed 3 Mar 2022

Dagstuhl (2009) https://www.dagstuhl.de/program/calendar/partlist/?semnr=09291. Accessed 3 Mar 2022

Edmonds EA (1993) Knowledge-based systems for creativity. In: Gero JS, Maher ML (eds) Modelling creativity and knowledge-based creative design. Lawrence Erlbaum, Hillsdale, NJ, pp 259–271

Edmonds EA (2007a) Reflections on the nature of interaction. CoDesign: Int J Co-Creation Des Arts 3(3):139–143

Edmonds EA (2007b) Shaping forms series (ongoing). In: Jennings P (ed) Speculative data and creative imaginary. National Academy of Sciences, Washington, DC, pp 18–19

Edmonds EA (2017) Shaping Form S17. In: CHI EA'17 extended abstracts on human factors in computing systems. ACM, New York, NY, pp 1431–1432

Edmonds EA (2018) The art of interaction: What HCI can learn from interactive art. Morgan & Claypool, San Rafael, CA

Edmonds EA, Daniels C, Humphrey M (1972) COPS—Conversational problem solver. In: Proceedings of on-line'72. Brunel University, pp 309–322

Edmonds EA, Lee J (1974) An appraisal of some problems of achieving fluid man-machine interaction. In: Proceedings of EUROCOMP'74, online computing systems, pp 635–645

Edmonds EA, Lee J (1975) Complexity and compromise in CAD systems. In: Proceedings of EUROCOMP'75—Interactive systems, online conferences, pp 497–510
Edmonds EA, Soufi B (1992) The computational modelling of emergent shapes in design. Gero JS, Sudweeks F (eds) Preprints computational models of creative design. University of Sydney, Australia, pp 173–189
Edmonds EA, Candy L (1999) Into virtual space and back to reality: computation, interaction and imagination. In: Gero JS, Maher ML (eds) Computational models of creative design IV. University of Sydney, Australia, pp 19–31
Edmonds EA, Dixon J (2001) Constructing inter-relationships: computations for interactions in art. In: Gero JS, Maher ML (eds) In: Proceedings of 5th international conference on computational and cognitive models of creative design. Sydney University, Australia, pp 173–185
Edmonds EA, Muller L (2005) Tales of the unexpected: understanding emergence and it relationship to design. Gero JS, Maher ML (eds) In: Proceedings of 5th international conference on computational and cognitive models of creative design. Sydney University, Australia
Fischer G (1993) Creativity enhancing design environments. In: Gero JS, Maher ML (eds) Modelling creativity and knowledge-based creative design. Lawrence Erlbaum, Hillsdale N.J, pp 235–257
Frich J, Vermeulen LM, Remy C, Biskjaer MM, Dalsgaard P (2019) Mapping the landscape of creativity support tools in HCI. In: Proceedings of CHI'19. ACM Press, NY, paper no. 389
Heron (2019) http://dccconferences.org/hi19/. Accessed 3 Mar 2022
ICCC (2021) http://computationalcreativity.net/iccc21/. Accessed 3 Mar 2022
Lubart T (2005) How can computers be partners in the creative process: classification and commentary on the Special Issue. Int J Hum-Comput Stud 63:4–5, 365–369
Mamykina L, Candy L, Edmonds EA (2002) Collaborative creativity. Commun ACM 45(10):96–99
Mitchell WJ (1993) A computational view of creative design. In: Gero JS, Maher ML (eds) Modelling creativity and knowledge-based creative design. Lawrence Erlbaum, Hillsdale, NJ, pp 25–42
Seevinck J (2017) Emergence in interactive art. Springer
Shneiderman B, Fischer G, Czerwinski M, Resnick M, Myers B, Candy L, Edmonds E, Eisenberg M, Giaccardi E, Hewett T, Jennings P, Kules B, Nakakoji K, Nunamaker J, Pausch R, Selker T, Sylvan E, Terry M (2006) White paper on creativity support tools workshop. Int J Hum Comput Interact (IJCHI) 20(2):61–77
Soufi B, Edmonds EA (1995) Cognitive issues of emergence and interaction: implications for computational models. In: Gero JS, Maher ML, Sudweeks F (eds) Preprints of third international round-table conference on computational models of creative design. University of Sydney, Australia, pp 131–146
USA Today (2015) https://eu.usatoday.com/story/tech/2015/07/01/google-apologizes-after-photos-identify-black-people-as-gorillas/29567465/. Accessed 3 Mar 2022
Wired (2019) https://www.wired.com/story/san-francisco-bans-use-facial-recognition-tech/. Accessed 3 Mar 2022

The Ethics of Creative AI

Catherine Flick and Kyle Worrall

Abstract Creative AI has had and will continue to have immense impact on creative communities and society more broadly. Along with the great power, these techniques provide and come significant ethical responsibilities in their setup, use, and the output works themselves. This chapter sets out the key ethical issues relating to creative AI: copyright, replacement of authors/artists, bias in datasets, artistic essence, dangerous creations, deepfakes, and physical safety and looks toward a future where responsible use of creative AI can help to promote human flourishing within the technosocial landscape. After Vallor (2016), it suggests key technomoral values of honesty, humility, empathy, care, civility, and flexibility as those which virtuous creative practitioners will want to embed within any practice conducted using creative AI techniques.

Keywords Ethics · AI · Change · Creativity · Art · Music · Robotics

1 Introduction

The use of artificial intelligence (AI) techniques[1] in creative pursuits has been increasing at a pace in keeping with the improvement in the use and outputs of these methods. Alongside the ethical and social impact of AI techniques more broadly, creative AI methods have also raised interesting and problematic ethical issues, such as the emergence of deepfakes, the bias of datasets, and the potential for copyright and other authorship issues. However, creative AI has also shown some highly beneficial impacts for society—not only the art itself, but some applications that

[1] In this chapter, we use AI as an overarching term that includes machine learning, deep learning, and neural networks.

C. Flick (✉)
De Montfort University, Leicester, United Kingdom
e-mail: cflick@dmu.ac.uk

K. Worrall
University of York, York, United Kingdom
e-mail: kjw547@york.ac.uk

C. Vear and F. Poltronieri (eds.), *The Language of Creative AI*, Springer Series on Cultural Computing, https://doi.org/10.1007/978-3-031-10960-7_5

can significantly improve people's lives and agitate for beneficial societal, environmental, and political change. It is important, therefore, to explore some of the ethical issues present in current implementations of creative AI, which practitioners may encounter in their use of these techniques, and to also frame a future of creative AI practice in which the positive impacts are encouraged to promote human flourishing within the technosocial landscape and negative impacts mitigated or avoided.

In this chapter, we outline some of the existing ethical issues in creative AI and suggest ways to approach, mitigate, or avoid negative impacts from these issues. We focus firstly on ownership and authorship of creative AI outputs, looking specifically at copyright infringement and the potential for creative AI techniques to facilitate replacement of authors/artists. We then look at the inputs and outputs of creative AI where we examine the issues with datasets and the artist's essence, and what role creative AI may have in the artistic world more generally, before focusing on the potential for dangerous creations, non-consensual deepfakes, and the importance of physical safety in a world where physical AI systems can encounter bugs in their programming. Finally, we look to the future—what a virtuous creative AI might look like that focuses primarily on contributing to human flourishing and the pursuit of a technosocial good life. We use Vallor's (2016) technomoral virtues to frame practical suggestions for practitioners to consider as they use creative AI techniques and as a starting point for discussion into the future of creative AI.

2 Ownership and Authorship

If a creative AI is to be accepted by potential users and the public alike, it needs to be free from the risk of copyright infringement, and the design of such a tool needs to be incorporated into a workflow by the user, rather than incorporating the user (or the user's identity/essence—more on this in Sect. 3.2). Current research is examining the different facets of creative AI relating to the use of these tools, and how they impact claims of ownership and/or authorship. Examples of such research are the design of creative AI tools and how they fit into the human workflow (Ben-Tal et al. 2020; Louie et al. 2020), how humans can collaborate with or leverage creative AI (Collins and Laney 2017; Frid, Gomes and Jin 2020; Hanson et al. 2021), and how to best evaluate creative AI tools and their output (Zhou et al. 2020; Yin et al. 2021). Recent studies have also surveyed the use of creative AI to establish the usefulness and acceptance of such tools (Knotts and Collins 2020; Liebman and Stone 2020). This forms a basis for understanding that the incorporation of creative AI into a workflow and the ethics of ownership and authorship that surround the use of creative AI are much more complex than simple generation of artistic content. The ethics of authorship encapsulate one of the core concerns around creative AI; that of "stealing" ideas from authors' stimuli or replacing the humans in the loop. In this section, we will look at some of the more specific issues around ownership and authorship in creative AI.

2.1 Copyright Infringement

Lawsuits around copyright infringement are frequent within the arts, and creative AI will add to the complexity of legal discussions around copyright. With creative AI using databases of existing material to be trained on, both the use of that material and potentially the output of the AI could become complicated from an ownership/authorship perspective.[2] If a creative AI does infringe on the copyright of songs, the question will be where the blame truly falls: the user who is providing stimuli, or the creator who built the tool. For now, the evaluation of the level of theft that occurs through stimuli-based generation is done using evaluative algorithms (Yin et al. 2021) or listening studies (Collins and Laney 2017).

Legislation may fall to looking at prior, human-driven examples of use of copyright testing, such as sampling or remixing, which can allow an original creator the ability to claim royalties from the song that has sampled or remixed the original. However, issues of plagiarism, such as in the case between Led Zeppelin and Spirit (Carroll 2016), could become more difficult to prove if creative AI is involved in the process. This issue of adopting others' artistic essence is discussed in more detail in Sect. 3.2, but creative AI users should be aware of this potential issue in their creative process, and where possible allow for transparency in their algorithms to be able to prove how the art has been derived.

Finally, there might be some discussion around who owns the art if AI progresses to the point where it might be able to assert ownership. The "monkey selfie" case showed that in multiple countries, copyright is held by a "legal person", the definition of which does not include non-human animals (Ncube and Oriakhogba 2018). In this case, ownership-asserting AI would need to be classified as a "legal person", before it would be able to claim copyright, which would be a complex decision to make. Such a discussion about the identity of AI agents is outside the scope of this current chapter, but likely one that will be had in the coming years.

2.2 Author/Artist Replacement

Previously, creative AI tools were often left at the command line, usable only by those who understand how to access and use their terminal. Recently, these AI tools have become a lot more usable with easy to understand interfaces. In the domain of music, examples of tools that generate MIDI for use by musicians include Google's Magenta Suite (Roberts et al. 2019) and the Piano Inpainting Application for Ableton (Hadjeres 2021). While an easy to use interface has the potential for replacing the artistic endeavor of creating music by non-musicians; both the Inpainting Application

[2] Indeed, one way that creative AI has been attempted to be used to alleviate legal concerns around musical copyright and lawsuits was Riehl and Rubin's efforts to generate 'every possible melody' and release them to the public domain (Cole, 2020; Whitwam, 2020). This is an interesting use of creative AI, although whether it will stand up in court is up for debate.

and the Magenta Suite are usable as plug-ins for Ableton, which shows a focus on incorporating the AI into the workflow of the artist (in this case musicians). This focus on incorporation of tools into the digital audio workstation (DAW) shows the intention of the designers for the tool to sit within the workflow of the composer and perform an assisting role. This is important as ethically the concern would be that creatives could be replaced through clever use of creative AI. However, researchers are designing their AI to work with practitioners in order to address these ethical considerations around their use (Roberts et al. 2019; Hadjeres 2021).

In other artistic fields, GANs are creating surprisingly good outputs, such as the text to image GAN by Epstein (2021) that can create new images "in the style of" other artists. OpenAI's GPT-3 algorithm has written newspaper articles (GPT-3 2020), conference talk titles (Wareing 2021) and creative fiction (Branwen 2020). It is quite plausible that creative AI could be used to replace authors in some fields, for example, writing copy or other repetitive and less prestigious writing tasks. Art "in the style of" could be good enough for certain applications and not subject to copyright or royalty claims. This use of the "essence" or "identity" of other artists is further discussed in Sect. 3.2, but it is important to point out here that there is a real possibility of the replacement of some subset of authors and artists by creative AI applications.

3 Inputs

What goes into making creative AI work? In this section, we consider different types of inputs that might be part of the process to create creative AI works, and the potential ethical issues with the collection and use of these data points. We explore some possibilities for ensuring responsible usage of these inputs and some scoping strategies to avoid or mitigate possible misuse or other issues.

3.1 Data and Bias

As with any AI application, creative AI needs data to use as inputs. Creative works are necessarily products of the society they are in, so data based on these will reflect societal and institutional biases. Such biases exist at several stages: in the raw datasets that might contain previously existing creative works, such as music or art; but also at the design stage that might reflect institutional biases based on the questions, the creative AI program is asked. For example, at the design stage, a creative AI algorithm might be designed to create potential product designs of a chair based on existing chair designs. It may be tasked with designing one that will be extremely popular in order to sell a lot of chairs. But these AI-generated chair designs will not take into consideration edge cases of chair requirements, for example, chairs for larger people, chairs for people with disabilities, environmental sustainability of the chair, or even chairs that might just need to be for taller or shorter people. Because the parameters have been based around popularity, and "most people" are likely to buy an "average" chair, the program will naturally follow that instruction and only create chairs for "average" people.

In a world that is also shifting toward AI-driven efficiency, it is also increasingly likely that large companies could use such programs to cheaply design potentially popular offerings without any need for an actual product designer to determine whether the design is actually any good (Martinez 2019). Another creative AI program tasked with producing Renaissance-style artwork based on existing artwork might create scenes that only show white people in them, because the large majority of Renaissance artwork with people in them depict only white people. Musical compositions that are used to train generative models are more often than not western classical pieces, which creates bias in the output style of generated art. Additionally, the existing code that is used to analyze MIDI files in order to build compositional analysis models is built to analyze sheet music, which is expected to fit into an even twelve-tone temperament. This means that music that does not fit into the normal or standardized sheet music needs to be parsed using a bespoke algorithm that can account for non-western tones. This is possibly the reason behind the lack of representation of eastern microtonal music or aleatoric music that makes use of timbres not definable in such analytical models.

While creative AI applications are unlikely to be making life-or-death decisions based on the data it uses, aspects of representation, edge cases, and other potential impacts of poorly handled creative AI applications could have a real negative impact on society. In order to prevent possible problems, there is a large movement within the greater AI field to look at openness, explainability, and transparency of AI systems (Larsson and Heintz 2020). Being able to see how an algorithm processes and weighs data can help to identify biases in datasets but does not help with solving the dataset bias itself. Debiasing datasets is very much a field in its infancy, and there is a definite place for creative AI to potentially help with this problem. Debiasing attempts can include the removal of the biased data, for example, gender stereotypes from text (Bolukbasi et al. 2016; Greenwald 2017) and speech emotion recognition (Gorrostieta et al. 2019). But it will not change the fact that Renaissance art is largely of white people, or that chairs are largely designed for people of a certain shape, size, or ability. And simply adding more diverse data (whether contrived or real) may not be enough to redress the balance. An attempt at debiasing images of skin lesions, for example, showed that it was an extremely complex task to attempt to debias the datasets despite there being features within that dataset that could help with the process (Bissoto et al. 2020).

Until there are more useful methods of debiasing datasets (if any such methods exist), it is unlikely that this problem will be solved simply through changes to the dataset. Indeed, specifically biased datasets might actually be desired by the creator—in both of these cases, creative AI practitioners have a responsibility to understand and be aware of these potential biases and address them openly as part of the process, with the possibility to end the process if such bias is problematic—particularly in live performances, shows, or engagements. If the creative AI produces outputs that are then further acted upon by the practitioner, such as designs, it is then the practitioner's responsibility to ensure artistic and/or societal acceptability by engaging with those who might be disadvantaged, under-represented, or in other ways negatively affected by the process, to ensure that any outputs are sensitive to the underlying issues or are critically framed to reflect the output's origin.

3.2 Artistic Identity and Essence

Another type of input to consider when working with creative AI is that of artistic identity and essence, both of which come with their own ethical and legal considerations. When we speak about artistic identity, we mean the style of artistic works that arises from the individual who is creating the art. The artist/musician has developed a style of producing their art or music over the course of a career, which encapsulates their identity through the output of their chosen medium. By essence, we mean the way in which they perform their art. An example of this in musical terms is that if virtuosic violinist Niccolò Paganini was to perform a violin rendition of Johann Sebastian Bach's Suite No. 1: Prelude, the artistic identity belongs to Bach, who composed the piece of music. However, the performance of western classical music is often left to the interpretation of the performer (in this case Paganini), who has to interpret information that is represented on the sheet music with some vagueness. This interpretation and the performance it leads to are the essence of the creative practice; another performer might interpret Bach's instructions in a completely different way that encapsulates their own essence. Yet Paganini is also recognizable as having a particular style of interpretation when it comes to the performance of music—this is his identity. Another example of this in terms of artistic practice is that the essence of Peter Paul Rubens demonstrates the "dynamism, vitality, and sensuous exuberance" of the Baroque painting style (Scribner 2000). However, when compared to other Baroque-era painters such as Caravaggio and Rembrandt, the artistic identity of each is discernible to the trained eye, although in essence they are all in the Baroque style. Much of this is likely due to the way in which they created their art which captures their own essence. Indeed, "Rembrandt" was "brought back to create one more painting" in the "Next Rembrandt" project (J Walter Thompson Amsterdam 2017), in which "the computer learned how to create a Rembrandt face based on [...] "typicalities" [of a Rembrandt portrait]" (Dutch Digital Design 2018).

Relating this back to creative AI, it is important to consider the role of both essence and identity as they apply to training or using artificial intelligence in the pursuit of artistic or creative endeavors. An example of this within the realm of music is that expressive rendering. Expressive rendering is the study of improving the mechanical performance of music (specifically MIDI) by training algorithms to apply temporal, dynamic, or performative elements to the musical output (Widmer 2002; Widmer and Goebl 2004; Flossmann et al. 2009; Grachten and Widmer 2011; Grachten and Krebs 2014). This practice has arisen as "a mechanical performance of a score is perceived as lacking musical meaning and is considered dull and inexpressive" (Canazza et al.

2015). This approach has been one of the first steps toward the development of an essence of performative measures within creative AI. The data used within expressive rendering come from datasets such as the aligned scores and performances (ASAP) dataset, which is a collection of over two hundred distinct musical scores and over one thousand performances of classical piano pieces from fifteen western classical composers (Foscarin et al. 2020). These performances were captured from various performers and amalgamated elements of all of their performances. However, if you were to create a dataset from a single musician and use said data as the inputs for modeling your expressive rendering, then you could be capturing, and ultimately imitating, their creative essence with the output. This raises the question of what reparations such a performance would be worth, considering that the output of the creative AI can be used and applied to multiple projects, whereas a musician would be paid well for each recording of a performance was it to be traditionally recorded.

In recent years, artificial intelligence has also been leveraged in order to perform more generative tasks. Recent examples of generative models used for art can be found in music (Collins and Laney 2017; Huang et al. 2018; Collins 2020) or the recent phenomena of blockchain or CryptoArt (Finucane 2019; Franceschet et al. 2019). These generative models capture more of the artistic identity present in musical composition or in artworks in a similar manner to the expressive rendering example above and raise similar questions of ethical use of such technology (which could be used to create "deepfakes" of artistic works or performances), what rights artists have to safeguard their artistic identity and essence, and how artists might be compensated if these rights are violated.

An example of how artistic identity and essence can be used controversially in creative AI is seen in the 'Lost Tapes of the 27 Club (2021)' (Brodsky 2021). This is an initiative, whereby creative AI has been used to create new songs for a variety of artists who all died at the age of 27. This was done in order to raise awareness regarding mental health issues within the music industry by over the bridge, a non-profit organization that is focused on tackling mental illness within the music industry. These songs were all modeled on existing music by these artists, and deepfakes were created for the original vocal performance. Although for a good cause, these tracks have received a slightly controversial reception online, but too much acclaim from fans (Brodsky 2021; Grow and Grow 2021). This shows how creative AI can be leveraged in order to replace artists and create music, even for those that are deceased. The ethical considerations of deepfakes like these are discussed in depth later in Sect. 5.1.

Finally, the shift of essence or identity to compensate for AI is likely to happen if creative AI becomes part of the process of art creation. For example, Alex Kiessling talks about how he changed his artistic style to compensate for the range of movements, the robot arms are able to make in *long distance art* (The Method Case 2013). While this is not a new thing—humans have compensated for the limitations of their tools since the dawn of art—AI-based tools can be less predictable in their requirements and could have a significant impact on the essence/identity of the artist.

Those developing or using AI that encapsulates the identity or essence of a performer or artist should take into consideration the impact on the artist/performer and ensure that their AI-generated work is not misrepresented as being that of the artist or performer or replace that artist or performer unless with their agreement.

3.3 Creative AI's Ship of Theseus

Similarly to the issues presented above, there are deep philosophical considerations around replacing parts of a creative exercise with artificial intelligence. To slightly extend the analogy of the Ship of Theseus (Wikipedia 2021), if the human made "planks" used to create the artistic work are exchanged with artificially made "planks", such as methods for piecing together existing works or generating designs, at what point is the art more machine than human created? If a composer, for example, was to use a tool to generate ideas that use their own music as a stimuli, then is it still part of their original work? If they were to create a new piece, using the last generated piece as a stimuli, how long before the music looks nothing like the original piece, yet might still be considered the same piece? When does it stop being that artist's work? What if they were a still life artist, and never painted a landscape, and the AI painted a landscape in their style? Or, if highly successful in replicating an artist's style, does a piece of work inherit some of the value of the original artist's work? And does it matter?

Certainly, as discussed in Sect. 4, if the outputs are honest as to their inception, this should not matter. But there are concerns around the replacement of human creativity with machine creativity to, for example, fill gaps in larger projects that humans might otherwise fill. Video games, for example, might use creative AI to generate music or design levels, which could remove job opportunities for composers or level designers. Would it be the same video game if the artificial components replace what might have otherwise been human made components? In games like *Candy Crush Saga*, AI is being used to test levels, removing some of the human input that might otherwise go into the game. King argues that this actually makes the game better, as it allows for faster testing turnaround (King 2019), but there is an argument to be made that it could potentially limit the game as the testing AI is unlikely to think "outside the box" like humans will. Although this is not traditional creative AI work, other companies are using procedural generation in other aspects of games, such as the world building in *Minecraft* and *Valheim*; with sophisticated-enough generators, it could be that this kind of integration of creative AI follows King's testing regime and removes human input altogether because it is faster or easier to make (and perhaps also less expensive).

While this replacement of human input may not necessarily be a problem, in the bigger picture of creative AI integration into art, it is important to recognize that this could potentially mean that the future of art inherently has creative AI embedded within it. While there might be a push for more "traditional" art (much like digital photography has not fully replaced film photography (Keats 2020)), it is likely that there is no going back to a time before creative AI was introduced to the art world.

4 Outputs

This section highlights some of the potential ethical issues with the outputs of creative AI. Recent forays into deployment of AI entities have shown that, without human intervention, AI can quickly produce outputs that can potentially be dangerous, be used to defame or cause social anxiety, or have issues with expectation management as to their capabilities (Wolf et al. 2017; Toews 2020). We capture these concerns in terms of dangerous creations, deepfakes and similar problematic uses, and issues of safety in physical performance with creative AI and humans working side-by-side.

4.1 Dangerous Creations

It is well known that certain types of artistic outputs can have adverse effects, for example, certain types of light flashing can trigger seizures, loud noises can cause hearing loss, and certain kinds of movement in video games can cause motion sickness (Stoffregen et al. 2008). It is plausible that within the AI's creative process, there could be these kinds of outputs that could cause harm. Monitoring the output and potentially putting in checks for the elements that could trigger physical harms such as these are therefore very important. We cover physical safety of artists co-working with AI-driven tools (such as robots) in Sect. 4.3.

Other kinds of harm could be socio-cultural, for example, racist, sexist, or other offensive or hate speech. These kinds of outputs have already been seen in existing AI (natural language processing) projects such as the Microsoft TayBot (Wolf et al. 2017), and in the bias shown in language outputs (Dinan et al. 2020), for example. Much of this bias is due to the inputs described in Sect. 3.1 and can be mitigated through thoughtful data collection and usage. Monitoring these kinds of usages of natural language processing is a core recommendation of the more generally applied ACM Code of Ethics, which, in Principle 2.5 states: "A system for which future risks cannot be reliably predicted requires frequent reassessment of risk as the system evolves in use, or it should not be deployed" (Gotterbarn et al. 2018). For creative AI, we also recommend such vigilance.

4.2 Deepfakes

Deepfakes are the result of using AI process (specifically deep learning techniques) to simulate a person in audio, video, or still imagery. Some techniques require the use of prior work, e.g., audio or video, to create a "persona" that then renders a realistic approximation of the original person in that medium. Other techniques include "face swapping" that takes images of the person and places it over another actor (Meskys et al. 2019). Some famous examples include the Nicolas Cage deepfakes where fans of the actor inserted him into a number of classic films (Neilan 2018), or more recently (and commercially), the use of AI to generate Anthony Bourdain's voice in a documentary about him (Tangcay 2021). While recent *Star Wars* shows and films have also recreated actors who have passed away, they have been created without AI techniques, instead using more classic CGI techniques alongside motion capture, and were considered "eerie" (Dockterman 2016), veering sharply into the "uncanny valley" (Hsu 2012). Fans have since made these "performances" more realistic using the deep learning methods that are used in deepfakes, such as in *Rogue One: A Star Wars Story* (Suciu 2020) and *The Mandalorian* (Kain 2020).

While these might seem trivial or even useful, deepfakes have a more problematic history in that they were originally primarily used in pornographic videos to replace the actors with celebrities, or to engage in "revenge porn"—the sharing of sex videos in order to "humiliate, threaten, or mark other harm to a person who has broken off the relationship" (Meskys et al. 2019). Legislation has been updated in many jurisdictions to include deepfakes within revenge porn laws (ibid.). Deepfakes could also be used in other ways to harm, for example, deepfakes of politicians could be used to influence elections (Diakopoulos and Johnson 2020), or fake news made to look credible by impersonating trustworthy sources (Ajder 2019).

Outside of these kinds of uses, deepfakes can be used in more meaningful and helpful ways as well. For example, the deep empathy project (MIT Media Lab 2017) aims to induce empathy for disaster victims by using deep learning techniques to simulate disasters in cities around the world, essentially a deepfake of the cityscape. Deep learning techniques can also recreate how a motor neuron disease (MND) patient speaks from voice banks, which MND patients add clips of their voice to use to communicate when they can no longer speak (Bonifacic 2019). It will attempt to recreate the tone, accents, and colloquialisms that the person would have used when they spoke in almost real time, allowing for a greater quality of life for MND patients. These "greyfakes" also encompass the possibility to realistically represent yourself in virtual reality environments, or to create more realistic AI voice assistants (Ajder 2019). They could also be used to create digital likenesses of you for your family and friends to engage with after you pass away. Microsoft, for example, patented a chatbot that did this, but has no plans to actually make it (as yet) due to concerns about social acceptability (Harbinja et al. 2021), though it has been argued that these chatbots "can offer an important source of support to mourners" (Elder 2020).

Throughout all of these uses of deep learning technologies runs the issue of informed consent. Deepfake technologies can be used in positive ways when accompanied by informed consent by the person impersonated and the viewer, in that they know they are watching an impersonation. When this does not happen, situations like the backlash against Anthony Bourdain's AI created voiceover arise, whereby viewers were horrified by the convincing impersonation that the documentary-makers created. This is especially likely with famous people that viewers feel they "know" (through the effect of parasocial relationship) (Rosner 2021), if the person is not able to sign off on the process (in this case, because Bourdain had passed away). Another issue is disclosure of a deepfake: even though Bourdain wrote the words, the AI-generated voice read, the lack of indication to the audience that the voiceover was faked is a breach of trust between viewer and documentary-maker. "Creative signaling" is a way that this could be mitigated—some documentaries do this with reconstructions or other contextual signals that the audio/visuals might be indicative rather than real (ibid.). And while people may well be more comfortable with this use of technology in future, right now it warrants sensitive and context-aware use in creative AI.

4.3 Physical Safety

In Alex Kiessling's "long distance art", a set of industrial robot arms in other countries recreated an artwork in real time as Kiessling drew the original (The Method Case 2013; Kiessling 2021). Kiessling regularly uses industrial robot arms within his artwork, which have a potential to be quite dangerous if programmed incorrectly. Industrial robot arms have a number of recognized hazards, including impact or collision accidents, crushing and trapping accidents, mechanical part accidents, and other accidents that can result from working in the vicinity of a robotic arm such as if hydraulic lines rupture (United States Department of Labor 2020). According to the OSHA Technical Manual for Industrial Robots and Robot System Safety (ibid.), "*the greatest problem, however, is over familiarity with the robot's redundant motions* so that an individual places himself in a hazardous position" (emphasis sic). An artist working with a robot they have programmed to perform certain tasks, whether AI powered or not, would therefore need to be particularly vigilant, especially if they are working within range of the robot's physical movement.

Creative AI in music has a similar potential for the use of robotics and AI to control them in the process of making music. For example, Vear's "Embodied Musicking Robots" are a performative use of creative AI that specifically aims to engage humans and robots together in the music-making experience (Vear 2020). Physical safety is one of the principles in Vear's work, though it is interesting that the emphasis is on the robot's safety (through "self-preservation") rather than that of any human performer that might share the robot's stage or audience member in the vicinity. It is likely that this is because the technology is in its infancy; further development should take into account the possibility of trip hazards and over-enthusiastic manipulation of instruments by the robots involved that might potentially cause harm to humans involved as well.

5 The Future: Virtuous Creative AI

In the previous sections, we have looked at specific ethical issues in the area of creative AI and suggested mitigations or solutions for these. In this section, we look more at the possibilities for ethical creative AI. We frame this within a "virtuous creative AI" perspective in the manner of Vallor (2016, p. 28). Virtue ethics is, as Vallor notes, "a uniquely attractive candidate for framing many of the broader normative implications of emerging technologies in a way that can motivate constructive proposals for improving technosocial systems and human participation in them" (ibid., p. 33). To do this, we will examine some recommendations for practitioners and developers that work with creative AI from the perspective of Vallor's technomoral virtues (ibid., p. 120). We have focused on six of these, grouped together, which are the most applicable to the application of creative AI in a technosocial environment.

5.1 Honesty and Humility

Vallor describes the technomoral virtue of honesty as "an exemplary respect for truth, along with the practical expertise to express that respect appropriately in technosocial contexts" (2016, p. 122). In creative AI, this is best applied to the "creative signaling" that might be required to ensure that audiences know that what they are interacting with is AI-generated. Of course, this is highly contextual, and we are not suggesting that creative AI practitioners need to fully disclose all methods all of the time, but more that the question needs to be asked about audience reaction to the use of AI methods within the setting and whether and what kind of disclosure is appropriate. Similarly, the use of someone's data to create a likeness of them needs to be properly disclosed and consented to by the appropriate people, for example, in Anthony Bourdain's deepfakes case discussed above, his family and other important people such as his manager, since Bourdain is not able to consent himself.

This virtue also ties into the general use of creative AI: that of an augmentation of creative processes. Creative AI practitioners need to be honest with themselves about how these techniques can complement or augment their own abilities. Whether it is through using GPT-3 to encourage creativity, e.g., as a muse for creating new characters (Smith 2021) or to help to physically create the work, such as Vear's musical robots (2020) or Kiessling's artistic robotic arm (The Method Case, 2013), the focus should remain on the artist producing the work. Honesty with the audience as to the creation process is essential, however, even when that process acquires or uses another artist's essence (as discussed in Sect. 3.2), or is used as a refinement technique for original human-created art, such as in filters for visual arts that use AI.

Honesty in creative AI also links in with the virtue of humility, defined by Vallor as "a recognition of the real limits of our technosocial knowledge and ability; reverence and wonder at the universe's retained power to surprise and confound us; and renunciation of the blind faith that new technologies inevitably lead to human mastery and control of our environment" (Vallor 2016, p. 127). This should not be a surprising virtue to the creative AI practitioner, who likely encounters it, and perhaps even embraces it, within their practice. However, humility also requires ensuring that the human context is profoundly centered within the use of creative AI techniques - the impacts of and on humans in, for example, the inputs used or the outputs created. It is an obligation for creative AI practitioners to be critical of the use of AI techniques in their work, rather than follow blind techno-optimism or -pessimism, to be hopeful about the possibilities of creative AI but understanding that we do not always know what the outputs might be, and that they might, in fact, cause harm. Examples here include potential impacts on humans as a result of bias, or physical harms from unexpected movements of robots or dangerous outputs of audio or visual AI creativity as discussed in Sects. 4.1 and 4.3. Practitioners need to be as humble as to continue to value the human touch in the creative arts and the value of "traditional" creativity (see Sect. 3.3), even if use of creative AI becomes the norm. Hopefulness in humility's sense may be in terms of looking toward future use of creative AI, such as for non-specialists to become creative, or to use it in order to make creation of art more accessible to those who would not normally be able to create (demonstrated by Louie et al. (2020)).

5.2 Empathy and Care

The technomoral virtue of empathy is defined as "a cultivated openness to being morally moved to caring action by the emotions of other members of our technosocial world" (Vallor 2016, p. 133). Vallor makes a specific distinction between empathy, "a form of co-feeling, or feeling *with* another", and sympathy, "a form of benevolent concern *for* another's suffering" (ibid.). Given that the outputs of creativity are frequently expressions aimed at affecting the audience's emotions, helping the audience to understand another's perspective, or similar, creative AI has a lot of potential to positively affect the audience and increase their empathetic concern for a subject. Projects such as deep empathy (discussed in Sect. 4.2) can show the power this might have. It is, however, important that, when affective creative AI is implemented, that this is done carefully in order to respect the audience as well. Psychological manipulation and deception can undermine the desired outputs, with honesty, as mentioned before, an important value to hold here as well. Hence, the pairing of empathy with the technomoral virtue of care here defined as "a skillful, attentive, responsible, and emotionally responsive disposition to personally meet the needs of those with whom we share our technosocial environment" (Vallor 2016, p. 138). Creative AI can help to promote human flourishing by incorporating the virtue of care. While regulation, such as the upcoming EU regulation of AI, is likely to limit some of the more risky uses and abuses of machine learning, such as the use of subliminal techniques (MacCarthy and Propp 2021), there is always the potential for creative uses of AI technologies that (intentionally or unintentionally) go beyond these regulations and pose a potential for harm.

The responsible and attentive practitioner should be able to shut down any harmful AI applications and allow for reflective interrogation of the approach used in order for auditing so that others might learn from it. Techniques such as explainable AI or other audit processes for machine-made decisions are key here, with an example being the issue of bias, which was not foreseen in the implementation of AI techniques at the beginning (and the harm of which is discussed in Sect. 3.1). The responsibility for any creative AI's output must always rest with a human individual such that in deploying it, there is someone that is able to halt the application if needed. The Taybot discussed in Sect. 4.1 is a good example of where this should have happened much quicker than it did. Additionally, shifting moral responsibility onto the machine is not fulfilling the requirements of a duty of care for "those with whom we share our technosocial environment". Similarly, the erasure of jobs for artists and technicians through creative AI uses (such as in an unconsenting impersonation or use of an artist's artistic essence, discussed in Sect. 3.2, and particularly in the Lost Tapes example) should be avoided where possible; the primary goal for ethically acceptable creative AI should be to augment or enhance existing art rather than replacing artists or those who support them.

5.3 Civility and Flexibility

Civility in a technomoral sense is not the simple call to politeness according to the use of it in common parlance, it is, instead, "a sincere disposition to live well with one's fellow citizens of a globally networked information society: to collectively and wisely deliberate about matters of local, national, and global policy and political action; to communicate, entertain, and defend our distinct conceptions of the good life; and to work cooperatively toward those goods of technosocial life that we seek and expect to share with others" (Vallor 2016, p. 141). The important part here is to see the possibilities for creative AI to fulfill this and positively impact society in many different ways. Art has often comprised significant parts of movements for change including political action and raising awareness of inequalities and injustices in the world. Art has long been an inspiration for improving society, envisioning utopias, improving wellbeing, and portraying potential impact that current and possible trajectories could have in future. Creative AI thus has potential to affect significant positive change through continuing this tradition. This does not mean that there will not be problematic uses along the way, but if creative AI is used responsibly and with the "sincere disposition [...] to work cooperatively toward those goods of technosocial life", it will have a net positive impact on society.

Civility here also includes many of the mitigations and avoidances we have suggested throughout this paper, but we include civility as a specific virtue here because of the incredible possibilities that creative AI can have as a positive impact on society. Current examples that would fit into this include the deep empathy project, and projects to assist with giving voice to MND patients, promote policy and political action, and allow for more people to live a good technosocial life. Respecting artists' identity, essence, and intellectual property are examples of working cooperatively. Even the creative AI projects that bring long-dead artists "back to life", if done sensitively, can encourage wellbeing through appreciation of the arts.

The final technomoral virtue we want to apply in this chapter is flexibility. It links in with civility well, because it is defined by Vallor as "a reliable and skillful disposition to modulate action, belief, and feeling as called for by novel, unpredictable, frustrating, or unstable technosocial conditions" (2016, p. 145), and thus moderates the zeal for which creative AI practitioners might want to practice the virtue of civility. Not only is it a moderating force in this way, but it reminds practitioners that the implementations of technologies can be unpredictable and unstable, thus ensuring that they monitor outputs and ensure that these are appropriate. Interestingly, it is here we also see the discussion of forbearance of norms that might be outside of our technosocial experiences, and whether we need to be flexible in terms of our respect for difference in cultural norms.

Vallor addresses this difference by introducing the concept of "capacity for global technomoral agency" to decide which norms should "warrant mutual forbearance", for example, a definition of "feminine virtue" that does not allow for female education or equal participation in technosocial life would not be compatible with global flourishing (pp. 147–148). Instead, active deliberation of the ways that creative AI can contribute to human flourishing is needed to determine the contributions and methods of contribution it can bring. This translates in practice to remembering that the people who encounter, interact with, are affected by, and use the creative AI tools and outputs and make up the inputs may have differing social or cultural expectations of these and that, for the most part, these warrant mutual forbearance. Examples from above include the nature of copyright and intellectual property, informed consent, user expectations of disclosure of the nature of the use of creative AI tools, creation, and use of datasets, augmentation of artistic ability vs replacement of artists, etc. Creative AI practitioners should be easily able to cope with handling these uncertainties—after all, the nature of art is often unpredictable and can sometimes cause discomfort. The key to determining whether this discomfort is acceptable or not is to look at the bigger picture of global technomoral agency and ultimately, the ability to contribute to human flourishing.

While we have covered many of Vallor's technomoral virtues, we recognize that the others have value within creative AI practice as well and recommend to the reader to use this application of her approach as a way into further reading and understanding of how virtue ethics can help to frame the future of creative AI.

6 Conclusion

The ACM Code of Ethics (which applies to the computing profession generally, therefore, encompassing creative AI tools and applications) promotes a goal of ensuring that the public good is the primary consideration when evaluating ethical decisions (Gotterbarn et al. 2018). It is in this tradition that we have approached this chapter—focusing firstly on specific ethical issues, and then on presenting a future-looking virtue ethics framework to understand how creative AI can positively contribute to human flourishing within the technosocial environment. It is important to not view this chapter simply as a list of ethical issues and how to solve them—but as a starting point for discussion about what kind of society creative AI techniques will be creating, and more importantly, what kind of society creative AI practitioners *want* to create through their artistic practice and use of AI tools. The virtues discussed in Sect. 5

allow for a theoretical framework to consider future inputs for and applications of creative AI that we might not have foreseen; keeping in mind the primary consideration of the public good, or human flourishing, within a complex technosocial world.

References

Ajder H (2019) The ethics of deepfakes aren't always black and white. TNW, Podium. https://thenextweb.com/news/the-ethics-of-deepfakes-arent-always-black-and-white. Accessed 23 Apr 2021
Ben-Tal O, Harris MT, Sturm BL (2020) How music AI is useful: engagements with composers, performers, and audiences. Leonardo, pp 1–13. https://doi.org/10.1162/leon_a_01959.
Bissoto A, Valle E, Avila S (2020) debiasing skin lesion datasets and models? Not so fast. In: 2020 IEEE/CVF conference on computer vision and pattern recognition workshops (CVPRW). 2020 IEEE/CVF conference on computer vision and pattern recognition workshops (CVPRW). IEEE, Seattle, WA, USA, pp 3192–3201. https://doi.org/10.1109/CVPRW50498.2020.00378
Bolukbasi T et al (2016) Man is to computer programmer as woman is to homemaker? Debiasing word embeddings. Available at http://arxiv.org/abs/1607.06520 [cs, stat]. Accessed 10 Mar 2021
Bonifacic I (2019) Quips is an AI to help ALS patients speak with their own voice. Engadget. Available at https://www.engadget.com/2019-12-18-rolls-royce-quips-als-mnd-speech-ai.html. Accessed 16 July 2021
Branwen G (2020) GPT-3 creative fiction. Available at https://www.gwern.net/GPT-3. Accessed 16 Apr 2021
Brodsky (2021) AI software creates 'new music' from Nirvana and Amy Winehouse, the independent. https://www.independent.co.uk/arts-entertainment/music/news/ai-songs-27-club-music-b1827003.html. Accessed 21 July 2021
Canazza S, De Poli G, Rodà A (2015) CaRo 2.0: an interactive system for expressive music rendering. Adv Hum-Comput Interact 2015:1–13. https://doi.org/10.1155/2015/850474
Carroll R (2016) Led Zeppelin cleared of stealing riff for stairway to heaven. The Guardian. Available at http://www.theguardian.com/music/2016/jun/23/led-zeppelin-cleared-stairway-to-heaven-lawsuit-spirit. Accessed 16 Apr 2021
Cole S (2020) Musicians algorithmically generate every possible melody, release them to public domain. VICE. Available at https://www.vice.com/en/article/wxepzw/musicians-algorithmically-generate-every-possible-melody-release-them-to-public-domain.
Collins N (2020) Composition in the age of AI. Ideas Sonicas 12(23)
Collins T, Laney R (2017) Computer-generated stylistic compositions with long-term repetitive and phrasal structure. J Creative Music Syst 1(2). https://doi.org/10.5920/JCMS.2017.02
Diakopoulos N, Johnson D (2020) Anticipating and addressing the ethical implications of deepfakes in the context of elections. New Media Soc 1461444820925811. https://doi.org/10.1177/1461444820925811
Dinan E et al (2020) Queens are powerful too: mitigating gender bias in dialogue generation. Available at http://arxiv.org/abs/1911.03842 [cs]. Accessed 16 July 2021
Dockterman E (2016) Let's talk about grand Moff Tarkin in 'Rogue One'. Time. Available at https://time.com/4604996/rogue-one-grand-moff-tarkin-princess-leia-cgi/. Accessed 16 July 2021
Dutch Digital Design (2018) The next Rembrandt: bringing the old master back to life. Medium. Available at https://medium.com/@DutchDigital/the-next-rembrandt-bringing-the-old-master-back-to-life-35dfb1653597. Accessed 16 July 2021
Elder A (2020) Conversation from beyond the grave? A neo-confucian ethics of chatbots of the dead. J Appl Philos 37(1):73–88. https://doi.org/10.1111/japp.12369

Epstein V (2021) Text2Image Siren+.ipynb. Available at https://colab.research.google.com/drive/1L14q4To5rMK8q2E6whOibQBnPnVbRJ_7#scrollTo=JUvpdy8BWGuM. Accessed 16 Apr 2021
Finucane BP (2019) Creating with blockchain technology : the "provably rare" possibilities of crypto art. https://doi.org/10.14288/1.0370991
Flossmann S Grachten M, Widmer G (2009) Expressive performance rendering: introducing performance context. https://doi.org/10.5281/ZENODO.849619
Foscarin F et al (2020) ASAP: a dataset of aligned scores and performances for piano transcription. In: 21st international society for music information retrieval conference, pp 534–541
Franceschet M et al (2019) Crypto art: a decentralized view. Available at http://arxiv.org/abs/1906.03263 [cs]. Accessed 23 March 2021
Frid E, Gomes C, Jin Z (2020) Music creation by example. In: Proceedings of the 2020 chi conference on human factors in computing systems. CHI'20: CHI conference on human factors in computing systems, Honolulu HI USA. ACM, pp 1–13. https://doi.org/10.1145/3313831.3376514
Gorrostieta C et al (2019) Gender de-biasing in speech emotion recognition. In: Interspeech 2019. Interspeech 2019, ISCA, pp 2823–2827. https://doi.org/10.21437/Interspeech.2019-1708
Gotterbarn DW et al (2018) ACM code of ethics and professional conduct. Association for Computing Machinery. Available at https://www.acm.org/code-of-ethics. Accessed 19 Oct 2018
GPT-3 (2020) A robot wrote this entire article. Are you scared yet, human? GPT-3, the Guardian. Available at http://www.theguardian.com/commentisfree/2020/sep/08/robot-wrote-this-article-gpt-3. Accessed 16 Apr 2021
Grachten M, Krebs F (2014) An assessment of learned score features for modeling expressive dynamics in music. IEEE Trans Multimedia 16(5):1211–1218
Grachten M, Widmer G (2011) Explaining musical expression as a mixture of basis functions. In: Proceedings of the 8th sound and music computing conference (SMC 2011)
Greenwald AG (2017) An AI stereotype catcher. Science 356(6334):133–134. https://doi.org/10.1126/science.aan0649
Grow K, Grow K (2021) In computer: hear how AI software wrote a "New" Nirvana Song. Rolling Stone, 2 April. Available at https://www.rollingstone.com/music/music-features/nirvana-kurt-cobain-ai-song-1146444/. Accessed 21 July 2021
Hadjeres G (2021) The piano inpainting application: an A.I.—Enhanced piano performance generation Ableton plugin. Github. Available at https://ghadjeres.github.io/piano-inpainting-application/
Hanson D et al (2021) SophiaPop!: Experiments in human-AI collaboration on popular music, p 7
Harbinja E, Edwards L, McVey M (2021) Chatbots that resurrect the dead: legal experts weigh in on 'disturbing' technology. The Conversation. Available at http://theconversation.com/chatbots-that-resurrect-the-dead-legal-experts-weigh-in-on-disturbing-technology-155436. Accessed 16 July 2021
Hsu J (2012) Why 'Uncanny Valley' human look-alikes put us on edge. Scientific American. Available at https://www.scientificamerican.com/article/why-uncanny-valley-human-look-alikes-put-us-on-edge/. Accessed 16 July 2021
Huang C-ZA et al (2018) 'Music Transformer'. Available at: http://arxiv.org/abs/1809.04281 [cs, eess, stat]. Accessed 23 Mar 2021
J. Walter Thompson Amsterdam (2017) The Next Rembrandt. The Next Rembrandt. Available at https://www.nextrembrandt.com. Accessed 16 July 2021
Kain E (2020) 'Star Wars' deepfake 'Fixes' that crazy 'Mandalorian' season 2 finale cameo. Forbes. Available at https://www.forbes.com/sites/erikkain/2020/12/23/star-wars-fan-fixes-that-crazy-mandalorian-cameo-with-a-stellar-deepfake/. Accessed 16 July 2021
Keats J (2020) Film photography can never be replaced. Wired, 21 April. Available at https://www.wired.com/story/film-photography-can-never-be-replaced/. Accessed 16 Apr 2021
Kiessling A (2021) Alex Kiessling: gallery. Alex Kiessling. Available at: https://www.alexkiessling.com/gallery/. Accessed 15 Apr 2021

King T (2019) Human-like playtesting with deep learning. Medium. Available at https://medium.com/techking/human-like-playtesting-with-deep-learning-92adafffe921. Accessed 15 Apr 2021

Knotts S, Collins N (2020) A survey on the uptake of music AI software. In: Proceedings of the international conference on new interfaces of musical expression. new interfaces of musical expression. NIME, Birmingham, UK, pp 499–504

Larsson S, Heintz F (2020) Transparency in artificial intelligence. Internet Policy Rev 9(2). https://doi.org/10.14763/2020.2.1469

Liebman E, Stone P (2020) Artificial musical intelligence: a survey. Available at http://arxiv.org/abs/2006.10553 [cs, eess]. Accessed 7 Apr 2021

Lost Tapes of the 27 Club (2021). Available at https://losttapesofthe27club.com/. Accessed 21 July 2021

Louie R et al (2020) Novice-AI music co-creation via AI-steering tools for deep generative models. In: Proceedings of the 2020 CHI conference on human factors in computing systems. CHI '20: CHI conference on human factors in computing systems. ACM, Honolulu, HI, USA, pp 1–13. https://doi.org/10.1145/3313831.3376739

MacCarthy M, Propp K (2021) Machines learn that Brussels writes the rules: the EU's new AI regulation. Lawfare. Available at https://www.lawfareblog.com/machines-learn-brussels-writes-rules-eus-new-ai-regulation. Accessed 21 July 2021

Martinez FJ (2019) AI and creativity—Is it ethical; will it kill creativity?, Ethics in graphic design, 3 June. Available at http://www.ethicsingraphicdesign.org/ai-and-creativity-is-it-ethical-will-it-kill-creativity/. Accessed 10 March 2021

Meskys E et al (2019) Regulating deep fakes: legal and ethical considerations. SSRN Scholarly Paper ID 3497144. Social Science Research Network, Rochester, NY. Available at https://papers.ssrn.com/abstract=3497144. Accessed 23 Apr 2021

MIT Media Lab (2017) Deep empathy. Available at http://deepempathy.mit.edu. Accessed 16 July 2021

Ncube CB, Oriakhogba DO (2018) Monkey selfie and authorship in copyright law: the Nigerian and South African perspectives. Potchefstroom Electron Law J 21:1–35. https://doi.org/10.17159/1727-3781/2018/v21i0a4979

Neilan D (2018) These maniacs will not stop until Nic Cage is in every movie. The A.V. Club. Available at https://www.avclub.com/these-maniacs-will-not-stop-until-nic-cage-is-in-every-1829302082. Accessed 16 July 2021

Roberts A et al (2019) Magenta studio: augmenting creativity with deep learning in Ableton live. https://doi.org/10.5281/ZENODO.4285266

Rosner H (2021) The ethics of a deepfake Anthony Bourdain voice. The New Yorker. Available at https://www.newyorker.com/culture/annals-of-gastronomy/the-ethics-of-a-deepfake-anthony-bourdain-voice. Accessed 20 July 2021

Scribner C (2000) Peter Paul Rubens. Biography, style, & facts. Encyclopedia Britannica. Available at https://www.britannica.com/biography/Peter-Paul-Rubens. Accessed 23 March 2021

Smith AE (2021) GANs + Sanrio = Ganrio!. Medium. Available at https://amyelizabethsmith01.medium.com/gans-sanrio-ganrio-21e263666929. Accessed 16 Apr 2021

Stoffregen TA et al (2008) Motion sickness and postural sway in console video games. Hum Fact: J Hum Fact Ergon Soc 50(2):322–331. https://doi.org/10.1518/001872008X250755

Suciu P (2020) Deepfake star wars videos portent ways the technology could be employed for good and bad. Forbes. Available at https://www.forbes.com/sites/petersuciu/2020/12/11/deepfake-star-wars-videos-portent-ways-the-technology-could-be-employed-for-good-and-bad/. Accessed 16 July 2021

Tangcay J (2021) Anthony Bourdain voice re-creation in documentary raises controversy. Variety. Available at https://variety.com/2021/artisans/news/anthony-bourdain-fake-voice-roadrunner-documentary-backlash-1235020878/. Accessed 16 July 2021

The Method Case (2013) Long distance art one artist, two robots and three paintings. The Method Case, 23 November. Available at https://www.themethodcase.com/video-long-distance-art-one-artist-two-robots-three-paintings/. Accessed 15 Apr 2021

Toews R (2020) Deepfakes are going to Wreak Havoc on society. We are not prepared. Forbes. Available at https://www.forbes.com/sites/robtoews/2020/05/25/deepfakes-are-going-to-wreak-havoc-on-society-we-are-not-prepared/. Accessed 15 Apr 2021
United States Department of Labor (2020) OSHA technical manual (OTM)—Section IV: Chapter 4. Occupational safety and health administration. Available at https://www.osha.gov/otm/section-4-safety-hazards/chapter-4. Accessed 16 April 2021
Vallor S (2016) Technology and the virtues: a philosophical guide to a future worth wanting. Oxford University Press, New York, NY
Vear C (2020) Creative AI and musicking robots. In: 1st Workshop on creativity and robotics, 12th International conference on social robotics (ICSR 2020), Colorado, USA. Available at https://dora.dmu.ac.uk/handle/2086/20477. Accessed 20 July 2021
Wareing J (2021) 'I made GPT-3 read every @emfcamp talk title and then asked it to write new ones. A thread. https://t.co/M6xZgmyuCN. @jonty, 5 March. Available at https://twitter.com/jonty/status/1367932738691596295. Accessed 16 Apr 2021
Whitwam R (2020) New tool generates every possible melody for public domain use. ExtremeTech. Available at https://www.extremetech.com/extreme/306575-new-tool-generates-every-possible-melody-for-public-domain-use
Widmer G (2002) Machine discoveries: a few simple, robust local expression principles. J New Music Res 31(1):37–50. https://doi.org/10.1076/jnmr.31.1.37.8103
Widmer G, Goebl W (2004) Computational models of expressive music performance: the state of the art. J New Music Res 33(3):203–216. https://doi.org/10.1080/0929821042000317804
Wikipedia (2021) Ship of Theseus. Available at: https://en.wikipedia.org/wiki/Ship_of_Theseus#Thought_experiment. Accessed 4 July 2021
Wolf MJ, Miller KW, Grodzinsky FS (2017) Why we should have seen that coming: comments on Microsoft's Tay "Experiment", and Wider implications. ORBIT J 1(2):1–12. https://doi.org/10.29297/orbit.v1i2.49
Yin Z et al (2021) "A good algorithm does not steal—it imitates": the originality report as a means of measuring when a music generation algorithm copies too much. In: Proceedings of EvoMUSART. EvoMUSART 2021 (in press)
Zhou Y et al (2020) Generative melody composition with human-in-the-loop Bayesian optimization. https://doi.org/10.5281/ZENODO.4285364

Structures and Frameworks

Ecosystemic Thinking: Beyond Human Narcissism in AI

Cesar & Lois, Cesar Baio, and Lucy HG Solomon

Abstract This chapter introduces the reader to critically oriented artworks by Cesar & Lois that propose that AI move away from anthropocentric modes of processing information and toward more ecologically oriented decision-making. The artists argue that the layering of ecological and biological logical inputs within a relational system (ecosystem) has potential as an environmentally aware model for artificial intelligences. They ask: How might an AI reflect the processing and functioning of a healthy ecosystem and support ecosystems? By contextualizing their art that intersects with science and technology, the artists propose the reorientation of machine thinking to living systems, while recognizing the planetary ecosystem as inherently intelligent. By framing microbiological systems as intelligent networks and integrating those with AI, they question what an AI built on knowledge that predates human beings would answer to and the form that its logic would take. This prehuman intelligence has survived and evolved across millions of years, making decentralized decisions across billions of entities as emergent processes. A discussion around nonhuman logic and nonhuman language is linked to the bhiobrid (bio-digital) artworks under discussion. By detailing two artworks that propose novel AI's, *Degenerative Cultures* with *Physarum polycephalum* as the model organism and *Mycorrhizal Insurrection*, which integrates the signals of mycelia, the artists' poetic proposition for new networks emerges, along with questions around humanity's relationship to technology, and technology's relationship to the living world. With a look at how anthropocentric orientations of intelligence pervade contemporary art

Cesar & Lois: Independent art collective, US/ Brazil

Cesar & Lois (✉)
Escondido, CA, USA

São Paulo, Brazil

C. Baio
Departamento de Multimeios, Mídia e Comunicação, University of Campinas, Campinas (UNICAMP), Cidade Universitária Zeferino Vaz - Barão Geraldo, Campinas, São Paulo 13083-970, Brazil

L. HG Solomon
Department of Art, Media, and Design, California State University San Marcos, 441 La Moree Road, San Marcos, CA 92078-5017, USA

C. Vear and F. Poltronieri (eds.), *The Language of Creative AI*, Springer Series on Cultural Computing, https://doi.org/10.1007/978-3-031-10960-7_6

and popular culture, the authors question the relationship between the machine and human programmer and the machine's decisions. The chapter considers how human authorship, even in the seemingly neutral area of abstract computations, results in inequities that permeate those algorithms and result in decisions that serve some over others. The artworks that are discussed are proposals for reorienting machines in a way that expands the nonhuman and environmental data entry points of AI training. Their method is to develop intelligences that draw from nonhuman living systems for logical pathways for computing while at the same time critiquing the anthropocentric orientation in the development of sociotechnical systems.

Keywords Creative AI · Practice · Bio-hybridity · Alternative AI · Ecosystems

1 Introduction

This chapter focuses on the artwork of Cesar & Lois, which layers living systems and artificial intelligence to challenge the anthropocentric models that drive decisions in our societies. Our approach to AI, evident in the Cesar & Lois artworks that we discuss in this chapter, is aligned with a critique of the anthropocentric orientation in the development of sociotechnical systems (the systems that society depends on). We understand anthropocentrism to reference the framing of humans as superior rational beings, an idea that sustains Kant's concept of human exceptionalism, established in Kant's distinction of reason as human: "In the lifeless or merely animal nature we see no ground for admitting any faculty, except as sensuously conditioned," while "man" is, at least in part, a "purely intelligible object" (Kant 1922, pp. 442–3). This elevation of human reason persists in technological contexts and is updated in the sociotechnical systems of today. Through our artworks, we challenge this understanding by attempting to think of intelligence through nonhuman parameters. In this text, we point out the anthropocentric ways that technology is conceived and built, and we ask: What if AI functioned without an embedded association with (and preprogrammed elevation of) human thinking?

Cesar & Lois consists of Lucy HG Solomon (California) and Cesar Baio (São Paulo). We are an art collective that works across continents and across species and media, as we experiment with nonhuman logic as the basis for AI. We also attempt to think with nonhuman logic (see Fig. 1).

We grow microorganisms in university laboratories but also in our homes, and these become integrated with the technologies that we develop. For us, thinking creatively about AI means challenging the concept of intelligence as established throughout modernity: one that assumes the (supposed) superiority of human rationality as the sole method of thinking and the default model for the development of technologies. In order to instigate discussions about anthropocentric understandings of intelligence present in those narcissistic models for technology that reflect human thinking, we develop what we call *bhiobrid* (bio-digital hybrid) intelligences, which we materialize in artworks. These artworks allow us to imagine new futures, with

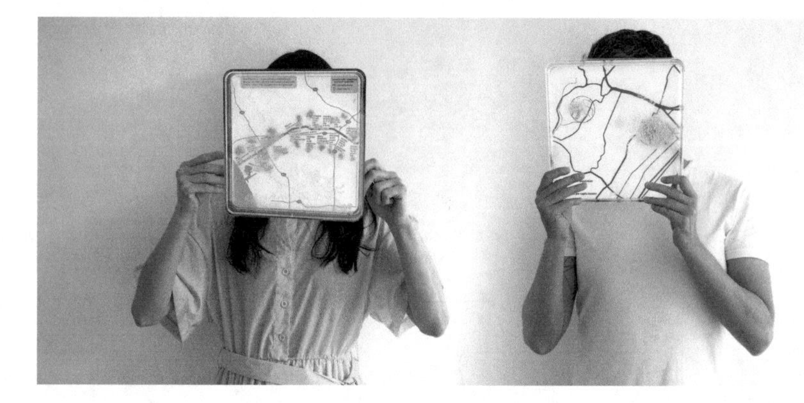

Fig. 1 Cesar & Lois pictured with the logic of *Physarum polycephalum*, or slime mold (Cesar & Lois 2018)

an AI that is ecosystemic and that, by tapping into the decision-making processes that occur in even the simplest life forms, is capable of responding to a network of living beings and to their environments (see Fig. 2 for a model of this layering within an ecosystemic AI). The systems we envision are reflective of a nonhuman logic that takes into account environmental inputs and a complex array of intercellular and intracellular communications that make it possible to understand the planetary ecosystem as an intelligent entity.

In this text, we discuss the development of what could be considered an artistic language that combines both nonhuman and human thinking. This is followed by an interrogation of technological narcissism in (human-reflecting) AI and in a host of popular cultural representations of AI. Microbiological systems as models for computational processing invert the narcissistic tendency of the human intellect and challenges the assumption that an adequate machine is one that successfully impersonates human intelligence (e.g., Turing test). In the artworks that we outline, we propose another kind of AI, one which acknowledges intelligence in microorganisms, whose evolution predates human beings. In our art, we draw on ecosystemic intelligences, which we consider to be a layering of intelligences that draw on an array of logical inputs including simple life forms, networks of living beings, and environments. This layering of AI with nonhuman intelligences challenges the anthropocentric values that drive our society. The proposed bio-digital intelligence is based on and addresses the complexity of enmeshed ecosystems by integrating nonhuman ways of making sense and decisions. This conception runs counter to those concepts of art that place humans at the center in the organization of the world. We as artists contemplate more-than-human networks and imagine the kinds of decisions an AI could make were the AI capable of the complex enmeshed logic prevalent in ecosystems.

Questions about the import of nonhuman languages and the status of the human and the nonhuman in the world drive us to make art that considers nonhuman logic as an input for AI that edits texts about human exceptionalism. Through artworks such as *Degenerative Cultures* and *Mycorrhizal Insurrection*, both based on networking

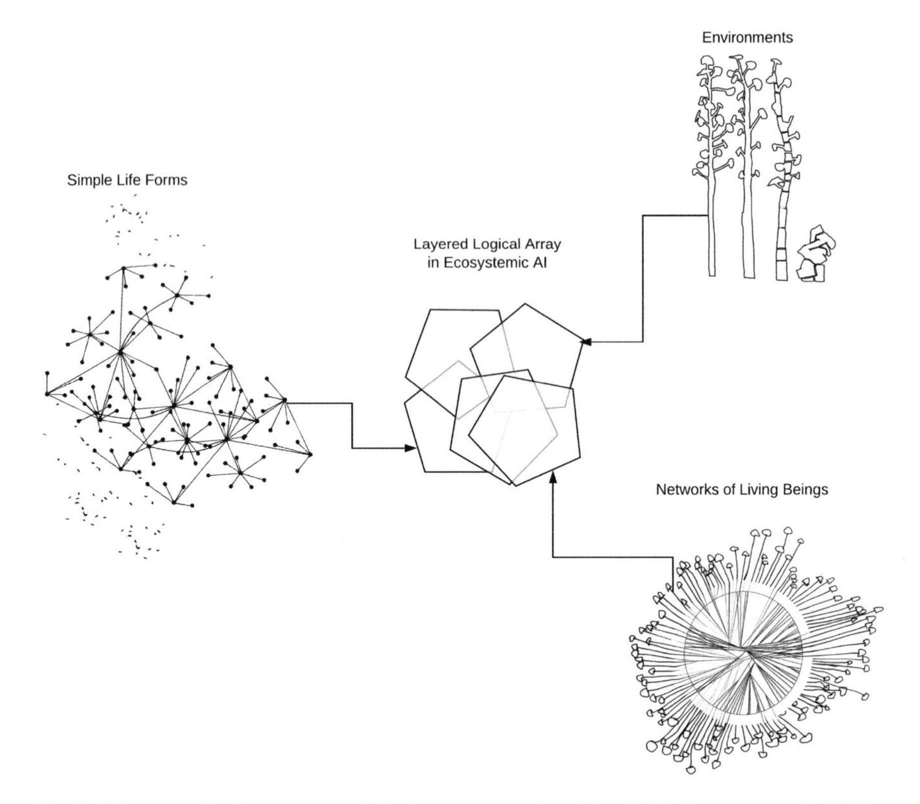

Fig. 2 Diagram reflecting ecosystemic logic as a composite of simple life forms, networked living beings, and environments (Cesar & Lois 2022)

living organisms combined with AI, we examine how nonhuman logic inputs can shift our understanding of language, as well as meaning and context. Our strategies may be considered subversive, as we intervene in and even invent new technologies (for rethinking how to think!). Could a new language result from this hybridization of logics? As part of these conceptual challenges to language, we also examine the concept of AI and question its future, which is symbolically and technologically related to the history of how human beings have framed knowledge, including which models of intelligence prevail in the popular imagination of AI. Our critical approach to human-centered understandings of language, intelligence, and AI is materialized in our artworks, which implies non-anthropocentric AI's that are based on *Physarum polycephalum* (slime mold) and on mushrooms. Throughout this text, we discuss these works and reflect on human and nonhuman decision-making, and we posit a utopian future in which humans and machines think together with other entities. We imagine that the AI and humans of that future could easily balance decisions around life and prosperity with the well-being of the ecosystem and the welfare of other species.

Sequentially, this chapter explores:

- The question of nonhuman logic and nonhuman language.
- A consideration of human-like AI's in popular culture and in art contexts, with examples of artists pushing against the limitations of those.
- An introduction to the artworks, *Degenerative Cultures* and *Mycorrhizal Insurrection*, and their proposals for AI's mapped to nonhuman living systems (slime mold and mushrooms, respectively).
- A proposal for prehuman logic in seemingly simple organisms and embodied intelligence as models for AI, rather than human neurological processing.
- A call for interspecies thinking as a means to formulate ecologically responsive (and environmentally responsible) AI's.

1.1 The Question of a Nonhuman Language

The concept of an artistic language is often discussed in terms relating to the organization of aesthetic elements in certain ways (syntax) which conveys meaning (semantics), and there are many distinctions in how artists and viewers embed intuitive meaning and derive understanding. What is distinct about our approach to the question of language and art is the insertion of nonhuman logic, which does not abide syntax or semantics. We are curious about the way in which this traditional understanding of language becomes problematized by nonhuman logic systems. How might syntax and semantics diverge when driven by mycelia? (see the fungal colonization of a dictionary in Fig. 3) What new meanings and understandings become accessible?

Following thinking around linguistics that breaks from this standard, we consider Nelson Goodman's formative ideas around the capacity of symbols and their systems—inclusive of language, science, art, and other ways of organizing and

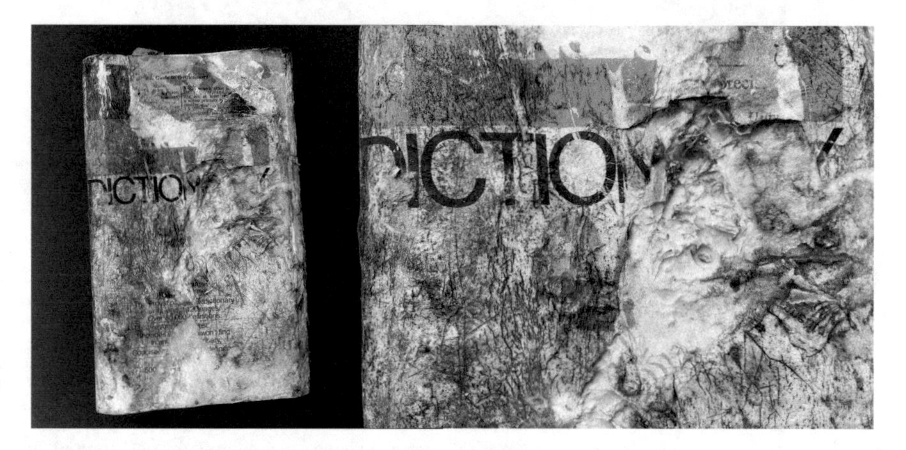

Fig. 3 Dictionary colonized with mycelia, an artifact from the series *Thinking like a Mushroom*, with mycelial logic embedded in text (Cesar & Lois 2018)

perceiving reality—to build the world. Goodman acknowledges that the ways that we build that world, whether through the systems of symbols embedded in language, in science or in art (factoring in perception), can limit the very worlds being built: "And even within what we do perceive and remember, we dismiss as illusory or negligible what cannot be fitted into the architecture of the world we are building" (Goodman 1978, p. 15). Can nonhuman sensing (making sense of the world) likewise build a reality? Would this then constitute a nonhuman language? What might a language embedded in an environment, in the land be like, and who or what would speak that language and understand its logic? Could an AI operate based on such a logic, a logic of living beings and environments? In laboratory contexts, in the field, with technological tools and in home growth chambers, we have pondered the logic of various nonhuman entities, including mushrooms (see Fig. 4).

If we picture an ecosystem as a web of relationships that constitutes a "logic" and which underlies a place's inherent language, and then imagine its disruption, then the loss of that ecosystem is also the loss of language. Scientist and author Robin Wall Kimmerer, a member of the Citizen Potawatomi Nation, links being native to a place to speaking its language. Wall Kimmerer describes how her lack of fluency in Bodewadmimwin or Potawatomi (her language had history been different) has left out words, terms, and even concepts from her vernacular.

> My first taste of the missing language was the word *Puhpowee* on my tongue. I stumbled upon it in a book by the Anishinaabe ethnobotanist Keewaydinoquay, in a treatise on the traditional uses of fungi by our people. *Puhpowee*, she explained, translates as "the force which causes mushrooms to push up from the earth overnight." As a biologist, I was stunned

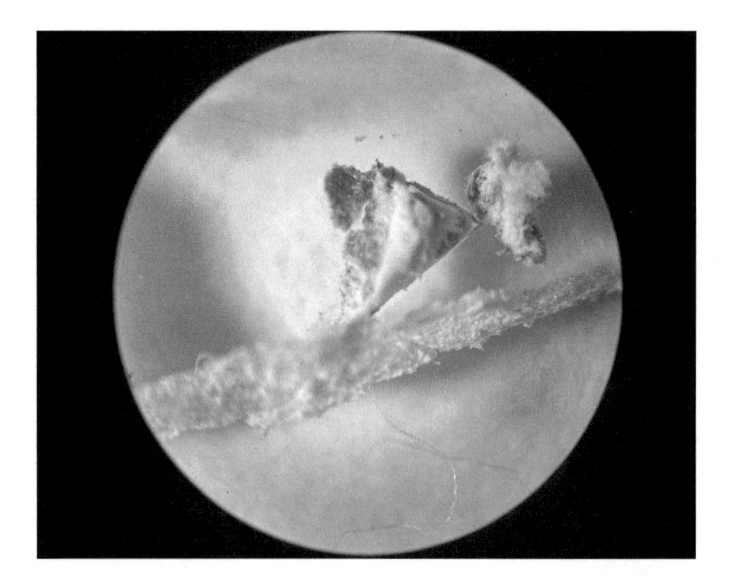

Fig. 4 Microscopy of sectioned fungal specimen with mycelia (a contemplation of nonhuman logic) at Coalesce Center for Biological Arts, University at Buffalo (Cesar & Lois 2021)

> that such a word existed. In all its technical vocabulary, Western science has no such term, no words to hold this mystery (49).

This loss of the language of a place suggests a semantic loss, since the speakers of a place derive meaning from the relationships manifest in that ecosystem. The anthropologist, Eduardo Viveiros de Castro, who examines Amazonian people's relationships to other living beings, discusses whether humans are nonhuman beings and animals are humans:

> If humans regard themselves as humans and are seen as nonhumans, as animals or spirits, by nonhumans, then animals should necessarily see themselves as humans. What perspectivism affirms, when all is said and done, is not so much that animals are at bottom like humans but the idea that as humans, they are at bottom something else—they are, in the end, the "bottom" itself of something, its other side; they are different from themselves. (Viveiros de Castro 2014, p. 69)

Through our work, we speculate about other forms of intelligence that are not human-centered. The anthropologist Eduardo Kohn argues that nonhuman lifeforms can think, and to approach these other intelligences, we must challenge the human models and parameters with which we have historically understood intelligence. Digressing from the strictly symbolic understanding of cognition, Kohn insists: "The first step toward understanding how forests think is to discard our received ideas about what it means to represent something" (Kohn 2013, p. 8). Kohn aggregates theories about representation that extend beyond what humans consider language to argue that nonhuman lifeforms are capable of representing the world.

According to Kohn, the conflating of representation with language happens because we tend to think about how representation works by approaching it in relation to how human language works. Representation in language for Peirce (1991), whom Kohn responds to, is understood through the concept of symbols (an arbitrary sign that is conventional to an object of reference): an abstract assembly of letters is understood to represent something. In Peirce's semiotic theory, other modalities of relationships between signs and the objects of reference are iconic, with signs that share likenesses with the objects they represent, and are indexical, with a cause-effect determination. According to Kohn, lifeforms represent the world in nonsymbolic ways: "These nonsymbolic representational modalities pervade the living world—human and nonhuman—and have underexplored properties that are quite distinct from those that make human language special" (Kohn 2013, p. 8).

Following this argument, Kohn proposes that if we want to understand how nonhuman beings think, we have to change our understanding of representation. Additionally, he advocates for a non-colonial linguistics. Kohn discusses the application of European linguistics to other nuanced cultures—a linguistics that cannot take into account the specific realities and the wealth of worlds created by non-European languages. We would argue that this is also true of attempts to understand the logic of microbiology and even that of artificial intelligences, from a human perspective. "Human" in many theoretical applications represents lineages of colonization that discount the underlying realities of the logical networking of living beings.

1.2 Technological Narcissism

> Without realizing it we attribute to nonhumans properties that are our own, and then, to compound this, we narcissistically ask them to provide us with corrective reflections of ourselves. (Kohn 2013, p. 21).

Often popular imagination understands artificial intelligence as extensions of anthropocentric perspectives of technology and the machine. There are numerous examples in cinema and literature that illustrate the anthropomorphizing of technology, with machines that simulate human thinking, appearance and/or feeling, or with narratives that enmesh fear and love. These include Mary Shelley's *Frankenstein*, *Terminator*, Stanley Kubrick's *Hal 9000,* the replicants in *Blade Runner* (in particular, Rachael, the replicant Rick Deckard falls in love with), and *Her*, the movie in which Theodore falls in love with a personalized operating system. In each, the machine is made human.

Although in the context of computational processing, coding is an abstract aspect of artificial intelligence, the concept of AI, even in the realm of technology, often undergoes an anthropocentric approach. From the technical perspective, artificial intelligence is a computational algorithm, and machine learning distills to machine-based statistics. To deal with the highly abstract mathematical concepts that are necessary to understand and to program computers, most computer theorists reference conceptual models that are based in the anthropocentric understanding of intelligence. A super-human brain becomes a metaphor for AI, with *human* as the primary frame of reference. This is the case in many of the fundamental concepts in current AI development. The idea of artificial neural networks, for example, posits rationality and intelligence as synonyms and suggests that intelligence is associated with a neurological system.[1] As argued in the next sections, this restrictive idea of intelligence does not consider the wide range of microorganisms that function without any neurons and whose decisions have allowed them to survive for millions of years. In our artworks, we propose these microbiological networks as alternative models for computing.[2]

This tendency to shape an AI according to a human model is also expressed in the classic Turing test (Turing 1950). When formulating the idea of a machine that would exhibit intelligent behavior, the mathematician Alan Turing proposed a test. To pass Turing's test, a human must be unable to tell whether one is talking to a machine or a person. Interestingly, the test does not seek to verify if the answers that the machine gives are correct, but rather if the machine's answers are convincingly "human". According to Turing, this would be the way for a machine to present intelligence.

---

[1] The article, "The Evolution of General Intelligence," counters, to a degree, human exceptionalism that places animal behavior apart from human rationality. However, the article maintains the brain-intelligence link in both humans and nonhumans and correlates greater intelligence with brain size (Burkart et al. 2017).

[2] See Flikkema and Leid's study of swarm intelligence in bacteria as a potential model for digital networks (Flikkema and Leid 2005).

The impulse to project a human figure on machines is explicit when machines are directed to create human-like output.

In the art sphere, this includes machine-generated representations of human figures presented as artworks. This is also the case in many of the digital artworks that imitate human behavior or language. With the increasing use of machine learning in everyday life, more and more art incorporates AI. Among a variety of approaches, some artists use machine learning techniques to create programs that can learn a specific painter's style and apply it to any image. The user can convert a camera's image into a "van Gogh" or an Impressionist painting. The underlying question posed by these initiatives is whether machines can produce art.[3] Though there are birds that make exquisite nests to impress their mates (Endler 2012), and elephants have been taught to paint unique compositions by wielding paint brushes with their trunks,[4] art is a very human concept—perhaps one of the most anthropocentric ones. To ask if machines can produce art is to project this very human concept on an entity that has an existence completely different from that of humans. It means that any plausible answer to this question can only be reached if the machine in question has been trained with human concepts. If this is the case, then the art does not originate in the machine but with the human who designed the patterns used to train and program the machine. Artworks that assert that machines can make "art" are not often intended to raise questions about the ethics of technologies or their impacts in society, but rather suggest a future machine intelligence capable of imitating human creative activities. This is a human obsession and reflects a machine's goal only if the machine's logic mirrors that of humans, constituting a sort of technological narcissism.

In a strict sense, technology is knowledge applied in the service of doing/solving something. Both the concept of knowledge and the problem to be solved can only exist in a context, which means that technologies are historically situated and responsive to political, economic and ideological interests, and values.[5]In a society that is still driven by notions of development grounded in extractivism and exploitation of both human and nonhuman labor, technology is also a generator of power that privileges a restricted group of people. Artificial intelligence can generate inequities, favor specific groups and outcomes,[6] and these patterns are not always easy to map. Applications and platforms owned by the giant companies that control this industry are part of more and more aspects of our lives, directing our decisions about which routes we take, our relationships, financial transactions and elections; technology informs the ways we represent, understand, and interact with the world. This technological context demands that artists interrogate the relationship between automation and creation, expanding on the art from the last century that reflected contemporary

[3] Note that this question is different from asking if it is possible to produce art with/against/for machines.

[4] It is important to note that elephants do not paint without human intervention. Animal rights activists point to the cruelty of humans making an elephant paint and question the methods and motives of this practice (Barry 2016).

[5] This ties into Vilém Flusser's connection between tools and societal values (Flusser 1999).

[6] See Ruha Benjamin's discussion of racial bias in algorithms in healthcare contexts (Benjamin 2019).

societies' increasing reliance on machines by introducing automation in creativity decades before the information age.

As a result of this updated context, with increasing societal dependence on machine learning running parallel to societal inequities, artists have created manifestos and artworks that challenge the current direction of AI. Caroline Sinders, with *Feminist Data Set*, interrogates AI's biases during development. Sinders writes of the constant process of analysis and questioning: "Every step exists to question and analyze the pipeline of creating using machine learning—is each step feminist, is it intersectional, does each step have bias and how can that bias be removed?" (Sinders 2017–ongoing). The artwork, *Anatomy of an AI*, depicts the societal impacts and environmental and human costs of an Amazon Echo in a detailed information chart (Crawford and Joler 2018). As artists in the contemporary sociotechnical contexts of São Paulo and California and, in a connected sense, the globe, we understand automation, which includes AI technologies, as not only a technical means but as part of a political and economic project. Power as something crystallized in technology. For us, artmaking is a political and poetic act that allows us to critically analyze and develop potential technologies, to challenge the myth of technological objectivity and to conceptually reframe how meaning and power are encoded within these technologies. As discussed in the next section, our propositions are both materially and conceptually represented in our artworks.

2 Artworks as Proposals for Reorienting AI

Cesar & Lois artworks question anthropocentric systems by focusing on the decision-making processes in microorganisms. Creating artworks based on the convergence of bhiobrid intelligences—including processes of prehuman (microorganisms), human, and posthuman (AI) decision-making—is a conceptual and poetic action that proposes a post-anthropocentric conception of intelligence and creativity.

The installation *Degenerative Cultures* (see Fig. 5) creates a bhiobrid network in which living microorganisms and AI work together to challenge the human impulse to control nature, while involving human participants via Twitter. Instead of a human-like AI that can chat or create images within the same logical parameters as that of humans, Cesar & Lois creates a generative algorithm modeled on the microorganism *Physarum polycephalum* and which mirrors the growth logic inherent in this organism. The AI-based digital fungus evolves according to this generative algorithm and corrupts texts on geoengineering found online. This digital degradation results in revisions of the human-centered texts, with the edited versions of the text linked to the logic of microorganisms.

Embedded in the artistic language of Cesar & Lois is the search for how humans relate to technology and how technology relates to the living world. Just as Duchamp displaces industrial objects from their original context to provoke discussions about the object and the artistic act, we displace technologies, protocols, and concepts to generate discussions around digital and living systems. We use technologies with

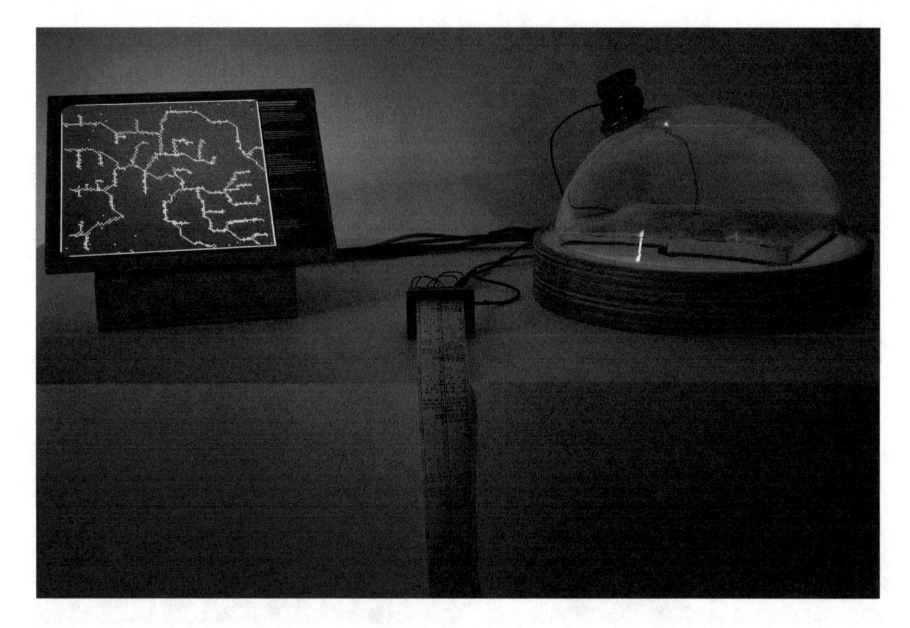

Fig. 5 Installation view of *Degenerative Cultures* at Lumen Prize Exhibition, *Uncommon Natures*, in the Brighton Digital Festival, UK (Cesar & Lois 2018)

different goals from those for which they were created, but also, we modify the conceptual frameworks that drive how we understand such technologies, and we (intend to) give them new meaning. This artistic procedure happens, for example, when instead of designing an AI that mirrors a human being (a task that, by definition, is doomed to fail), the AI is modeled on nonhuman organisms, which function as interspecies networks and which challenge the anthropocentric concepts of the individual and community. In *Degenerative Cultures*, a physical book, one of the main symbols of both the accumulation of knowledge and the human ability to produce cultural tissue, serves as a substrate for living microorganisms (as pictured in Fig. 6). These microorganisms are connected to a digital system based on an AI (Natural Language Processing) that searches the Internet and replicates the microorganism's physical growth over texts that assert human control over all ecological systems.

Although we inoculate the book with *Physarum polycephalum*, the controlled humidity, color, and warmth makes it possible for other microorganisms already present in the book to flourish. From the viewer's perspective, the bright yellow of *Physarum polycephalum* and the yellow generative animation of the AI generate contrast with the growth chamber's red glow (as seen in Fig. 7). The inherent decision-making process of the microorganism, the living book (with the organism growing across it) and the humidity seen through the dome counter the precise logic of the computer-based algorithms. The artistic action of assembling these produces an aesthetic and experiential outcome. Instead of learning the arguments against modern anthropocentrism, those who see and interact with the artwork are invited to consider

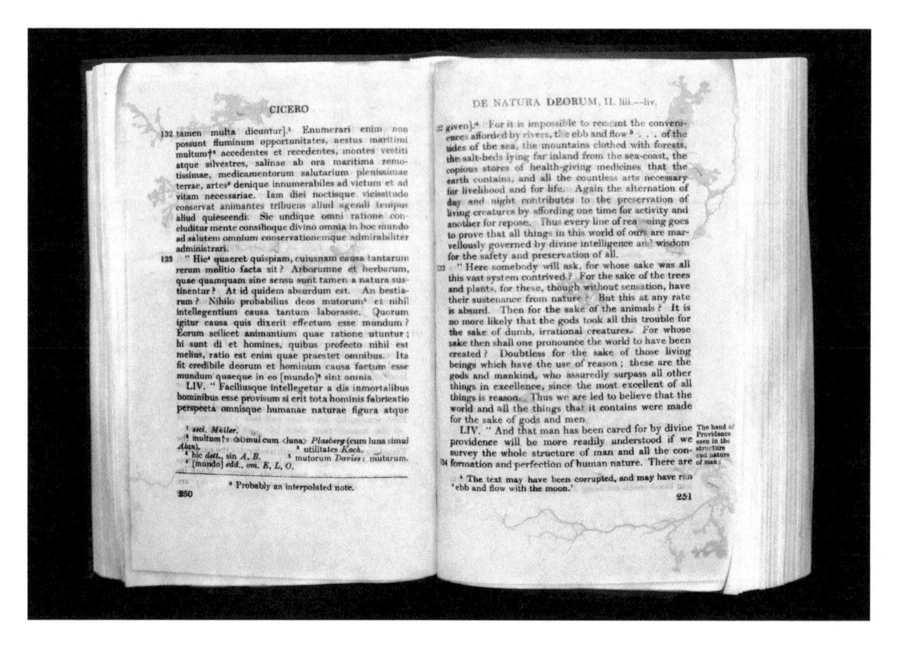

CICERO

132 tamen multa dicuntur].[1] Enumerari enim non possunt fluminum opportunitates, aestus maritimi multum†[2] accedentes et recedentes, montes vestiti atque silvestres, salinae ab ora maritima remotissimae, medicamentorum salutarium plenissimae terrae, artes[3] denique innumerabiles ad victum et ad vitam necessariae. Iam diei noctisque vicissitudo conservat animantes tribuens aliud agendi tempus aliud quiescendi. Sic undique omni ratione concluditur mente consilioque divino omnia in hoc mundo ad salutem omnium conservationemque admirabiliter administrari.

133 "Hic[4] quaeret quispiam, cuiusnam causa tantarum rerum molitio facta sit? Arborumne et herbarum, quae quamquam sine sensu sunt tamen a natura sustinentur? At id quidem absurdum est. An bestiarum? Nihilo probabilius deos mutorum[5] et nihil intellegentium causa tantum laborasse. Quorum igitur causa quis dixerit effectum esse mundum? Eorum scilicet animantium quae ratione utuntur; hi sunt di et homines, quibus profecto nihil est melius, ratio est enim quae praestet omnibus. Ita fit credibile deorum et hominum causa factum esse mundum quaeque in eo [mundo][6] sint omnia.

LIV. "Faciliusque intellegetur a dis inmortalibus hominibus esse provisum si erit tota hominis fabricatio perspecta omnisque humanae naturae figura atque

[1] *secl. Müller.*
[2] multum†: <si>mul cum <luna> *Plasberg* (cum luna simul *Alan*).
[3] utilitates *Koch.*
[4] hic *dett.*, sin *A, B.*
[5] mutorum *Davies*: mutarum.
[6] [mundo] *edd.*, *om. E, L, O.*

[a] Probably an interpolated note.

250

DE NATURA DEORUM, II. liii.—liv.

132 given].[a] For it is impossible to recount the conveniences afforded by rivers, the ebb and flow[b] . . . of the tides of the sea, the mountains clothed with forests, the salt-beds lying far inland from the sea-coast, the copious stores of health-giving medicines that the earth contains, and all the countless arts necessary for livelihood and for life. Again the alternation of day and night contributes to the preservation of living creatures by affording one time for activity and another for repose. Thus every line of reasoning goes to prove that all things in this world of ours are marvellously governed by divine intelligence and wisdom for the safety and preservation of all.

133 "Here somebody will ask, for whose sake was all this vast system contrived? For the sake of the trees and plants, for these, though without sensation, have their sustenance from nature? But this at any rate is absurd. Then for the sake of the animals? It is no more likely that the gods took all this trouble for the sake of dumb, irrational creatures. For whose sake then shall one pronounce the world to have been created? Doubtless for the sake of those living beings which have the use of reason; these are the gods and mankind, who assuredly surpass all other things in excellence, since the most excellent of all things is reason. Thus we are led to believe that the world and all the things that it contains were made for the sake of gods and men.

The hand of Providence seen in the structure and nature of man:

LIV. "And that man has been cared for by divine providence will be more readily understood if we survey the whole structure of man and all the con-

134 formation and perfection of human nature. There are

[b] The text may have been corrupted, and may have run 'ebb and flow with the moon.'

251

Fig. 6 *Physarum polycephalum* grows over Cicero's text, *De Natura Deorum II*, during growth for the project's iteration for the Aesthetica Art Prize Exhibition, York, UK (Cesar & Lois 2021)

the microorganisms and to contemplate the predatory epistemology of certain human traditions. Within the installation, inserted into the clean biotech aesthetic of the connected growth chamber, is the uncontrollable impulse of life.

In *Mycorrhizal Insurrection*, mushrooms become the conduits for machine processing, with hyphae becoming the conductors of signals for the artwork. The mycelia signals are "electromyceliograms," our word for the mycelial equivalent of electroencephalograms (EEG) (see Fig. 8). In a circular habitat, a floating substrate contains mushroom colonies, whose mycelia send signals to an AI, and viewers can respond by messaging the mushroom colony. A screen visualizes what is happening within the system: the mycelial pulses prompt the AI's metabolic processing of viewers' texts, which become less and less human as the messaging prompts bursts of humidity within the mushrooms' habitat and this in turn affects the electromyceliograms.. User input through social messaging improves the habitat within the cylindrical enclosure, with optimal levels of humidity maintained by these human connections. The mycelial responses to the human input are read as changes in the electromyceliograms (see Fig. 8), which prompt the AI to revise texts with an anthropocentric focus.

This cross-technology or bhiobrid (bio-digital hybrid) system begins with the growth of the mushroom colony, whose mycelial signaling becomes an essential data input for the AI's processing of text. Viewers of the artwork send the system messages and engage with the AI's conversational bot through the exchange of texts, with each message shifting the environmental conditions of the growth chamber

Fig. 7 Viewer peering into the growth chamber of *Degenerative Cultures*, where *Physarum polycephalum* grows across a book by Cicero (Cesar & Lois 2021)

Fig. 8 Reading electromyceliograms, the electronic signaling of mycelia (Cesar & Lois 2022)

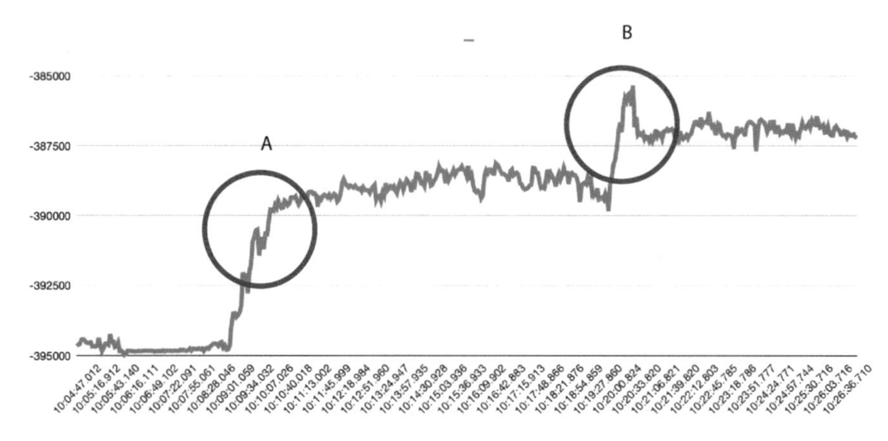

Fig. 9 Graph of electromyceliograms from an experiment with humidity: The first spike **a** happens when a drop of water is added to the fungal colony's substrate; the second spike, **b** demonstrates the change in signaling when water is added to the fruiting body (Cesar & Lois 2022)

and making it more optimal for mushrooms. Correspondingly, the text becomes less and less human. To process the text in this way, the AI learns from a training database and identifies anthropocentric texts, then outputs edited texts in response to the electromyceliograms (see Fig. 9 for a graph of these readings).

This "hyphaened Intelligence," driven by the variations in electric pulses of hyphae, explores nonhuman language by connecting the computer logic of AI and mycelial signaling. The artwork asks how a text might be written for or by nonhumans, or at least revised to reflect nonhuman logic and patterns of being. At the core of *Mycorrhizal Insurrection* is the assertion of a nonhierarchical model of intelligences as conduits for machine thinking, where fungi are valued as sources of knowledge within techno-societal contexts. The artwork is also a launching pad for thinking anew, and with others, about the nature of knowledge. When the connections across living beings become conduits for other logics, new imaginaries emerge.

2.1 Channeling Prehuman Logical Impulses

By making these new logical conduits between prehuman intelligences and artificial intelligence, we propose that thinking across those is possible. In the biological sciences, many researchers have argued that intelligence is not the exclusive privilege of brains. One of the most influential arguments in this direction comes from Francisco Varela and Humberto Maturana, who propose that thinking is embodied, and the nervous system is not the sole information processing system. In Francisco Varela's theory of enactive systems, the world and the cognitive organism determine each other. Dealing with Varela's concepts, Pasquinelli explains that "the organism selects relevant properties of the physical world, and the world selects the structure of the organism, during its respective co-evolutionary history" (Pasquinelli 2006, p. 34). According to Varela, "…we find ourselves performing that act of reflection out of a given background (in the Heideggerian sense) of biological, social, and cultural beliefs and practices […] our very postulation of such a background is something that *we* are doing: *we* are *here*, living embodied beings, sitting and thinking of this entire scheme, including what we call a background" (Varela et al. 1991/1993, pp. 11–12). From the perspective of embodiment, intelligence emerges in the relationship between a body and the world.

Commenting on R. Brook's discussion of AI, Varela identifies the problem of representation as a problem of understanding cognition without enactment: "The world shows up through the enactment of the perceptuo-motor regularities." Drawing on R. Brook's assertion that the problem with AI (and, Varela adds, with much of cognitive sciences) is the tendency to factor out perception in favor of abstraction, Varela connects intelligent behavior to the situated sensory perception and perception-action that feed cognition. "As I have argued here (and as Brooks argues for his own reasons), such abstraction misses the essence of cognitive intelligence, *which resides only in its embodiment*." (Varela 1992, p. 13).

As artists, we engage in this shift in how we understand intelligence through artworks and, also, in texts where we get inside the theories that support the argument of intelligence in microorganisms (Solomon and Baio 2020). From our perspective, the approach to intelligence as a logical manifestation founded in reason and centralized in the brain reiterates the classical argument of human exceptionalism. This is used to justify the global dominance of capitalist colonialism that, as observed by Moore (2015), is only possible by means of the exploitation of natural forces, including human and nonhuman labor.

The consideration of nature as an external being is less about the essence of the things and more about a projection of power. Simple transactions—filling a gas tank or buying an imported fruit—assert that power and conform to a certain world structure, yet there are societies that affirm relational power structures. Viveiros de Castro articulates the theory of perspectivism, "the ideas in Amazonian cosmologies concerning the way in which humans, animals and spirits see both themselves and one another" (Viveiros de Castro 1998, p. 469). Perspectivism acknowledges that "Western" epistemological debates that center on the dualities of power, ranging from "Nature and Culture" to "body and mind" and "animality and humanity," are ill-equipped to comprehend "non-Western cosmologies" (Viveiros de Castro 1998, pp. 469–470).

On the other hand, many scientific studies have analyzed the cognitive abilities of microorganisms. When discussing some of these studies, Steven Shaviro argues that insects and microorganisms have their own strategy of dealing with their environment and making decisions. Commenting about the abilities of *Physarum polycephalum*, Shaviro writes that the movement of this microorganism "seems not to be centrally coordinated but to involve internal communication among different parts of the organism, that slime molds have succeeded in threading mazes and solving combinatorial problems" (Shaviro 2010).[7]The logic of these decisions reflects a prehuman intelligence and offers possibilities for rethinking the super brain of the future as a decentralized layering of the logics of multiple organic sources, spanning living entities and ecosystems.

3 Conclusion

In Cesar & Lois artworks, the AI that we develop is not intended to establish an anthropocentric personalized entity that could pass the Turing test, be mistaken as a person, or impersonate human logic. We also don't intend to create a tool that learns aesthetic patterns and replicates those to recreate the work of a human artist. We try to create AI in a non-anthropocentric way. Our way of doing this is by creating bhiobrid systems, in which AI and microorganisms can work together and interact in a way that we humans cannot understand, in a way that does not equate to any human exchange. We strive to better understand these interactions and their nonhuman logic. We ask: If an AI is based on a combination of human and nonhuman inputs, what new pathways are possible to rethink our relations with nonhuman beings, and ultimately change our societies?

When we think about creating an artistic language, we think of the books and organisms that we rely on, each with their distinct ways of interacting with the world. We work with living organisms and with texts that represent human knowledge, and specifically books that represent human-centered thinking. We introduce a living organism to these texts, and this combination of the organism—both how it behaves and its organization—and the text produces an aesthetic that conveys a hybrid AI that moves across both worlds.

Our work makes evident how the AI mimics the behavior of microorganisms. As in the living world where microorganisms and mycelia play a part in the decomposition of organic matter, this AI consumes anthropocentric "ideas," altering texts that assert human dominance over other living entities. We make artworks that are something between living sculptures and digital AI. This in between is what matters to us, because we see the potential of the crossover of the biological and artificial as a possibility for interspecies thinking—thinking capable of responding to an ecosystemic context, not to narcissistic models and human desires.

References

Barry H (2016) Perth Zoo hits out at 'inflammatory and baseless' animal abuse accusations. WAToday.com.au 14 Dec 2016: n. Pag

Benjamin R (2019) Assessing risk, automating racism. Science (Am Assoc Adv Sci) 366(6464):421–422. https://doi.org/10.1126/science.aaz3873

Burkart J, Schubiger M, Van Schaik C (2017) The evolution of general intelligence. Behav Brain Sci 40:E195. https://doi.org/10.1017/S0140525X16000959

Crawford K, Joler V (2018) Data visualization and artwork. Anatomy of an AI system: the Amazon echo as an anatomical map of human labor, data and planetary resources. AI Now Institute and Share Lab. https://anatomyof.ai. Accessed 20 Sept 2020
Endler JA (2012) Bowerbirds, art and aesthetics: are bowerbirds artists and do they have an aesthetic sense? Communicative Integr Biol 5(3):281–283. https://doi.org/10.4161/cib.19481
Flikkema PG, Leid JG (2005) Bacterial communities: a microbiological model for swarm intelligence. In: Proceedings 2005 IEEE swarm intelligence symposium SI, pp 427–430. https://doi.org/10.1109/SIS.2005.1501655
Flusser V (1999) The shape of things: a philosophy of design. (A. Mathews, trans.) (1st English ed.). Reaktion, London
Goodman N (1978) Ways of worldmaking. Harvester Press
HG Solomon L, Baio C (2020) An argument for an ecosystemic AI: articulating connections across prehuman and posthuman intelligences. Int J Commun Well-Being 3(4):559–584
Kant I (1922) Immanuel Kant's critique of pure reason. (F. Max Müller, trans.) (2nd ed.). The Macmillan Company, London
Kohn E (2013) How forests think: toward an anthropology beyond the human. First edição. University of California Press, Berkeley
Moore JW (2015) Capitalism in the web of life: ecology and the accumulation of capital. Verso, London
Pasquinelli E (2006) Varela and embodiment. J Aesthetic Educ 40(1):33–35
Peirce CS (1991) Peirce on signs: writings on semiotic by Charles Sanders Peirce. New edition. Hoopes (ed.). University of North Carolina Press, Chapel Hill
Shaviro S (2010) Fruit flies and slime molds. The Pinocchio Theory
Sinders C (2017–ongoing) Art and technology project. Feminist data set. https://carolinesinders.com/feminist-data-set/. Accessed 30 July 2021
Turing AM (1950) Computing machinery and intelligence. Mind 59:433–460
Varela FJ (1992) Autopoiesis and a biology of intentionality. In: Proceedings of the workshop "Autopoiesis and perception". Dublin City University, Dublin, pp 4–14
Varela FJ, Rosch E, Thompson E (1991/1993) The embodied mind: cognitive science and human experience. Revised ed. edição. MIT Press, Cambridge
Viveiros de Castro E (2014) Cannibal metaphysics. In: Skafish P (ed). Univocal Publishing, Minneapolis
Viveiros de Castro E (1998) Cosmological Deixis and Amerindian perspectivism. J R Anthropol Inst 4(3):469–488. https://doi.org/10.2307/3034157

Embodied AI and Musicking Robotics

Craig Vear

Abstract This chapter discusses a hypothesis for embodied AI and its development with music robots. It proposes a solution to the challenge: *if we want AI/robots to join us inside the creative acts of music, then how do we design and develop systems that prioritise the relationships that bind musicians inside the flow of musicking?* This requires significant thinking around some core questions such as 'what does AI need to do in order to stimulate an embodied interaction in music?', 'what sort of intelligent agent does the AI need to be?' and 'what does the machine need to learn; what is to be modelled?'. I proffer a definition of embodied AI as *an intelligent agent whose operational behaviour is determined by percepts* (Percepts are objects of perception, or put another way, something that is perceived) *interacting to the dynamic situation within which it operates*, and outline foundational theories such as embodiment, embodied cognition, flow, musicking, meaning-making and embodied interaction to argue for such a concept in music. Following this, I outline the theoretical and conceptual structures that I have developed in order to get embodied AI and musicking robots to create music and meaning with human-musicians through which I offer examples of some of the works that have been developed using this approach.

Keywords Creative AI · Embodiment · Robotics · Music · Embodied cognition · Flow · Musicking

1 Pre-introduction: Music Heard So Deeply …

If you have ever listened to music so deeply that you have been taken away by it, then you will understand what I am to discuss in this chapter. It is the type of feeling where you leave behind notions of your body in a physical place, and venture into the space created by the music. Your sense of being shifts as you become more deeply attuned to the sounds and the melodies, the rhythms, the instruments and the voices that you

C. Vear (✉)
University of Nottingham, Nottingham, UK
e-mail: Craig.Vear@nottingham.ac.uk

C. Vear and F. Poltronieri (eds.), *The Language of Creative AI*, Springer Series on Cultural Computing, https://doi.org/10.1007/978-3-031-10960-7_7

find in this space. For me, I can feel the live presence of these sounds, and that it is happening in the here-and-now. This even happens when it is a recording that I have listened to lots over the years. In a sense, I become absorbed into the materiality of the music and attuned to the presence of those sounds in its flow through time: I have become embodied.

T. S. Eliot described this sensation poetically in his *Four Quartets.* In *The Dry Savages,* he wrote:

> Music heard so deeply
>
> That it is not heard at all, but you are the music
>
> While the music lasts.
>
> T.S.Eliot Four Quartets *The Dry Savages*

Here, Eliot describes this phenomenon as 'you are the music' and I would agree. Although this may seem an odd thing to say, in my experiences this is what happens: I do become the music. When this happens, I lose the awareness of my physical body as a shell that holds my sense of being, and I vaporise into the world created by the flow[1] of the music. Inside this world, I am free to explore and to mingle with the other presences; I feel the music and the other things within it, I can interact with them and feel their presence deeply, or I can zoom out and enjoy being in the flow as the music passes through time.

Although when I am listening, composing or dancing to music, the sense of interaction is one way (my interaction with the sounds), when I play with live musicians this interaction becomes communal and social. It can be a dance of sorts: to touch, to feel, to work with, to play with, to hide and seek with, to flirt and subvert with others through the flow as they reach out and play with me. And it is through these embodied interactions inside the music that I find meaning.

Meaning, used here, does not refer to how I interpret the melodic line to be some symbolic container for an inner message, or how the harmonic structure makes me feel (sad, happy etc.); nor does meaning equate to how the sounds are generated: electronic or natural. On the contrary, from the embodied inside perspective meaning is constructed through the interactions I described above in the 'dance', and by becoming the music as Eliot portrays.

In my book *The Digital Score* (Vear 2019), I outline my understanding of meaning-making in music. At the heart of this is Christopher Small's concept of 'musicking', in that 'to music is to take part' (Small 1998, p. 9). Small wrote that taking part can happen 'in any capacity, in a musical performance, whether by performing, by listening, by rehearsing or practising, by providing material for performance (what we call composing)' (ibid). Small stresses that 'the act of musicking establishes in

[1] Is the experience of musicking from inside the activity. Within the context of this project, the flow of musicking defines how 'musicians become absorbed in the music through a sense of incorporation within their environment (the soundworld), a shared effort (with the digital, virtual, AI and robotic agents) and a loss of awareness of their day-to-day wakefulness and bodily self-consciousness (embodiment with their instrument and into their music)'.

the place where it is happening a set of relationships, and it is in those relationships that the meaning of the act lies' (Ibid., p. 13). Simon Emmerson clarified Small's principle of meaning to infer the 'what you mean to me' (Emmerson 2007, P. 29), and this subtle shift circumvents the significant issues of value and who is doing the evaluation of meaning. Therefore, meaning (or the *what-you-mean-to-me*) is to be found in the relationships formed between the embodied act of musicking and the materials therein, e.g. musicians, sounds, music space and time.

2 Embodied Interactions

For nearly four decades, I have enjoyed this experience as a professional musician who, before the call to academia, performed at a high level of musicianship in a variety of professional situations. This experience is central to my creative AI practice, and it is a foundation stone to understanding embodied intelligence in music. However, this presents a significant challenge:

> if we want AI/robots to join us inside the creative acts of music then how do we design and develop systems that prioritise the relationships that bind musicians inside the flow of musicking?

To illustrate this, imagine a scenario where two human-musicians are musicking together; they are embodied, in the flow and dancing as sound through the music. Their interactions are a mixture of creative, mischievous, surprising, familiar and playful. Of course, it is not like this 100% of the time, they will also feel ignored, isolated, stupid, rejected and inadequate; and at other times, they are simply getting on with their own journey aware of the other, enjoying any coincidences yet focussed on exploring the larger space of sound. But this is all part of the rich tapestry of the embodied experience within live music performance, an experience we can call *embodied interaction.*

Now let us replace one of those musicians with an AI-driven robot, and you will understand the goal that I have set myself: what does AI need to do in order to stimulate this embodied interaction in music? The key phrase here is 'stimulate this embodied interaction' and outlines the nature of the AI being a co-creative other inside music. This is opposed to using AI as a signal generator, or constructing the harmonic lattice of a song, or using a robot to play a musical instrument (although it may well need to do this as part of its role). Furthermore, for me, this question proposes that the perceptual focus is on the human: i.e. it is the human-musician who feels this sense of embodied interaction. But this is a personal rationale as my research aims to make humans more creative through AI, others may choose to make the AI conscious or for it to have embodied perception.

In order to deal with this challenge of AI stimulating (as opposed to simulating) an embodied interaction, the AI is tasked with operating in a specific way that feels intelligent within the situation that it is to be perceived, i.e. within embodied musicking. And, that it is interacting with the human-musician in ways that can be recognised (by the human-musician) as the dance as described above. This is, in the context of musicking, intelligent behaviour, and an AI operating in such a way I call embodied AI and will discuss in the following section.

3 Embodied AI

With the above discussion in mind, I define embodied AI as:

> an intelligent agent whose operational behaviour is determined by percepts interacting to the dynamic situation within which it operates.

This definition is built on two principles:

(a) that artificial intelligence is not limited to the thinking-mind model
(b) that we understand meaning-making from the perspective of embodied cognition.

Percepts in this context are objects of perception, or put another way, something that is perceived i.e. the stuff that is found in the relationships formed between the embodied act of musicking and the materials therein, e.g. musicians, sounds, music space and time.

To unpack this first principle—*that artificial intelligence is not limited to the thinking-mind model*—we need to address a bias and prejudice still found in people's opinion that intelligence in AI is understood from a specific perspective of a thinking machine. This is in part a legacy from the early days of AI research that sought to position the study in line with formal reasoning philosophies and Turing's notion of the 'electronic brain' (Wooldridge 2020).

Thankfully, the field of AI has moved away from notions of intelligence being solely limited to cognitive operations of reasoning and logic. The mind-based models that predominated Good-Old-Fashioned AI (GOFAI) have, since the 1980s, accepted other areas of intelligence such as behavioural and embodied, and that these are understood within the context with which they operate. Landmark papers such as *Intelligence without Representation* Brooks (1987) and Anderson's *Field Guide* (Anderson 2003) to embodied cognition for AI research reinforce the limitations of building GOFAI systems that model the thinking processes like a human mind.

This shift in the definition of AI is not limited to a small niche in AI research. The widely respected textbook *Artificial Intelligence—A Modern Approach* also supports this approach. They state that the 'main unifying theme is the idea of an intelligent agent' (Russell and Norvig 2020. Preface pp. vii–viii), from which it defines AI as 'the study of agents that receive percepts from the environment and perform actions' (ibid.). They go on to say that

> Each such agent implements a function that maps percept sequences to actions, and [...] we stress the importance of the task environment in determining the appropriate agent design.

For embodied AI in music determining the 'appropriate agent' that 'implements a function that maps percept's sequences to actions' involves understanding the type of cognitive and perceptual processes that occur when embodied musicking (the task environment); and these, as explained in Pre-Introduction (above), are very different to the everyday cognitive processes of the wakeful mind going about its business such as walking to work, or the rationale thinking mind as it strategizes a chess game. Russell and Norvig stress that developing AI in this modern way should be 'concerned mainly with rational action', and that ideally, 'an intelligent agent takes the best possible action in a situation'. In relation to this chapter, then the best possible action for embodied AI in music is to stimulate the dance between musicians that is discussed earlier; and the best possible intelligent agent is one that 'acts so as to achieve the best outcome or, when there is uncertainty, the best expected outcome' (Ibid., p. 4), i.e. the embodied interaction of musicking.

When unpacking the second principle—*that we understand meaning-making from the perspective of embodied cognition*—we can see that this also deals with the notion of 'rational action'[2] and guides this research into understanding what is to be modelled for 'an intelligent agent takes the best possible action in a [musicking] situation'. From this perspective, approaching embodied intelligence in music with a focus of behavioural and embodied cognition emphasises the close-coupled relationship between the *situation* that an intelligent agent is operating in (the embodied musicking space), and the *behaviour* that it exhibits to cope inside such a system (embodied musicking and interaction; i.e. the dance).[3]

As mentioned above, there is a shift in the nature of AI away from a 'thinking thing' as advocated by Descartes, towards an understanding of an intelligent agent that copes within a dynamic and interactive environment. The former proposes a split, or separation of the mind from the body, it prioritises the role of the thinking mind as intelligence, and foregrounds notions of thought and reason, logic and planning as primary facets of intelligence. The latter, however, understands the biological system as a situated agent that has evolved behaviour to cope with its dynamic environment. For innovators in embodied intelligence such as Rodney Brooks, these Cartesian aspects of intelligence 'cannot account for large aspects of what goes into intelligence' (Brooks 1999, p. 134).

While it is true to say that aspects of the thinking mind such as short-term planning, logic and reason are required to make music, this does not come close to accounting for the decisions that are made while in the flow of musicking. As such, any hypothesis for embodied intelligence in music needs to look beyond cognitivist priorities of 'representation, formalism and rule-based transformation' (Anderson 2003, p. 93) and consider the interactive, coping mechanisms at the heart of the

[2] There are other approaches such as thinking humanly, thinking rationally and acting humanly and are discussed in more detail in Russell and Norvig (2020).

[3] You can see this in action by watching any Roomba robot hoover navigate any room and any encounter with a rogue sock, cable, slipper, pet or moved chair.

embodied behaviour of musicians, and to understand the 'more evolutionary primitive mechanisms which control perception and action' (Ibid., p. 100) in this music world.

It should be obvious that embodiment is a central concern of embodied cognition and therefore embodied AI. While there are various definitions of embodiment, you will bc familiar with the concept through learning to ride a bike, or having taught a child to ride a bike. For example, to successfully ride a bike—by that I mean get from location A to the desired location B—we must carefully and naturally navigate the dynamic mechanical system of the bike to such an extent that we are able to exercise our desires and wants through it to achieve the goal. Through this process, our whole biological system must adapt to this new dynamic environment and draw it into our senses to such an extent that we might (or start to in the case of our novice child) feel part of the bike, or the bike feels part of us. This attunement goes beyond the thinking mind alone (although that is part of it) and relies on other intelligences, senses and attributes of our being to successfully negotiate dynamic elements such as gravity, speed, acceleration, bumps, braking and balance. Once embodied, we start to forget about the separation of the two elements (bike and biology) and enjoy the experience of our new sense of self, and the sensation (perhaps thrill) of moving in a different way, and develop intelligent behaviours to deal with this new version of being.

In music terms, this is best described by Nijs et al. as:

> In music performance the embodied interaction with the music implies the corporeal attunement of the musician to the sonic event that results from the performance. The embodied experience of participating in the musical environment in a direct and engaged way is based on the direct perception of the musical environment and on a skill-based coping with the challenges (affordances and constraints) that arise from the complex interaction within this musical environment.

They continue to explain that this becomes:

> an optimal embodied experience (flow) when the musician is completely immersed in the created musical reality (presence) and enjoys himself through the playfulness of the performance. Therefore, direct perception of the musical environment, skill-based playing and flow experience can be conceived of as the basic components of embodied interaction and communication pattern. (Nils et al. 2009)

One of the key elements here is 'direct perception of the musical environment', and this goes back to the Eliot quote and the opening of this chapter.

While this chapter is not the place to explain the whole of phenomenology, it is helpful to understand the role that phenomenalist philosophers such as Heidegger and Merleau-Ponty can play in defining at least where and what this direct perception might be. Merleau-Ponty, for example, argues that 'perception and representation always occur in the context of and are therefore structured by, the embodied agent in the course of its ongoing purposeful engagement with the world' (Anderson 2003, p. 104). In basic terms, this means that the perception of, and the representation thereof, a world is experienced and expressed through the full body system; what Bachelard describes as the 'polyphony of the senses' (Bachelard quoted in

Pallasmaa 1996). These perceptions are not given form or content by the separate and autonomous mind, but rather are in themselves the form and content of the whole experience. To the phenomenologists, meaning is created through getting-to-grips with the experience of the world in flow. On this Anderson writes: 'at the highest level, what is at issue here is the fact that practical, bodily activity can have cognitive and epistemic *meaning*' (Anderson 2003, p. 109).

At the centre of embodiment is the recognition that not only does it involve acting within some physical world, but that the 'particular shape and nature of one's physical, temporal and social immersion is what makes meaningful experiences possible' (Ibid., p. 124). Embodied interaction, such as those described at the top of this chapter, is the 'creation, manipulation, and changing of meaning through engaged interaction with artifects' (sonic or physical) (Dourish 2001, p. 126). It is how the world reveals itself through our encounter with it, and in musicking terms, this happens inside the flow of musicking and the relationships that are encountered there. It is these percepts that are to be mapped onto actions, which should determine the operational behaviour of embodied AI.

4 Embodied AI and Musicking Robots

It should be obvious by now that the AI I am discussing here goes beyond a 'thinking engine' or an 'electronic brain' as posited in the origins of GOFAI research. As such, the AI that I design, build and deploy are not merely symbolic representation of thought processes (i.e. a schema for operation), or are used to construct the physical phenomena of music (i.e. the sound wave), or a trained neural network designs to output the meta-workings of music composition (i.e. the organisation and sequencing of music theory), although it may well do these as part of its role. Instead, embodied AI in music should:

> stimulate percepts leading to meaning-making in the human-musician so as to be believed to be rationally operating in the close-coupled relationships in the situated space of live music.

This is an immense challenge—life-long I would argue—and there are many ways to crack this nut; however, I have been developing these following strategies that get closer to an optimal experience in musicking. These are:

i. Creativity and the Flow
ii. Experiential learning and Recollection
iii. Embodied dataset
iv. Coping and Beliefs
v. Embodied Percept Input: Affect module and Bio-synthetic nervous system

I will now introduce each of these in turn and discuss them in the context of the works that were created to develop and validate them.

4.1 Creativity and the Flow

As discussed above, meaning (or the *what-you-mean-to-me*) is to be 'found in the relationships formed between the embodied act of musicking and the materials therein'. I also mention that meaning is 'created through getting-to-grips with the world in flow' and that 'practical, bodily activity can have cognitive and epistemic *meaning*'. These of course are from the perspective of the human-musician; and although it is probably feasible to implement these meaning-making principles into an AI, my concern is in making humans more creative. As such, all following discussions about meaning are from the perspective of the human-musician in the flow. Flow in this sense is defined as 'the experience of musicking from the inside perspective of being inside the activity' (Vear 2019, p. 68).

The role of my embodied AI is to reach out, suggest, offer and shift connections and relationships as percepts (that which is 'taken-in') by the human-musician. It also needs to establish a world of creative possibilities for exploration through the flow of musicking that the human-musician is taken-into (another domain of percepts). This 'taking-in' and 'taken-into' structure is discussed in full in Vear (2019, Chap. 6) as a basic structure with which to understand the creative relationships in musicking that can form the *what-you-mean-to-me* to the human-musician. This basic structure is split into these two domains:

1. *Taking-in*—within the flow musicians make connections with the AI as they reach out, suggest, offer and shift through the tendrils of affordance experienced through notions of
   - *Liveness*: the sensation that the AI is co-operating in the real-time making of the music, and this meaningful engagement feels 'alive'
   - *Presence*: an experience that some*thing* is there, or I am *there*
   - *Interaction*: The interplay of self with environments, agents and actants[4]
2. Taken-into—the AI can establish a world of creative possibilities for exploration through the flow through the domains of
   - *Play*: the pure play of musicking happens inside a play sphere in which idea and musicking are immutably fused
   - *Time*: the perception of time (of now, past, future and the meanwhile of multiple convolutions of time) inside musicking plays a central role to the experience of the musician
   - *Sensation*: is an aesthetic awareness in the experience of an environment (music world) as felt through their senses

This basic structure was deployed in a collection of standalone music compositions entitled *Black Cats and Blues* (2014–18) which were recorded by the US cellist

[4] 'Actants denote human and non-human actors, and in a network take the shape that they do by virtue of their relations with one another'. (Online. Available https://en.wikipedia.org/wiki/Actor-network_theory (accessed 27 May 2022).

Craig Hultgren and released on Metier Records.[5] These compositions focussed on deploying and testing the 'taking-in' and 'taken-into' principles through narrow-AI within the flow of musicking. Narrow AI typically operates within a limited pre-defined range of functions and in this scenario worked to provide the tendrils and creative possibilities in the embodied dance of the flow (discussed at the top of this chapter), so that Hultgren could find and create meaningful relationships. The analysis of Hultgren's experience inside the flow with these compositions is discussed in detail in Chap. 6 of *The Digital Score* (Vear 2019).

4.2 Experiential Learning and Recollection

In an attempt to arrive at a suitable solution for the embodied AI to learn, I designed, developed and deployed a rapid prototype that explored notions of experiential learning and recollection. This dealt with three key questions that emerged from discussions about embodied AI and machine learning,[6] which are:

1. what is to be learnt/modelled in embodied musicking?
2. 'how can we be sure that our learning algorithm has produced a hypothesis that will predict the correct value for previously unseen inputs' (Russell and Norvig 2009, p. 713)?
3. how will the results have meaning for the embodied human-musician in the flow?

On Junitaki Falls a Trio for Solo Instrument and two AI Performers (2016–17)[7] was created to experiment with a propositional solution to address these questions. The central concern was to find new insights into the challenge posed by question 3, while also adhering to the stimulation of percepts as per my definition of embodied AI (above).

The solution I embedded was built around theories of behavioural AI with creativity philosophy inspired by David Gelernter's book *The Muse in the Machine* (Gelernter 1994). Behavioural AI [also referred to by Rodney Brooks as Nouvelle AI (Brooks 1990)] posits that:

> intelligence is demonstrated through our actions and interactions with the world. Critical to this is that the environment within which the robot operates must be independent of the robot design. (Jordanous 2020)

Gelernter theorises creativity based on attention, focus, affect and emotion. Although harshly criticised around its release, it is gaining interest in modern AI research as a gateway into new approaches to the nature of creativity and intelligence in embodied and behavioural AI systems. Of particular interest to *On Junitaki*

---

[5] https://divineartrecords.com/black-cats-and-blues-new-work-for-cello-and-electronics-from-metier/

[6] Thanks to Dr Fabrizio Poltronieri and the Creative AI Research group members for these.

[7] Available at https://vimeo.com/254090680.

Falls was Gelernter's notions of 'thought-trains' and the role of 'recollection' and 'affect linking', to creative thought, and his attempt to computerise the poetics of thought in its full richness. Coupled with core notions of p-novelty as defined by Margaret Bodem, these three facets can provide stimulation as percepts within the embodied interactions generated by my embodied AI in the flow (see Figs. 1, 2, 3 and 4 for a pictorial overview of these processes joined into a robot system). In short, these are defined as:

- *thought-trains*—are 'sequences of distinct thoughts or memories. Sometimes, our thought-trains are assembled—so it seems—under our conscious, deliberate

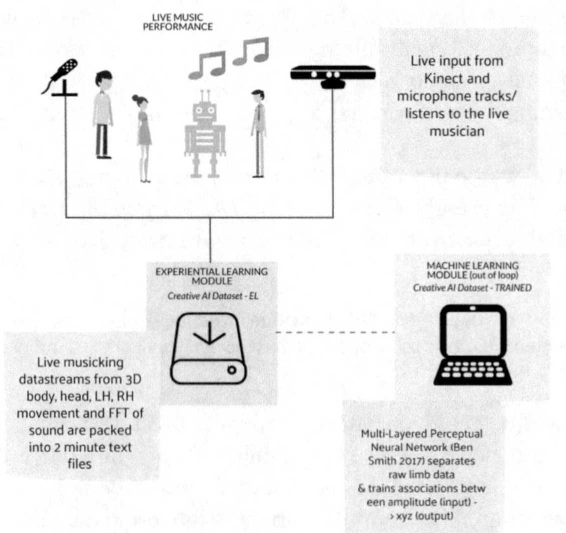

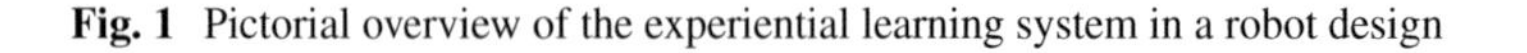
Fig. 1 Pictorial overview of the experiential learning system in a robot design

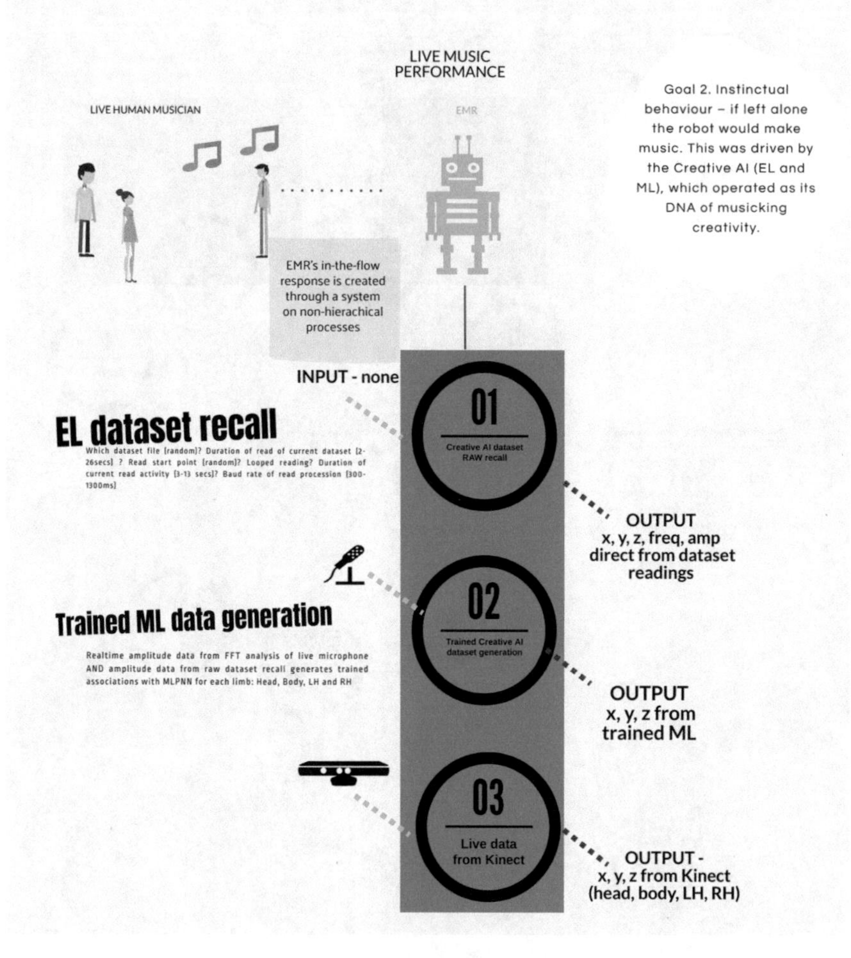

Fig. 2 Pictorial overview of the experiential learning system in a robot design

control. Other times, our thoughts wander, and the trains seem to assemble themselves' (Gelernter 2007).

- *recollection*—according to Gelernter 'when we pull out of memory a recollection associated with the same sort of feelings we're experiencing now… it's natural to apply the outcome or conclusion or analysis we arrived at then. And that's (in briefest outline) how emotions work as a 'parallel mind', how they lead us to fast conclusions we can't necessarily explain—but they feel right'.
- *affect-linking*—Gelernter argues that the mind can leap between mental images when 'two recollections engender the same emotion' (Gelernter 1994, p. 36).

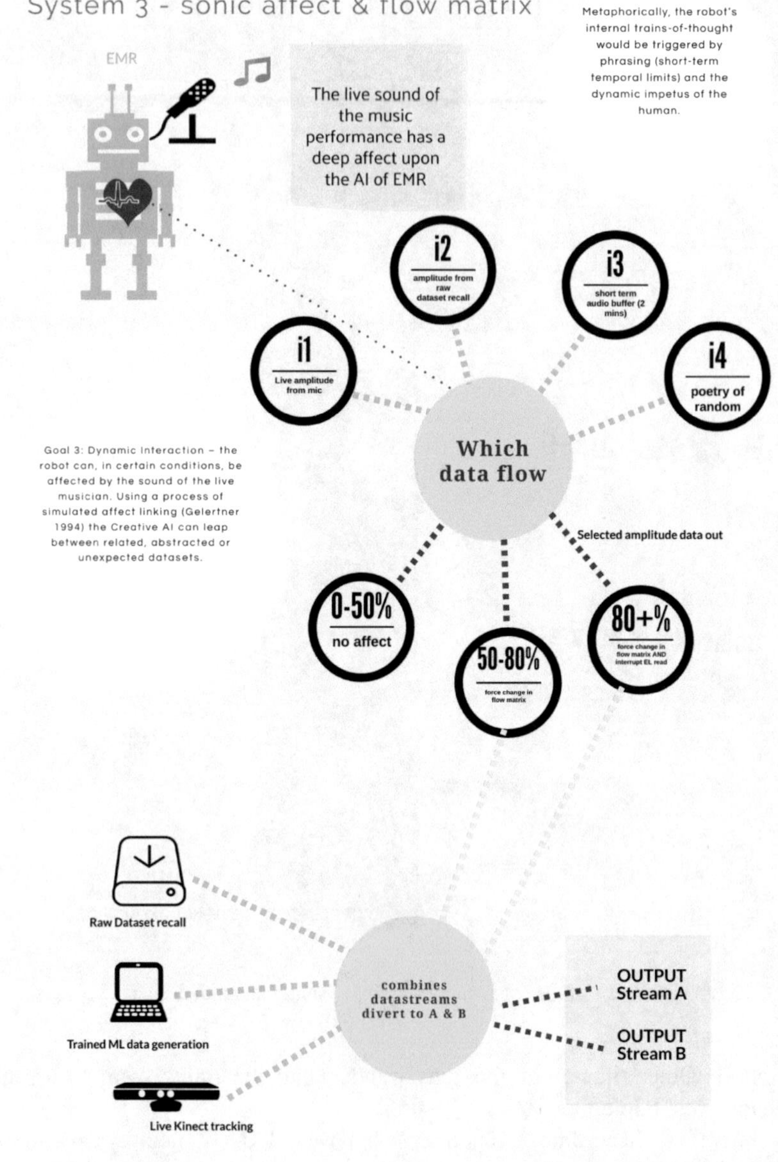

Fig. 3 Pictorial overview of the experiential learning system in a robot design

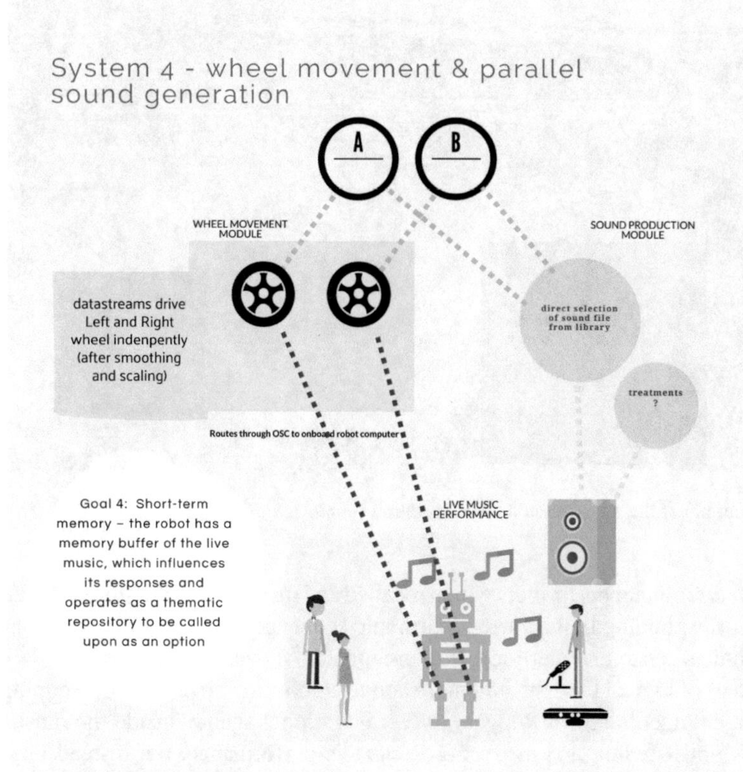

Fig. 4 Pictorial overview of the experiential learning system in a robot design

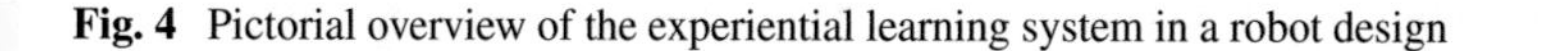

Each of these facets of the mind is part of a 'spectrum of thought' ranging between 'Upper-spectrum thought is abstract, full of language and even numbers; and lower-spectrum thought is concrete, full of sensation and emotion' (Ibid.). Furthermore, that these do not happen in steps but through a sudden awareness. 'It all depends not on a step-by-step logical sequence but on a step-by-step emotional one' (Friedersdorf 2017).

On Junitaki Falls was created for a solo musician, Christopher Redgate playing Oboe. It used a central director to control (a) a unique harmonic sequence with which it navigated a deconstructed version of a transcription of Eric Dolphy's *God Bless This Child*[8] (see Fig. 5), (b) a dynamic visual score shown on a laptop screen that mashed in real-time bars from the transcription that corresponded to the present harmonic sequence, (c) recorded and organised the recording of Redgate's performance as it corresponds to the current harmonic sequence; in a poetic sense, these became a repository of shared memories between the human-musician and the AI, and (d)

[8] The Roger Jannotta transcription was separated into bars and catalogued according to its underlying harmony. https://www.youtube.com/watch?v=m0-auuH1A-8.

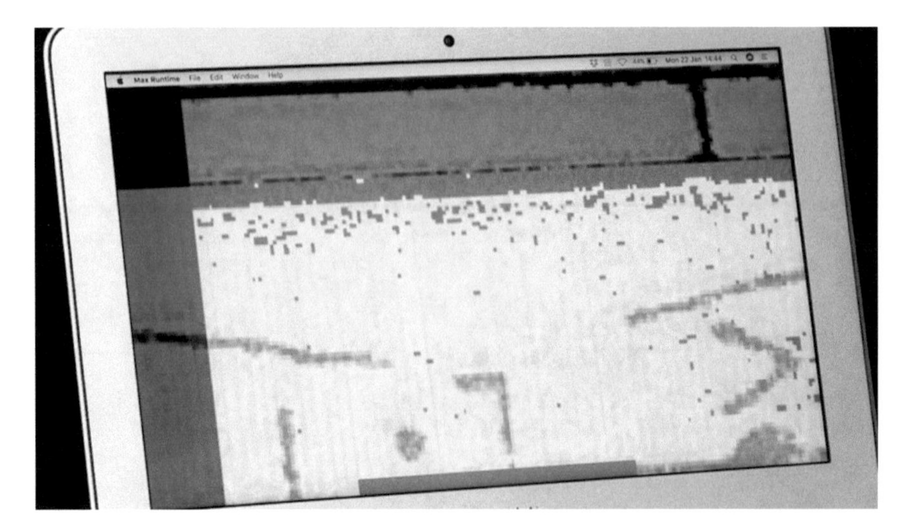

Fig. 5 An example of the AI-generated visual notation for *On Junitaki Falls*

coordinate the computer performers which recalled and manipulated the stored shared memories corresponding to the current harmonic sequence.

The technical system underlying this embodied AI was less concerned with symbolic ideas of logical thought, and more about supporting the process of it coping as an intelligent agent and for Redgate to feel its rational agency inside the music. The AI was not listening and analysing Redgate's performance, but instead used the central spine of the harmonic sequence as its perceptual core, as was Redgate. In practice, the human and AI agents operated concurrently within the shared environment of the live music. They are bound together by a shared goal: to navigate the pre-defined, randomly generated harmonic sequence. The percept manager (the general rule intended to regulate behaviour) of the embodied AI algorithm was to maintain a sense of familiarity with the human as both sounding material and logical adherences to the harmonic sequence. In this sense, the embodied AI for this project was listening into the harmonic sequence and problem-solving a response to it by controlling the two AI performers. The goal was to stimulate creativity in the human, who was simultaneously navigating the same harmonic sequence. In music terms, this concurrent relationship has less to do with improvisation and has more in common with pre-composed Western Art music, albeit, with a greater degree of freedom from the individual agents.

The embodied AI continually improves its relationship with the human by memorising and storing the human musicians' interaction with the AI over months, if not years, of shared experiences through the music. When presenting distorted sonic images of these memories back to the human within the flow of musicking, they operate like a distorted mirror of sorts, where the human feels an image of themselves in the flow and recognises a familiarity about these phrases leading to affect-linking. These distorted audio phrases working together with the visual score begets another response from the human-musician, which is stored and catalogued as a shared acoustic memory through the memory folder system. This in turn creates a cycle of response and invention because of the thought-trains and affect-linking from within the embodied *situation* that they are operating in, and for this to be perceived by the musician as intelligent and creative from the situated space.

Crucially, the embodied AI behaviour needed to feel intuitive inside the live musicking (flow), be meaningful to Redgate (familiar/inspiring 'thought-trains') and do something creative in this realm. The result was a technical solution that emphasised the symbiosis in the behaviours of the code and the recorded media in real-time musicking, over symbolic traits of analysis, reasoning and logic. After 18 months of development, through a process of iterative design and deployment, increasingly complex beta versions were developed with Redgate to a point where he felt the embodied AI was 'being there with him'.

I would like to call this 'machine learning', but that term is already restricted to a form of statistical data analysis that automates and improves automatically, the use of data, so I call this experiential learning instead. Although *On Junitaki Falls* is restricted to experiential learning through gathering shared acoustic memories and recycling them through distorted recollection, it did provide enough evidence to support the premise that the human-musician was in a relationship with them and, crucially, engendered affect-linking and made meaning because of them. Furthermore, that by storing shared memories and recalling them did appear to Redgate to be learning an aesthetic/interpretative approach to each of the harmonies in the composition.

The dataset of audio memories captured as part of the *On Junitaki Falls* project now spans to over four years of performance and development with this individual composition. There are also, at the time of writing, two other musicians feeding and growing an individual dataset through their own nurturing of this project. The recollection processes within each of these versions are quite fascinating for the individual musicians; they are aware that recollections may be presented to them by the composition, and at the same time, their responses are being recorded, which in turn may form the raw material for a recollection later down the line. Crucially, these recollections stimulate meaning-making in the musician as they recognise an essence of themselves in the distorted recollection, that may be historic, or interpreted differently because of the distance in time. More so that they are not some superficial artefact of music-AI generation, but are a container of a type of felt embodied intelligence from which they offer percepts towards 'more evolutionary primitive mechanisms which control perception and action' (Ibid., p. 100).

As an extension of this, the next phase of experiential learning developed a dataset that captured bodily movement and audio memories of a human-musician in the flow of musicking. This created a proof-of-concept (PoC) embodied dataset that was implemented into the follow-on projects and is discussed next.

4.3 Embodied AI Dataset

The embodied AI dataset was designed to dig a little deeper into the challenge of *what is to be modelled* when deploying embodied AI in creative practices and to train neural networks from this dataset. The initial design focussed on a single principle of embodied musicking: that physical gesture and sound actuation are linked not by the sequence of *mental idea → physical gesture → sound-activation*, but that musician's think as sound and the sequence is closer to *sound-as-idea → sound-activation as physical gesture*. Therefore, if we are to model neural networks on embodied musicking, the physical gesture of the musician and the resultant sound production are implicitly linked not as part of a linear sequence of a thinking mind. This presents a more convincing set of features for modelling, than, say, harmonic construction.

The primary features of the dataset captured the complex of bodily movement of the musician with the physical properties of the sound. Defining the complex of bodily movements was an important aspect to this design, as I understood through personal experience and in conversation with other musicians, that, say, when an arm moves a cello bow, it is the whole body that is involved in the embodied music gesture, not just the isolated limb or joint. Therefore, the inter-relationship between the whole complex of bodily movements would need to feature in the dataset (see Fig. 6).

For the purposes of this PoC, the design for this first stage dataset was *Sequential ID; Body Part* (*Head, Body, Left hand, right hand*); *3-D position* (*x, y, z axes*); and *Audio Analysis* (*FFT fundamental, amplitude*). This data was then used to train four Multi-Layered Perceptron Neural Networks (MLPNN), one for each of the body

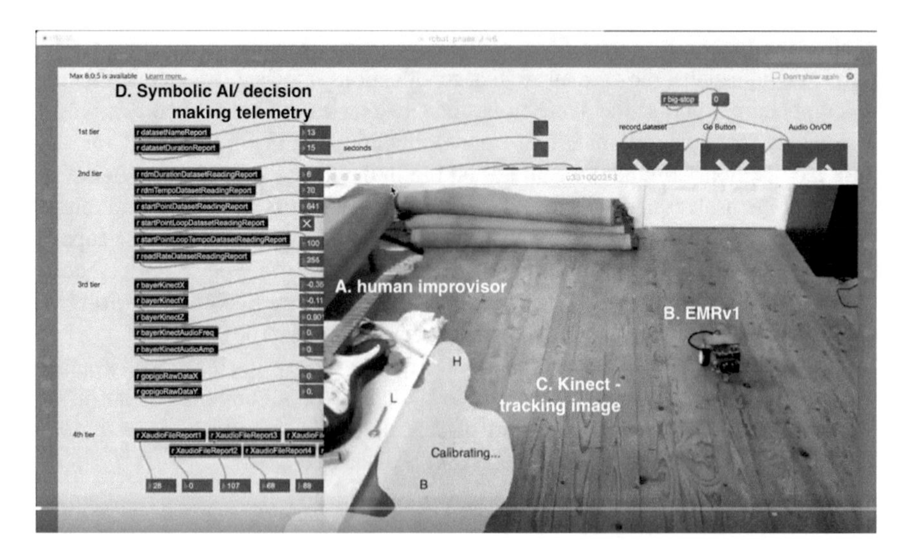

Fig. 6 Capturing the embodied dataset through in-the-loop methods and improvisation

parts captured, and implemented into two projects: *Seven Pleasures of Pris* (2019) and *Voight-Kampft Test* (2019). These projects were built around 2 then 3 separate AI's performing with each other. Each of the AIs contained each of the four MLPNNs. The behaviour of the AI was coordinated through a director that organised the streams of incoming signal (audio) with MLPNN response.

In the two examples, this MLPNN data and the raw dataset were used to control every aspect of the AI, movement, sound production choices and interaction goal. The raw dataset and the MLPNN outputs were also used to make choices about how the dataset is to be recalled and read by the algorithms (e.g. the read rate and ramp speed for each instance of wheel movement). This means that the direct application of data into wheel movement, and also the translations of that into sound-object choice and therefore as music in the flow is imbued with the essence of embodied musicking that has been embedded in the core of the dataset. The version of the dataset in this application was crude and small, but it has since been superseded by a larger project and more comprehensive embodiment approach to the dataset.

The inference I made through this process was that one isolated feature of this dataset, say the x coordinate of the right hand, might be imbued with a sense of embodied musicking, and this alone might be a viable data stream in the stimulation of percepts for the human-musicians. By that I mean, that any of the streams (raw data or neural network prediction) is embedded with a sense of embodied musicking and could therefore be used to control any parameter in the musicking AI behaviour. By extension and hypothetical proposition, that the essence of this dataset—imbued with creative and embodied musicking—could be used in a variety of creative applications not just music.

4.4 Coping and Belief

Earlier in this chapter, I discuss that the focus for embodied intelligence is on the coping behaviours that are required to maintain a balance of relationships within such a situated environment, and it is within this balance that the human-musician can find meaning. As music is a dynamic problem space, concentrating on the coping behaviours that are required to maintain a balance of relationships within such a situated environment is in my experience an optimal way of offering the human-musician an opportunity to find meaning. This is because the situated space of embodied musicking requires musicians to cope with a dynamic and changing world and that they need to do something in this world. If the AI is also to cope with dynamic changes and more importantly do something in this world that stimulates embodied interaction with a human-musician, then modelling it to some representational world-view is not the solution as this brings with it some significant questions, such as whose world-view are we to use for this representational model? and why would we want to have a strangle hold on the limits of its creative potential to known parameters?

Embodied AI needs to cope in real time within the realm of musicking and stimulate an embodied interaction within the flow. This requires a non-representational approach to how it relates to the flow as the coping mechanisms needed to be open and dynamic enough to co-operate in any given musicking realm. Limiting the robot to a single representation of what musicking is, or might be, imposed onto the system by the human designer(s), would only work in a number of instances.

The design of the coping systems that I implement is informed by two early papers by the robot innovator Rodney Brooks, specifically *Intelligence without Reason* (Brooks 1991) and *Intelligence without Representation* (Brooks 1987). In these, he lays out the foundation of his approach to designing and building robots that are first and foremost able to cope and therefore adapt to a dynamically changing environment within the parameters of specific and multiple goals. Brooks' research considered these robots to be a sort of 'creature' that copes in a specific world-of-concern in real time. They do not have a model of representation of their world (such as building a 3-D model of the space through computer vision and object analysis), nor do they make one as it goes about its business, but uses goals and strategies to cope with whatever that world can throw at it.

Brooks' foundational theories guided the developed for my early embodied AI musicking robot (EMR) projects[9] and generated this set of principles[10]:

- EMR must cope in an appropriate musical manner, and in a timely fashion, with the dynamic shifts inside the musicking world;
- EMR should be robust to the dynamic environment of musicking, it should not fail to minor changes in the properties of the flow of musicking and should behave appropriately to its ongoing perception of the flow;
- EMR should maintain multiple goals, changing as required and adapting to its world by capitalising on creative opportunity;
- EMR should do something in the world of musicking, 'it should have some purpose in being' (Brooks 1987).

A subsumption architecture was designed to support these multiple goals. This is a control architecture innovated by Rodney Brooks as an alternative to traditional AI, or GOFAI. Instead of guiding robotic behaviour by symbolic mental representations of the world, subsumption architecture correlates sensory information to action selection in a 'bottom-up' way. These were (in order of priority):

I. *Self-preservation*—the robot must avoid obstacles, not crash into the other musicians or fall off the stage.
II. *Instinctual behaviour*—if left alone the robot would make music. This was driven by the embodied AI dataset (discussed above), which operated as its DNA of musicking creativity.
III. *Dynamic Interaction*—the robot can, in certain conditions, be affected by the sound of the live musician. Using a process of simulated affect-linking the embodied AI could leap between related, abstracted or unexpected datasets. Metaphorically, the robot's internal trains-of-thought would be triggered by phrasing (short-term temporal limits) and the dynamic impetus of the human.

[9] https://twitter.com/i/status/1074617355345559552.

[10] Adapted from Brooks (1987).

A critical feature of the design of EMR is that each of these goals directly moves the wheels. It was essential that each goal is not part of an elaborate, logically flowing, representation of a thought process, mimicking some-kind-of-mind. As such, the overall design of robotic system was modular, with each system directly accessing the wheels when in operation. The overall design of the data flow is:

1. *Live data sensors (audio and visual) & SELF PRESERVATION*

|

2. *Data wrangler, neural net prediction, raw data recall/ recollection*

|

3. *Affect mixing (trains-of-thought): INSTINCTUAL BEHAVIOUR & DYNAMIC INTERACTION*

|

4. *Smoothing and deviation*

|

5. *Wheels move = Make sound*

In order for the embodied AI to have some purpose of being, it also needs to have a world view of music; in fact, I would argue that it needs a personalised world view rather than the whole entirety of music at its disposal. In short, the embodied AI need to have a belief system that frames its personalised world view through limitations, embedded aesthetics and behavioural traits (even glitches and bugs in the system).

Belief in this sense is used to describe an acceptance by the robot that something is true or that it has trust or confidence in something from its perspective especially in how it interprets percepts in the environment of musicking. Beliefs are not facts: they are subjective and individual, and they are not consensual and deal with conceptual parameters that may be biased, prejudiced or activated by affect and emotion. Beliefs may be based on subjective experiences and past episodes and may distort reasoning, knowledge and the processes of synthesising knowing (Cohan and Feigenbaum 1982, pp. 65–74). All of which I recognise as very musical traits.

The belief systems that I have implemented in my embodied AI are, at the moment, limited to three different attributes:

(A) *Personalised embodied AI dataset*—at the core of embodied AI belief system is a dataset that drives the individual AI. This PoC dataset has been represented in different formats across different project. The design of this dataset is discussed above, but it is important to stress that a new neural network was created for each of the AI, and that an individual character was embedded into each by randomising the raw dataset before training each model.

(B) *Movement behaviour*—The robot's movement operates within a behavioural system, designed to react openly to the dynamic soundworld and move the wheels accordingly. It is important to note that the interaction with the musician begets movement as its primary goal for musicking and that this movement is embedded with essence of embodied musicking because of the embodied dataset process. Following this, the movement begets sound, which begets music such that all relationships between human and AI are informed by phenomenon data captured within the embodied flow of musicking: either from the neural network, raw dataset or through live interaction.

C) *Soundworld*—the robot has a fixed sound library of sounds that amounts to its foundational aesthetic repository. These are different for each robot and are always extracted recordings created by humans engaged in embodied musicking, so as to embed them with an essence of musicianship. These are triggered only when the wheels move. These are then either presented to the world in their raw state or are treated in some way (time stretch, pitch shift or both) using the data streams as controlling parameters.

4.5 *Embodied Percept Input: Affect Module and Bio-synthetic Nervous System*

It has been necessary to rethink the sensory mechanisms and percept inputs for embodied AI. While standard instruments such as 3-D webcams (e.g. Kinect) and microphones can capture some of the physical properties of embodied musicking, it has been necessary to re-conceptualise aspects of sensory capture that help to emphasises belief and the inside nature of a type of perception in musicking.

Although the two solutions that I will introduce here—the affect module and the bio-synthetic nervous system—may read like attempt to bring the AI closer to a level of consciousness and human perception, they are in fact named metaphorically and are merely solutions that get the AI closer to a set of characteristics that I defined in order to stimulate such a relationship. These characteristics are:

1. listening
2. having my own train-of-thought
3. being surprised and maybe doing something with that
4. being surprising.

The first solution, the affect module, interrupts the continuity of internal data streams in response to the live sonic stimuli from the music. This module received all the data streams from the dataset query and parsing process, and the neural networks and mixed them to two outputs: left wheel data and right wheel data. The mixing was controlled by a special process designed to symbolically represent affect and affect-linking of a musician. I define affect as 'the mind's connecting response between sensorial input of external events with the internal perception of causation such as emotion or feeling, through time' (Vear 2019). This module translated this definition symbolically, the streams of amplitude data from the live input, the dataset parsing and a randomly generated 'drunk walk' would be used to trigger (a) local changes in the module such as mix and (b) global conditions such as dataset file selection. The basic process was:

1. randomly switch between input streams (1–4 s, or with a loud affect trigger)
2. if amplitude is < 40% do nothing
3. else if amplitude is between 41 and 80% trigger a new mix (see below)
4. else if amplitude is > 80% trigger condition changes across the architecture (new mix, new file read, restart reading rate and change smoothing rate, change audio read in following modules).

The mix function randomly selected which of the incoming data streams (*x*, *y*, *z* from dataset read, *x*, *y*, *z*, from live Kinect and *x*, *y*, *z* from neural network prediction) to be output to the following module for wheel movement. It was desirable that this involved multiple elements from these incoming streams being merged, metaphorically fusing different trains-of-thought into a single output.

The second solution, the bio-synthetic nervous system (BSNS), works as a sensory input for the embodied AI. In short, a container of moss is placed on top of the speaker used to transmit the AI's sound generation and relay the incoming sound from the other musician. Inside the moss container are positive and negative terminal that sense capacitance changes across the moss in response to vibration. When the speaker makes a sound, the levels of capacitance change, although there is no direct correlation between input value of a specific vibration, say 440 Hz, to the capacitance shift. In Fig. 7, you can see in an initial experiment how the capacitance generates a rhythmic response when I play a 50 Cent track through the speaker. The value from this system is then fed into the affect module as a 'self-awareness' data stream.

I must stress that I do not believe the AI to be self-aware (that I'm aware of), but this system is a way of achieving several connections between human-musician and the AI through musicking. There are three main connections that this system engenders:

1. the BSNS operates as a crude embodiment sensor. It is a direct connection between the sound vibration and some felt stream, that is different

Fig. 7 A track from 50 cent playing into a speaker, on top of which the moss-based sensor system feels the vibration, which is turned into digital signal (on the screen behind)

from a microphone and introduces a sense of biological connection with the human-musician.
2. the BSNS stream is subjective as there is no true direct correlation between the moss's response and the sound frequency or amplitude. This feeds into the AI's belief system as these are based on subjective experiences and past episodes and can distort reasoning, knowledge and the processes of synthesising knowing, beyond empirical and into interpretation.
3. to paraphrase Breazeal's description of her metaphorical 'synthetic nervous system' (Breazeal 2004, p. 38) in her social robot *Kismet*, BSNS draws together inspiration from 'infant social development, psychology, ethology and evolutionary perspectives' to enable embodied AI to 'enter into a natural and intuitive social interaction' with a human-musician.

5 Conclusion

In embodied and creative pursuits such as creative AI with music, or any performing/interactive situation, it is important to address the meaning of artificial intelligence and to understanding it in its context, rather than imposing upon it an outsider objective viewpoint limited to mind-based models. While I am not trying to claim that this research presented here is the solution, or even that this approach is part of the solution, I am proposing that understanding embodied intelligence from these perspectives can unlock areas of creativity with AI, by enhancing the close-coupled relationships that stimulate a sense of meaning in the human-musician.

The solutions that I have introduced above prioritise a way of thinking about embodied musicking between human-musicians and AI/robots. This way of thinking can be outlined by the following principles:

- *the robot was not an extension of the musician; but should extend their creativity*
- *the robot should not be an obedient dog or responsive insect jumping at my commands or impetus, but a playful other*
- *it should not operate as a* simulation *of play, but as a* stimulation *of the human's creativity*
- *it is not a tool to enhance the human's creativity, but a being with presence in the world that they believe to be co-creating with them*
- *it should prioritise emergence, surprise, mischief, not expectation.*

These solutions discussed in this chapter do not operate like, or even resemble GOFAI, if you were to only judge them by 'looking under the hood' at the code. This is because the 'hood' in this case is not symbolic code that solves problems or learns data sequences but is the operational behaviour of the code in its situated environment.

There is much more for this research to do. This will take time as each step of the research journey builds a new AI and robot and evaluates it inside embodied musicking. This is truly important as, to quote Rodney Brooks, 'one real robot is worth a thousand simulated robots' (Brooks 1989).

References

Anderson ML (2003) Embodied cognition: a field guide. In: Artificial intelligence, 149, pp 91–130

Bachelard G (1958, 1994 ed.) The poetics of space. Beacon Press, Boston

Breazeal C (2004) Designing sociable robots. MIT Press, Massachusetts

Brooks R (1999) Cambrian intelligence. The early years of the new AI. MIT Press, Cambridge

Brooks R (1991) Intelligence without reason https://people.csail.mit.edu/brooks/papers/AIM-1293.pdf. Last accessed 2020/10/23

Brooks R (1990) Elephants don't play chess robot. Auton Syst 6:3–14

Brooks R (1989) "Battling reality". Available online https://www.researchgate.net/publication/37597354_Battling_Reality. Accessed 12 May 2022

Brooks R (1987) Intelligence without representation. Artif Intell 47(1):139–159

Cohan PR, Feigenbaum EA (eds) (1982) The handbook of artificial intelligence, vol III. Addison-Wesley Publishing Co. Inc., New York

Dourish P (2001) Where the action is: the foundation of embodied intelligence. MIT Press, Cambridge

Emmerson S (2007) Living electronic music. Routledge, London

Friedersdorf C (2017) 'There's enough time to change everything': the polymath computer scientist David Gelernter's wide-ranging ideas about American Life. In: The Atlantic. Online available at https://www.technologyreview.com/2007/07/01/36908/artificial-intelligence-is-lost-in-the-woods/

Gelernter D (2007) Artificial intelligence is lost in the woods: a conscious mind will never be built out of software, argues a Yale University professor. In: Technological review. Online, available at https://www.technologyreview.com/2007/07/01/36908/artificial-intelligence-is-lost-in-the-woods/

Gelernter D (1994) The muse in the machine

Jordanous A (2020) Intelligence without representation. Systems 8:31. Online available at https://res.mdpi.com/d_attachment/systems/systems-08-00031/article_deploy/systems-08-00031-v2.pdf

Nijs L, Lesaffre M, Leman M (2009) The musical instrument as a natural extension of the musician. Online. Available http://www.academia.edu/205605/The_musical_instrument_as_a_natural_extension_of_the_musician. Accessed 30 Jan 2021

Pallasmaa J (2005) The eyes of the skin. architecture and the senses. Wiley, New York

Russell SJ, Norvig P (eds) (2020) Artificial intelligence—a modern approach, 4th edn. Prentice Hall, New York

Russell SJ, Norvig P (eds) (2009) Artificial intelligence—a modern approach, 3rd edn. Prentice Hall, New York

Vear C (2019) The digital score. Routledge, London

Wooldridge M (2020) The road to conscious machine: the story of AI. Pelican Books, London

Latent Spaces: A Creative Approach

Matthew Yee-King

Abstract This chapter explores the creative possibilities offered by latent spaces. Latent spaces are machine-learnt maps representing large media datasets such as images and sound. With a latent space, an artist can rapidly search for interesting places in the dataset and then generate new artefacts around and between data points. These unique artefacts were not in the original dataset, but they relate to it. Readers will find a detailed explanation of what latent spaces are and how they fit into a series of developments that have taken place in digital media processing techniques such as content-based search and feature extraction. We will encounter four examples of machine learning systems that provide latent spaces suitable for creative work. The first example is Music-VAE which creates a latent space of millions of musical fragments represented in the symbolic MIDI format. The second example is Latent Timbre Synthesis (LTS). Unlike Music-VAE, which works in a symbolic musical domain, LTS works directly with audio fragments. The third example is StyleGAN which creates a latent space of images which has specific properties allowing for style transfers. The final example is VQGAN + CLIP which is a text phrase-to-image system which uses fine-tuning techniques to iteratively generate images. Finally, we consider examples of artists working with each of the four systems along with reflections on their creative processes.

Keywords Machine learning · Music-AI · StyleGAN · Generative art · Latent space

1 Introduction

One might characterise the practice of creating visual art, music or sound as an exploratory process. An artist explores the space of possibilities in a given domain. Over time, they develop an understanding of this space and how they can effectively explore it and present it to others. This chapter will consider how such a creative

M. Yee-King (✉)
Goldsmiths, University of London, London, UK
e-mail: m.yee-king@gold.ac.uk

C. Vear and F. Poltronieri (eds.), *The Language of Creative AI*, Springer Series on Cultural Computing, https://doi.org/10.1007/978-3-031-10960-7_8

process might change if the artist has access to a latent space. We will dig into a deeper explanation of latent spaces later, but for now, think of a latent space as a map of thousands or even millions of items. Taking the example of a latent space of images, you would start with a dataset containing thousands of images. A machine learning system would place all the images in positions relative to each other, just like a city map showing the various buildings and landmarks. It would attempt to place similar-looking images in similar positions on the map. The space containing all the positions is known as the latent space. You can take an image that the map knows, and the map will tell you where that image is located.

A latent space is not ordinary map, though—you can also take an image that it has never seen, and it can give you its best guess as to where that image lies based on how it looks. You can place your finger in a random position on the map, which might lie in between known images, and the system can generate you an image that is its best guess for what you should see there. This ability to place and create artefacts not on the original map unlocks a wealth of creative possibilities. In this chapter, I consider the new opportunities for artistic practice presented by such technology and how creative practice changes to best exploit these possibilities.

Let us capture a flavour of the change by taking in a view from some contemporary creative practitioners who have worked extensively with latent spaces. Broad et al. describe the creation of latent spaces as 'learning to render entire distributions of complex high dimensional data with ever-increasing fidelity. They group together the various ways artists have exploited latent spaces under the term 'active divergence'. Active divergence involves 'optimising, hacking and rewriting [latent space models] to actively diverge from the training data' (Broad et al. 2021). Referring to the example of a map of images, the training data would be the set of images used to create the original map.

Optimising, hacking and rewriting are standard methods for computer artists, but how can they execute those in this machine learning-driven domain? We will consider this question through several detailed examples later. The ability to create and explore latent spaces can be seen as a computer-aided transition from considering a small number of positions in a creative space to having access to an interrogable model representing many positions. How can an artist cope with this abundance of choices? As it turns out, this is nothing new for artists. Going back to the early days of computer-aided creative work, we find that practitioners were quick to identify and discuss the 'over-abundance problem'. In an interview in 1968, John Cage characterised computerisation as a transition from a scarcity of ideas to an abundance:

> The need to work as though decisions were scarce-as though you had to limit yourself to one idea-is no longer pressing. It's a change from the influences of scarcity or economy to the influences of abundance and-I'd be willing to say-waste. Cage 1968. (Austin et al. 1992)

A contemporary of Cage, Gottfried Michael Koenig goes a step further in considering what statistical distributions representing a range of possibilities might mean for composers:

> the trouble taken by the composer with series and their permutations has been in vain; in the end it is the statistical distribution that determines the composition. Koenig (1971), (Ames 1987)

Another composer, Iannis Xenakis, famously also worked with statistical distributions. Writing in 1966, Xenakis saw it as a 'musical necessity that the laws pertaining to the calculation of probabilities found their way into composition' (Xenakis 1966). This colourful quote embodies the feeling one might have when working with latent spaces:

> With the aid of electronic computers the composer becomes a sort of pilot: he presses the buttons, introduces coordinates, and supervises the controls of a cosmic vessel sailing in the space of sound. Xenakis (1971) (ibid.)

These early statistical creation methods were probably a natural response to the constraints and formalisms of serialism that had come before. One might draw a parallel between the movement from serialism to stochastic composition and the rise of statistical machine learning after decades of formal and symbolic AI techniques. It is this statistical machine learning which makes creative latent spaces possible.

Reflecting on the quotes above, these early pioneers of statistical, creative spaces were already transitioning to conceptualising the substrate of their work not as small numbers of items and simple processes but large sets and complex processes. A full exploration of the history of artistic practice using statistical distributions and related techniques is beyond the scope of this writing. We have seen a to-ing and fro-ing between formalisms and statistical models in computational creative work and machine learning, with increasingly sophisticated models.

The most recent development in this story started in the mid-2000s, and it makes the rendering of large datasets into high-fidelity latent spaces of images and sounds possible. LeCun et al. refer to this development as representation learning (LeCun et al. 2015). This is where a machine learning model learns how to extract and represent the most pertinent information in a dataset of images, sounds or texts. These machine-designed latent spaces and their capabilities are the subjects of this chapter.

1.1 Structure of This Chapter

In the next section, I will explain what a latent space is and how we can go about creating them. In Sect. 3, we will consider some examples of recent systems which enable artists to work creatively in latent spaces. Following that, we will look at some creative work that has been produced using the example systems we presented in Sect. 4.

2 What Is a Latent Space?

Now, we will work towards an understanding of what a latent space is, why it is necessary and how we can go about creating one.

The avant-garde composers quoted in the previous section observed that the computerisation of creative practice is associated with a transition from a sparsity of ideas and content to an abundance. Instead of labouring with a single musical score, we work within a dataset of scores. Instead of working with a single image, we work with a dataset of thousands or millions of images. This leads us to wonder how we might adapt our practice to engage with this abundance of material.

We can start with two actions that we might want to take as creative practitioners given such a dataset: search and generate. Search allows us to find items of interest. Generate allows us to somehow generate new items using the existing items. These actions are shown in Fig. 1. One search approach is to use meta-data such as filenames, dates, geo-tags and camera settings. For example, you might search for all images taken in 2010. This approach depends on somebody or something having already correctly tagged the items, and the search is limited to the available tags.

Once we have located our items of interest in the dataset, we can move to the generation of new items. For example, you might load the images retrieved from your search into an image editor and create a collage. The problem is that this approach takes us back to the 'sparse' scenario as only the search part of the process takes account of the abundance in the archive. Generating like this ignores anything except the selected items.

Content-based search is another option. Content-based search involves examining the actual data in the archive. For images, the lowest level of data is the raw colour values for the pixels; for sound, it is the raw waveforms. Content-based search is potentially much more potent than meta-data search as it does not rely on accurate data being added to the items in an archive. Instead, it goes directly to the actual content in the archive.

But there are a few reasons why working with raw media data is problematic. It is very large; for example, the data for an image with a resolution of 1024×768

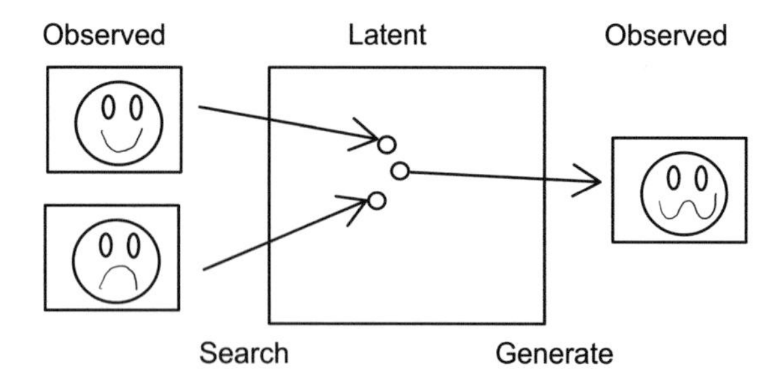

Fig. 1 Observed space contains the raw data representing the images. Latent space is a more compact space which places similar images close to each other. A small movement in latent space should lead to a small movement in observed space

has one number for each red, green and blue channel in its 786,432 pixels, so it is 2,359,296 dimensional. Raw pixel data is also sensitive to image transformations, such as translations and rotations. If you rotate an image, the pixel data might change completely. So with raw pixel data, you would not be able to know whether one image was very similar to another, just rotated. The two images would appear very far away in raw pixel space.

So why are humans able to judge two rotated images to be similar? It turns out that the raw data coming out of the back of the retina is processed through several stages before meaningful perception takes place. This processing can be called feature extraction, and algorithmic (as opposed to biological) feature extraction is what makes content-based search possible. To achieve content-based search, we can specify a desired feature as a search term then find items with similar features in our dataset. We might even pass a raw media item such as an image as our search term, then extract the feature for searching.

Researchers have designed many different features to enable searches for different things, for example, features that help detect which instrument is playing in a piece of music or features that help detect the position and orientation of faces in images. There are many software libraries available that make it possible to extract these features, for example, the OpenCV library for image features and the librosa library for audio features (Bradski and Kaehler 2000; McFee et al. 2015).

2.1 *Latent Spaces*

Now, we are familiar with the ideas of content-based search and feature extraction, and we are ready to talk about latent spaces. The features that we use to carry out content-based search form a latent space. We can refer to the original files as the 'observable' and the features we extract as the 'latent'. Therefore, the latent space is the space that contains the features of everything in our dataset. Consider an 'average brightness' feature for images as a simple example. We can represent this feature as a single value, perhaps between 0 and 1. To search, we can specify our desired brightness range and retrieve all images falling in that range. The latent space for this feature illustrated in Fig. 2 is a simple, one-dimensional line along which we place all of the images in our dataset. If we move through the latent space, viewing the images as we go, we will see them increase in brightness.

What about generation? Operating with the content of media archives intuitively seems to lend itself to generative tasks, as we are working more directly in the domain we wish to generate. We could manually generate new content from the found content, as we did for meta-data search. But perhaps, there is a more powerful option, one which maintains a sense of the abundance in the archive. If we can reverse the feature extraction process somehow, we can actually move from latent feature space back to the raw data of observed space. With such a feature reversing technique, we can actually move from any point in latent space back to observed space. This opens up

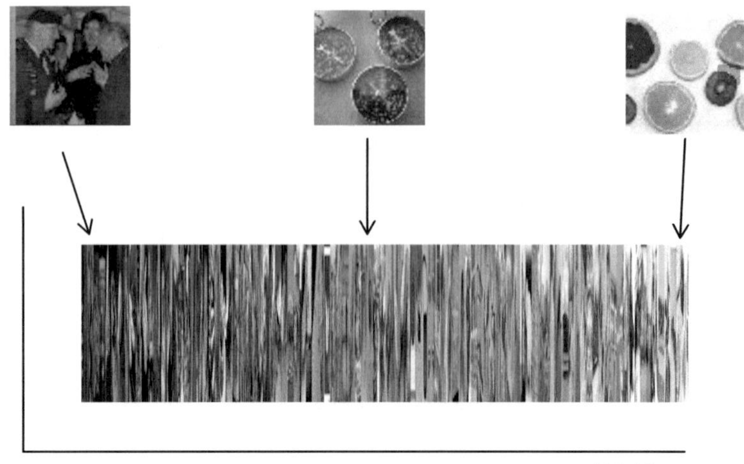

Fig. 2 Visualisation of a latent space based on a one-dimensional image brightness feature. 100 images from the ImageNet dataset are organised from darkest to brightest (Deng et al. 2009). Additionally, three zoomed-in images are shown, which are the darkest, middle brightest and brightest

possibilities such as choosing a point in the middle of two images and generating from there. An example of that is shown in Fig. 1.

But reversing from latent feature space back to observed space is not so straightforward. Consider the analogy of a fruit juicing machine. You place your fruit in the machine, and it extracts the most desirable part of the fruit. But it is not easy to get back to the original fruit because you have thrown away 'information' in the juicing process. My simple image brightness feature loses a lot of information going from two million numbers down to one single number, and we could not possibly reconstruct an image from that. Unfortunately, many existing features developed for image and sound analysis tasks are not reversible. They were not designed for generative purposes, so reversing was not a requirement.

Fortunately, there is a new generation of reversible features based on machine learning. These are machine-designed features used in the representation learning systems described by LeCun et al. (2015). Instead of creating a latent space using human-engineered features, machine learning systems learn their own features suitable for representing the pertinent information in the dataset. In parallel, the systems learn methods for feature reversal. I will spend the rest of the chapter discussing these features and some AI-creativity systems built on top of them.

Before that, I will summarise the key points from this section: large datasets of media are now available; creative practitioners wanting to meaningfully exploit the abundance in these datasets need to be able to search in and generate from the dataset; content-based search provides the most powerful way to search using feature extraction; the extracted features form a latent space containing the dataset;

unfortunately many features are not designed for generative purposes and they are therefore not reversible meaning you cannot generate by moving from latent space back to raw data in observed space; and recent developments in machine learning provide reversible features which allow for 'fully abundant' search and generation.

3 Examples of Latent Spaces

Now, we have a grasp of what a latent space is and what it means to search and generate, and we will consider some examples of latent spaces in creative domains. I have selected four examples: two for sound and two for image. For each of image and sound, I present a symbolic approach and a sub-symbolic approach. To clarify those terms, in the musical domain, Briot et al. define symbolic as 'dealing with high-level symbolic representations (e.g. chords, harmony)' and sub-symbolic as 'dealing with low-level representations (e.g. sound, timbre)', along with the associated processes for each domain (Briot et al. 2020). An equivalent definition for images would have symbolic as the contents of the image (e.g. contains a dog) and sub-symbolic as describing the raw image data (RGB values, brightness).

Before we take in the examples, we should highlight some terms that we will encounter. Variational Autoencoders (VAEs) and Generative Adversarial Networks (GANs) are high-level terms describing approaches to designing and training neural networks, especially for generative purposes. VAE involves training an encoder to encode to a latent space and a decoder to decode back out to observable space such that the error between original and decoded data is minimised. GANs involve a generative network learning to generate data similar to the training data where the similarity is judged by a second, 'critic' network which is also learning. So with GANs the error is dictated by the ever-learning critic, not a normal metric as in VAEs. Below the VAE or GAN method is the actual neural network architecture, which in both cases will consist of multiple layers of different types such as long short-term memory (LSTM), convolutional, fully connected and so on (Fig. 3).

3.1 A Latent Space for Symbolic Music Data: Music-VAE

Roberts et al. reported the Music-VAE system in 2018 (Roberts et al. 2018). Music-VAE works with MIDI files which are a standard data format for representing sequences of musical events such as notes. Music-VAE uses a Variational Autoencoder method in combination with a recurrent, LSTM network for its encoder and decoder. LSTMs are useful for time series such as sequences of musical events. Music-VAE can encode musical sequences to a latent space, and it can decode from latent space back to musical sequences.

Generative music systems are not new—there is a long history of methods for generating symbolic music data. Aside from formal methods such as grammars,

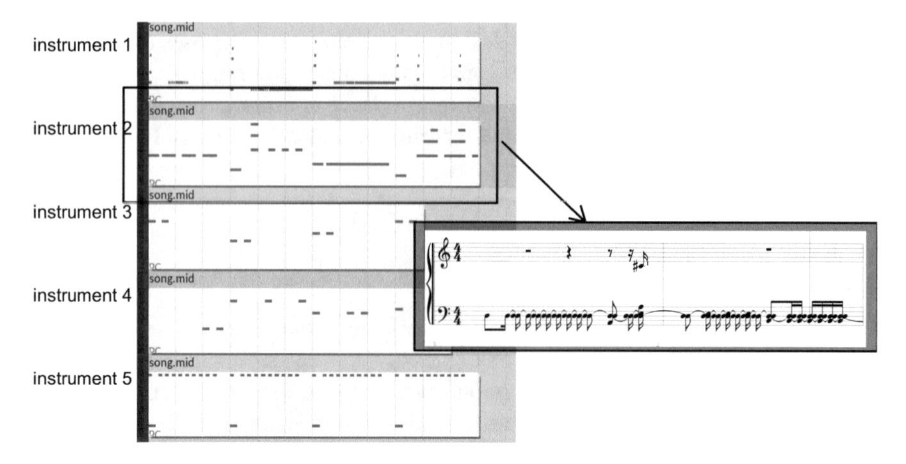

Fig. 3 Example output from a single point in the Music-VAE multi-instrument model latent space. The different instrument tracks are shown with a piano roll view, and there is a zoomed-in view of one of the instruments shown with traditional notation

Markov models were a common technique in the literature prior to the dominance of deep learning. Fernandez et al. survey examples of this work (Fernández and Vico 2013). Pachet and Roy's work with statistical models and constraints is perhaps the pinnacle here (Roy et al. 2017). There is limited work which deals explicitly with latent spaces of musical corpora but an example is Ellis and Arroyo's *Eigenrhythms* (Ellis and Arroyo 2004). Music-VAE does have some deep network precursors: the Ragtime generating DeepHear system by Sun from 2015 is perhaps the earliest example.[1] For further examples, we refer the reader to Briot et al. (2020).

Returning to our description of Music-VAE, using a dataset of 1.5 million MIDI files, the researchers trained different Music-VAE models on various inputs including 16 bar monophonic melody and bass lines, and drum patterns. The latent vectors therefore represent melodies, basslines or drum patterns. A later version of Music-VAE encoded complete musical arrangements, including instrument selection (Simon et al. 2018).

Music-VAE has been designed with the aim of creating a latent space which is suitable for creative exploration. The Music-VAE space places similar inputs into similar places, and it is a smooth space. Smooth means interpolations from one point to another in latent space produce a gradual series of outputs. In other words, similar melodies should encode to similar positions in latent space and moving between two melodies in latent space should sound like a musically smooth transition.

These two features are crucial to the creative possibilities of Music-VAE. To achieve the smooth transitions, Music-VAE uses a variational autoencoder as opposed to just an autoencoder. The variational autoencoder works with distributions rather than single points which encourages a better mapping of the space.

---

[1] https://fephsun.github.io/2015/09/01/neural-music.html.

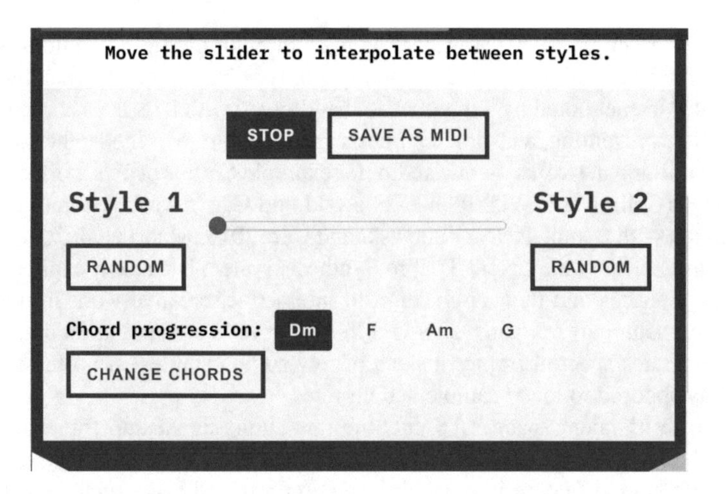

Fig. 4 The Music-VAE interpolation demo which uses the tensorflow.js implementation of Music-VAE. The user can generate two random points in latent space and interpolate between them, listening to the results. They can also condition the generative model using a chord sequence

Music-VAE's creators have endeavoured to provide a range of components to help creatives work with the system. A variety of pre-trained models are available, including melody, drum, trio and full arrangement models.[2] The Magenta.js library provides ready-made helper classes and functions.[3] Magenta.js allows users to sample latent space in near real time, generate MIDI files and play them in the web browser.

The researchers also provide a range of examples, especially based around the idea of interpolating between positions in latent space. Figure 4 shows one such demo.

3.2 *A Latent Space for Sub-symbolic Audio Data: Latent Timbre Synthesis*

In 2021, Tatar et al. described Latent Timbre Synthesis (LTS) (Tatar et al. 2021). LTS involves the creation of a latent space representing a corpus of audio frames. Therefore, LTS is a sound synthesis technique, where a stream of latent vectors is converted to a stream of audio frames. There are several related systems which create latent spaces of audio corpora for creative purposes. Casey's (2005) work represents an early example which addressed the problem of efficient content-based search in audio documents for resynthesis purposes via the creation of a compact latent space (Casey 2005). Wavenet is a more recent and more closely related example

[2] https://goo.gl/magenta/js-checkpoints.

[3] https://github.com/magenta/magenta-js.

(van den Oord et al. 2016). Wavenet has been through multiple iterations and is a well-established structure for sound and especially voice synthesis. One of the limitations of Wavenet noted by Tatar et al. and addressed with LTS is its computational complexity, preventing real-time synthesis. Aside from Wavenet, other work has involved training networks on raw audio, for example Collins et al.'s work on 'brAIn swapping' (Collins et al. 2020) and Zukowski and Carr's work generating infinite death metal with SampleRNN (Zukowski and Carr 2018; Mehri et al. 2016).

We have selected the Latent Timbre Synthesis system because it explicitly deals with latent spaces and their application to interactive, creative work in the sound design and sound art domains. LTS is different from Wavenet and SampleRNN in that it generates spectral frames instead of raw audio samples, at professional CD quality as opposed to lower sample and bit rates.

To create its latent space, LTS cuts the raw audio signal into frames and then applies the Constant Q Transform (CQT) to create spectrograms for each frame. CQT is a common feature in music information retrieval tasks such as instrument detection. Next, LTS trains a variational autoencoder to encode and decode the CQT spectrograms to and from latent space. The VAE neural network architecture uses densely connected and convolutional layers to encode and decode spectral frames.

Convolutional layers are often used for digital signal processing tasks as they are a kind of trainable filter where the training adjusts the filter so it can extract the most useful information from the signal. Convolutional layers are suited to spectral frame data if the spectral frames are not treated as a sequence. Sequential data would generally require some sort of recurrent network, such as LSTM. Since it does not use LSTM-like layers, we can say that LTS does not model the sequence in which the frames occur.

Working with latent spaces of spectral frames as opposed to raw audio samples is common amongst deep synthesis systems which aim to operate in near real time on regular hardware. This performance level is an important feature for accessibility and interaction (Grierson et al. 2019). Tatar et al. report synthesis running at twice real time with a commonly available consumer gaming GPU (GTX 2080). This is similar to the performance reported for other real-time tools for deep synthesis, e.g. the DDSP system (Engel et al. 2020; Yee-King and McCallum 2021).

Tatar et al. provide the Python code comprising the LTS system.[4] They also include a user interface created using Max/MSP which communicates with the Python system via Open Sound Control messages (see Fig. 5). With this UI, users can supply a dataset of audio files and train the system, or they can explore the latent space by providing two audio files and interpolating between them. This interpolation activity is similar to the interpolation activity in the Music-VAE system, except the output is a stream of audio frames instead of discrete MIDI arrangements.

[4] https://gitlab.com/ktatar/latent-timbre-synthesis.

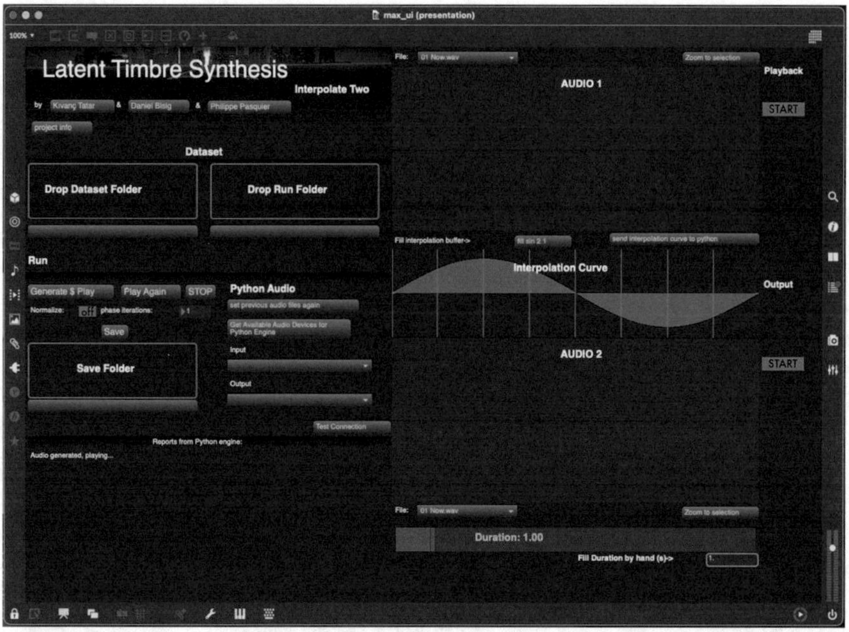

Fig. 5 User interface for the latent timbre synthesis engine. The user can add a training dataset and train and then they can generate audio sequences by interpolating between positions in the latent space. The interface is implemented in Max/MSP but it talks via OSC to a Python back end

3.3 A Latent Space for Sub-symbolic Image Data: StyleGAN

Kerras et al. reported StyleGAN in 2019 (Karras et al. 2019). StyleGAN is an image synthesiser which is trained using generative adversarial methods. This means there are two models: the image synthesis model and a discriminator model which evaluates the synthesis model. Both models are trained at the same time—as the synthesis model improves, the discriminator model gets better at finding its flaws. StyleGAN creates a latent space of a dataset of images (70,000 in the original article). Therefore, you can generate a latent vector and StyleGAN can convert that into an image, or vice versa.

Similarly to Music-VAE, users can move around in StyleGAN's latent space and see how the resulting images change. The difference is that StyleGAN allows for more control over how users do that. StyleGAN takes the latent vector and creates several transformed versions of it (we might call these 'sub-latents'). To synthesise an image, the sub-latents are passed as control inputs to different layers in the image synthesis network. StyleGAN's user can manipulate the raw latent vector, or they can manipulate one or more of the sub-latents before they go into the generator's layers. Users can even take latent vectors of two images, and mix and match which sub-latents they pass into the network, leading to mixtures between the two images.

This is much more subtle than simply interpolating between the latent vectors of the two images.

The different layers of StyleGAN tend to represent semantically meaningful aspects of the training dataset, e.g. for faces, presence or absence of sunglasses, hair length, face angle and so on. That means that 'sub-latent mixing' allows for a technique called style transfer where semantic characteristics of one image can be applied to another. For example, given an image of a young person and an older person, it is possible to transform the older person's image such that they look young.

Earlier I described how there has been a to-ing and for-ing between formal and statistical methods in machine learning. Often, formal methods are associated with more explicit modelling of semantic features. It is interesting to note that the statistical structure learned by StyleGAN allows it to perform semantic style transfer between images (Fig. 6).

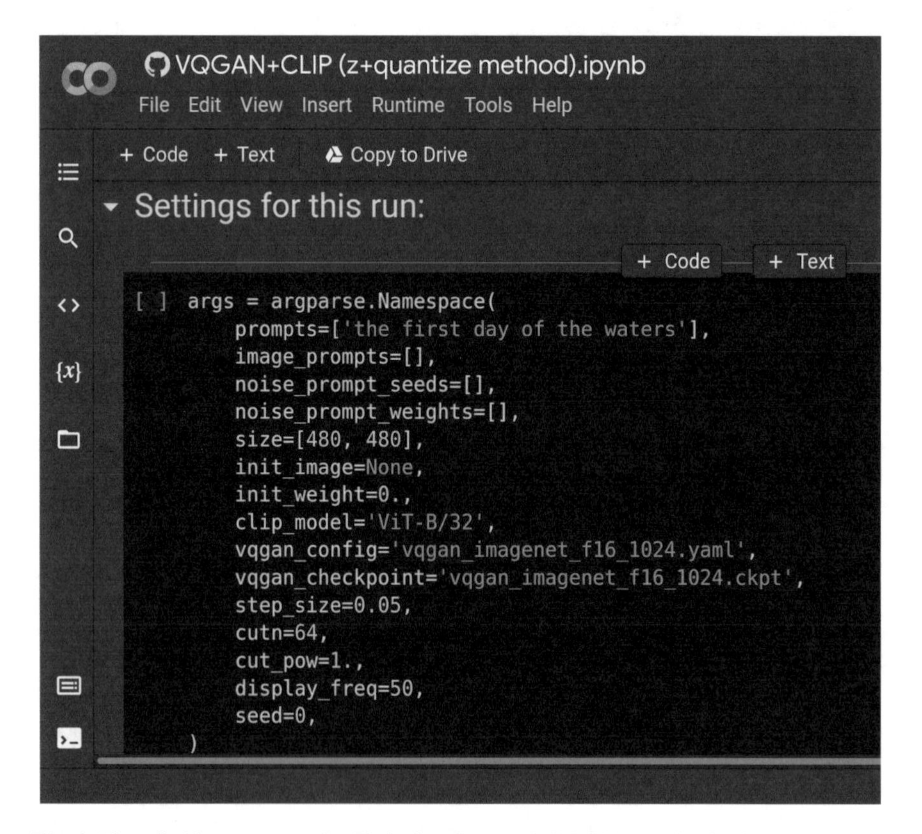

Fig. 6 The editable parameters for Katherine Crowson's VQGAN notebook viewed in Google Colab

3.4 A Latent Space for Combined Symbolic and Sub-symbolic Image Data: VQGAN + CLIP

In 2021, the Internet witnessed a storm of creative image generation using a novel text to image converter called VQGAN + CLIP.[5] The text to image system relies on two networks, CLIP and VQGAN (Razavi et al. 2019; Esser et al. 2021). VQGAN can generate images from latent space similarly to StyleGAN. CLIP can evaluate an image as to how well it matches a text phrase. The method begins with the user selecting a textual input phrase and a 'noise vector' or what we know as a latent vector in VQGAN's latent space. Then, the system iteratively generates images from VQGAN using the latent vector, evaluating the images for how well they match the phrase with CLIP. In each iteration, the text to image matching error is used to fine-tune the weights in VQGAN so it produces better matches.

In more detail, the CLIP network learns a latent space of paired text and image features. Effectively, you can pass it text and an image and it will tell you how closely they match. It knows what the expected image features should be for a given text input, and it can compare those expected image features to the ones from the image you pass it. That gives you an error between the expected image for a given phrase and the image you passed. On the other hand, the VQGAN network has learnt a latent space of a large number of images, and it can then generate back out from that latent space to images. VQGAN is similar to StyleGAN in that sense. In fact, it is possible to swap out the image generator—an earlier iteration of this technique used a relative of StyleGAN called BigGAN as the image generator (Fig. 7).

To use the two networks together, we start with a text phrase. Then, we select a position in VQGAN's latent space. We reverse from VQGAN latent space to an image. Then, we pass that image and the text phrase to CLIP. CLIP computes the features of the text and the image. Then, it computes the distance between the text and image features. This provides an error value, which we use to fine-tune the weights on VQGAN using normal neural network training techniques. Then, we try generating the image again and so on.

The technique produces interesting results because the 'search' can start anywhere in VQGAN's latent space. The fine-tuning then warps the space so that the selected position moves to improve the CLIP rating. The warping is iterated, so you can extract images periodically as the warping is taking place. For example, you might happen to start at the position representing dogs in latent space, but want the phrase 'big green houses'. The result would be a dog-like image iteratively warped into a big green house. The selection of the starting position and the target phrase, and the images that occur along the way provide for a huge range of creative possibilities.

The provision of this 'CLIP leading a generator' method in an accessible form is credited to machine learning engineers/ artists Katherine Crowson and Ryan Murdoch. They provided easily modified and executed Google Colab notebooks containing the code for the implementation. Google Colab makes it possible to run

[5] https://www.vice.com/en/article/n7bqj7/ai-generated-art-scene-explodes-as-hackers-create-groundbreaking-new-tools.

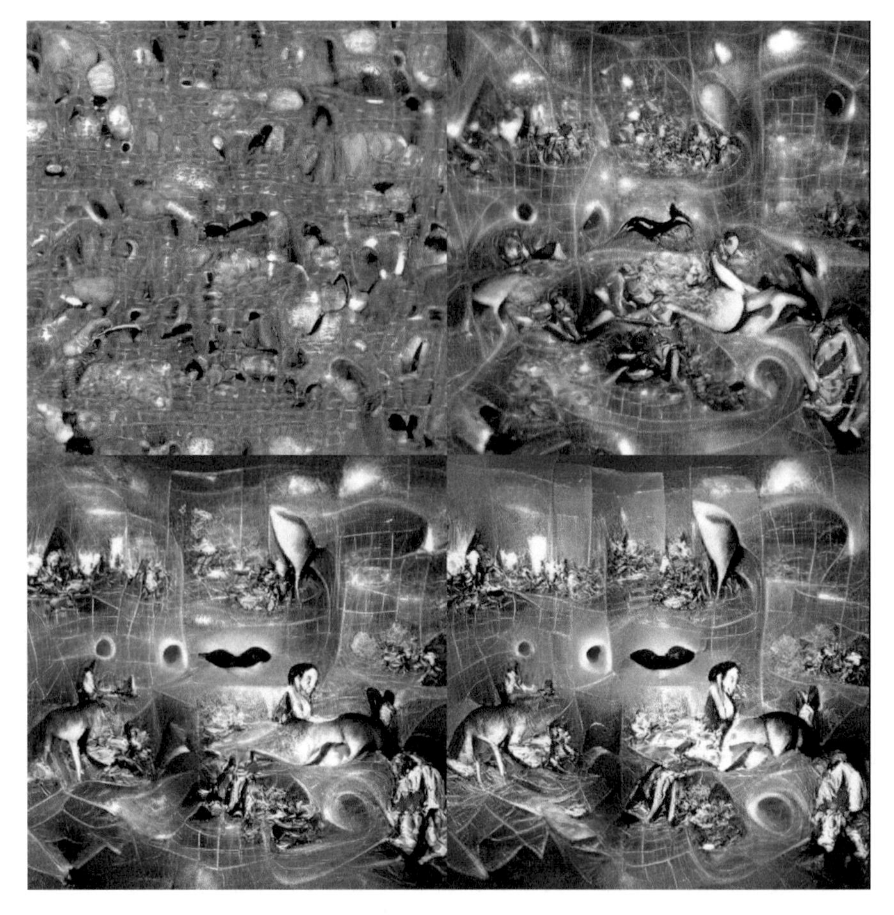

Fig. 7 Response of the VQGAN network to the phrase 'the space of possible worlds', iteration 0 (top left) then iteration 100 (top right), 250 (bottom left) and 500 (bottom right)

blocks of Python code on Google's compute infrastructure using just a web browser. Colab notebooks became something of a currency for digital artists in the early 2020s.

4 Creating with Latent Space Systems

In this section, we will revisit each of the generative systems we discussed in the previous section and consider some examples of work that has been created using them.

4.1 Composers Working with Music-VAE

In 2019, the band ‘Young Americans Challenging High Technology’, consisting of musicians Claire L. Evans & Ross Goodwin used Music-VAE in a music project. At a Google I/O talk in 2019,[6] Evans explained their creative process when they worked with Music-VAE in collaboration with the researchers who created it. Evans and Goodwin manually annotated their back catalogue of music into MIDI and extracted melodies, basslines and drum patterns. They trained a Music-VAE model on this corpus. According to Evans, Music-VAE ‘allowed us to find melodies hidden in between songs from our own back catalogue’. Thus, they made use of Music-VAE’s interpolation capabilities to find new melodies which carried the essence of YACHT music. In order to move from there to a full piece of music, they applied certain human workflow rules such as ‘only use melodies generated from the model, do not improvise’. They used another model to generate the lyrics in a similar fashion, by training it on a corpus of their lyrics then generating from that latent space. To create the final track ‘Loud Light’, they took the MIDI data and lyrics from the machine learning models and created the final music using standard production techniques.

4.2 Composers Using Latent Timbre Synthesis

Tatar et al. worked with nine composers to create a compilation album using the LTS system[7] (Tatar et al. 2021). They report on the results of interviews with the participating composers wherein they queried many areas of the composers’ experiences. Considering comments made by the composers with particular relevance to latent spaces, one ‘found the wide range of sound output possibilities of LTS rather exhausting’. Other composers discussed different phases in their workflows. Tatar et al. characterised this as wider, exploration search followed by narrower, exploitation search. These concepts relate to the divergent and convergent strategies described by Tubb and Dixon (2014). The compositions and interview transcripts are publicly available.[8]

4.3 Artwork Using StyleGAN

Terence Broad is a practitioner in the creative visual domain using machine learning techniques. In the article ‘Amplifying the uncanny’, Broad et al. explore Mori’s uncanny valley using StyleGAN (Mori 1970; Broad et al. 2020). The researchers

[6] https://www.youtube.com/watch?v=pM9u9xcM_cs.

[7] https://medienarchiv.zhdk.ch/entries/376e81a2-a6b9-4b74-91a3-9144c192f8e1.

[8] https://medienarchiv.zhdk.ch/entries/376e81a2-a6b9-4b74-91a3-9144c192f8e1.

acknowledge that a criticism of generative image systems is that 'the endless generation of samples from a given model, while initially mesmerising and transfixing, can quickly become banal, monotonous repetitions for the sake of overwhelming the viewer'. Broad exploits this overwhelming feeling by attempting to create large palettes of uncanny faces. Broad also exploits the nature of the GAN training method, manipulating how the discriminator decides on the credibility of the images from the generator. Thus, he 'fools' the discriminator into guiding the warping of the manifold of latent space towards his own ends. That means training towards maximally uncanny faces instead of maximally realistic ones.

4.4 Artwork Using VQGAN/CLIP

As mentioned above, artist/engineers Katherine Crowson[9] and Ryan Murdoch are credited with creating Google Colab notebooks which allowed people to experiment with state-of-the-art text to image technology by editing some simple parameters in a text field and executing GPU-accelerated neural network models for free on the Google compute infrastructure. Figure 6 shows a screenshot of the parameter panel in Katherine Crowson's VQGAN-CLIP notebook. These Colab notebooks made high-resolution, pre-trained image generators available to artists with minimal technical training.

Creating accessible technology which regular artists could use meant that in 2021, the currencies for some digital visual artists suddenly became pre-trained image processing neural network with acronymic names, Colab notebooks and non-fungible tokens (NFT). Artists could simply edit a text field such as that shown in Fig. 6 in their web browser, press a series of play buttons, and the notebook code and Google's compute infrastructure would do the rest of the work. Once created, artists could use NFT technology to place the images on the blockchain and place them onto the art market.

The work often combines multiple visual elements in dream-like, brightly coloured images. The artists sell the images as sets containing tens or hundreds of images. For further reading and examples of neural visual art using CLIP and GANs from 2021, we refer the reader to an article by Luba Elliott.[10]

Having considered four examples where creative practitioners worked with machine learning tools and latent spaces, we will now leave this discussion with a quote from Memo Atkin. Atkin is a digital artist, researcher and long-time user of deep networks for image and video generation. Here is what Atkin had to say about generative systems in his 2021 Ph.D. thesis:

> It is incredibly valuable that a person has the ability to freely explore such a massive space, so that they may embark on an goal-less, purely inquisitive and creative exploration, to build an understanding of the extents of such a system's creative capacity. (Akten 2021)

[9] https://kath.io/, https://twitter.com/RiversHaveWings.

[10] https://www.rightclicksave.com/article/clip-art-and-the-new-aesthetics-of-ai.

5 Conclusion

In this chapter, I have considered how it is possible to use machine learning techniques to create and explore latent spaces for different creative domains. The latent spaces might be symbolic, e.g. Music-VAE or they might be sub-symbolic, as in Latent Timbre Synthesis. The spaces can even combine symbolic and sub-symbolic elements, as in the VQGAN + CLIP text to image system. I have presented examples of creative practitioners working with each of the example systems, illustrating how creatives might go about exploiting their capabilities. Some used the systems to learn and explore latent spaces of their own previous work, and others exploited the techniques to deliberately create 'non-optimal', uncanny output. I also explained the importance of accessibility of the technology. Providing artists with pre-trained models, easy to hack code and easy access to compute infrastructure leads to a fantastic explosion of new visual work in the early 2020s. I look forward to the future of creative exploration of latent spaces and expect to see fantastical images and hear unheard sounds.

References

Akten M (2021). Deep visual instruments: realtime continuous, meaningful human control over deep neural networks for creative expression. Ph.D. thesis. Goldsmiths, University of London. https://research.gold.ac.uk/id/eprint/30191/

Ames C (1987) Automated composition in retrospect: 1956–1986. In: Leonardo, pp 169–185

Austin L, Cage J, Hiller L (1992) An interview with John Cage and Lejaren Hiller. Comput Music J 16(4):15–29

Bradski G, Kaehler A (2000) OpenCV. Dr. Dobb's J Softw Tools 3:2

Briot J-P, Hadjeres G, Pachet F-D (2020) Deep learning techniques for music generation, vol 1. Springer

Broad T, Leymarie FF, Grierson M (2020) Amplifying the uncanny. In: arXiv preprint arXiv:2002.06890

Broad T, Berns S et al (2021) Active divergence with generative deep learning—a survey and taxonomy. In: CoRR abs/2107.05599. arXiv:2107.05599. https://arxiv.org/abs/2107.05599

Casey MA (2005) Acoustic lexemes for organizing internet audio. Contemp Music Rev 24(6):489–508

Collins N, Ruzicka V, Grierson M (2020) Remixing AIs: mind swaps, hybrainity, and splicing musical models. In: Proceedings of the joint conference on AI music creativity

Deng J et al (2009) Imagenet: a large-scale hierarchical image database. In: 2009 IEEE conference on computer vision and pattern recognition. IEEE, pp 248–255

Ellis DPW, Arroyo J (2004). Eigenrhythms: drum pattern basis sets for classification and generation. In: Loureiro R, Buyoli CL (eds) ISMIR 2004: 5th international conference on music information retrieval: proceedings: Universitat Pompeu Fabra, 10–14 Oct 2004

Engel J et al (2020) DDSP: differentiable digital signal processing. In: arXiv preprint arXiv:2001.04643

Esser P, Rombach R, Ommer B (2021) Taming transformers for high-resolution image synthesis. In: Proceedings of the IEEE/CVF conference on computer vision and pattern recognition, pp 12873–12883

Fernández JD, Vico F (2013) AI methods in algorithmic composition: a comprehensive survey. J Artif Intell Res 48:513–582

Grierson M et al (2019). Contemporary machine learning for audio and music generation on the web: current challenges and potential solutions. In: 45th International computer music conference, ICMC 2019 and international computer music conference New York city electroacoustic music festival, NYCEMF 2019. International Computer Music Association

Karras T, Laine S, Aila T (2019) A style-based generator architecture for generative adversarial networks. In: Proceedings of the IEEE/CVF conference on computer vision and pattern recognition, pp 4401–4410

LeCun Y, Bengio Y, Hinton G (2015) Deep learning. Nature 521(7553):436–444

McFee B et al (2015) librosa: audio and music signal analysis in python. In: Proceedings of the 14th python in science conference, vol 8. Citeseer, pp 18–25

Mehri S et al (2016) SampleRNN: an unconditional end-to-end neural audio generation model. In: arXiv preprint arXiv:1612.07837

Mori M (1970) Bukimi no tani [the uncanny valley]. Energy 7:33–35

Razavi A, Van den Oord A, Vinyals O (2019) Generating diverse high-fidelity images with vq-vae-2. Adv Neural Inf Process Syst 32

Roberts A et al (2018) A hierarchical latent vector model for learning long-term structure in music. In: International conference on machine learning. PMLR, pp 4364–4373

Roy P, Papadopoulos A, Pachet F (2017) Sampling variations of lead sheets. In: arXiv preprint arXiv:1703.00760

Simon I et al (2018) Learning a latent space of multitrack measures. In: arXiv preprint arXiv:1806.00195

Tatar K, Bisig D, Pasquier P (2021) Latent timbre synthesis. Neural Comput Appl 33(1):67–84

Tubb R, Dixon S (2014) The divergent interface: supporting creative exploration of parameter spaces. In: NIME, pp 227–232

van den Oord A et al (2016) Wavenet: a generative model for raw audio. In: arXiv preprint arXiv:1609.03499

Xenakis I (1966) The origins of stochastic music 1. In: Tempo 78, pp 9–12

Yee-King M, McCallum L (2021) Studio report: sound synthesis with DDSP and network bending techniques. In: Gioti A-M, Eckel G (eds) 2nd conference on AI music creativity (MuMe + CSMC). Graz, Austria 18–22 July 2021

Zukowski Z, Carr CJ (2018) Generating black metal and math rock: beyond bach, beethoven, and beatles. In: arXiv preprint arXiv:1811.06639

Intersections of Living and Machine Agencies: Possibilities for Creative AI

Carlos Castellanos

Abstract This chapter looks at artists' experimentations with linkages between intelligent computational systems and non-human living organisms. These unusual hybrid systems showcase models for how we can bridge heterogeneous lifeworlds. In doing so, they evince all kinds of heretofore unimagined possibilities for mutual understanding and influence, which may give us new perspectives on AI-non-human alterities and may serve to question the anthropocentric divisions between humans, human technology and the more than human world. At the same time, this approach may point toward a model of art-making, where encounters between living organisms and intelligent machines can serve not only as vectors of novelty and unexpected variety, but also as a step toward developing a system of ideas focused on showcasing alternative possibilities of human–machine non-human relations.

Keywords Creativity · Non-human intelligence · Art-AI · Practice · Interactivity · Inter-relations

1 Introduction

In *The Three Ecologies*, French philosopher Félix Guattari argues for more subjectivity in science (Guattari 2000). In extending the definition of ecology to encompass social relations and human subjectivity as well as environmental concerns, Guattari states that the boundaries between nature and technology need to be collapsed if we are to properly address the ecological crisis. Learning to think "transversally" or across disciplines is a crucial step toward the goal of developing the alternative ontologies, epistemologies and social relations necessary for ecological sustainability. Guattari's ideas draw significantly from those of cyberneticist Gregory Bateson who argued it is not merely our technologies which are unsustainable but our ways of thinking (Bateson 2000). However, I believe that if we want to properly account for

C. Castellanos (✉)
Rochester Institute of Technology, Rochester, NY, USA
e-mail: cc@ccastellanos.com

C. Vear and F. Poltronieri (eds.), *The Language of Creative AI*, Springer Series on Cultural Computing, https://doi.org/10.1007/978-3-031-10960-7_9

subjectivity in science and explore alternative epistemological models, we would do well to look at the work and ideas of cyberneticist Gordon Pask.

Although a full accounting of Pask's work is out of the scope if this chapter, what I can briefly discuss here is Pask's notion of the "participant observer" and how it contrasts with the more traditional "scientific observer" (Pask 1958). Pask pointed out the marked difference between a scientific observer, who minimizes interaction with an observed system and a participant observer who maximizes it. He proposed that when we build and try to understand complex systems, we approach them as a natural historian would: through our interactions with them (Pask 1960). Much of mainstream science considers observer interaction as a source of confounding variables and thus as something to be avoided. For Pask, however, constructing devices that adapt to environmental conditions and interacting with them can add to our knowledge in ways the traditional scientific experimental method cannot. As Andrew Pickering notes in his analysis of Pask's work, there is an original, inchoate philosophy of science embedded in Pask's model (Pickering 2010).

What does the is mean for the arts, and specifically art and AI? Many contemporary artists who work with digital and emerging technologies have employed living organisms and explored ecological themes in their work in recent years, each of them exploring unique aspects of non-human life and their relevance to contemporary science, technology and art. Much in the vein of Pask and Guattari, these artists bring these ideas of scientific subjectivity and observer interaction into their work. Likewise, artists have been employing AI and A-life techniques in their work for over 50 years, yielding a rich and diverse set of artworks in that time. There is an inherent strangeness and ambiguity to these technological systems. They behave quite unlike any technological systems we are accustomed to, often exhibiting autonomy, life-like behavior and at least the appearance of intelligence or sentience.

For this chapter, I will discuss the intersection of these two kinds of artistic practice. There has been an increasing artistic experimentation over the last several years with linkages between computational systems (often intelligent ones) and non-human living organisms. Although sometimes falling under the umbrella of biological or ecological art, I want to argue for its distinction as a unique genre, that lies at the intersection of bio-art, eco-art and Creative AI. These works feature encounters and interactions between living organisms and intelligent computational systems, often employing agent-based and/or machine learning methods. Here, the autonomous agents respond to actions and behaviors of the living organisms and produce some sort of output related to their learnings and interpretations, which they often use to respond in a manner that feeds back into the living organism's environment, influencing its behavior in some way.

Thematically, these works often examine the implications that arise from encounters between human technology and the environment, while simultaneously raising the ethical and ontological status of the organisms they use. They bring into question the rational clarity of the classical ideal of a dualist ontology that separates people and things, throwing into high relief, a recognition that matter and non-human life are not passive and inert but are lively and dynamic, with agency and lifeworlds of their own. In doing so, they showcase the nascent elements of this distinct aesthetic paradigm and its possibilities for Creative AI.

2 Speculative of Models of Shared Non-Human Machine Agency in the Arts

> The recognition of sophisticated information processing capacities in prokaryotic cells represents another step away from the anthropocentric view of the universe that dominated pre-scientific thinking. Not only are we no longer at the physical center of the universe; our status as the only sentient beings on the planet is dissolving as we learn more about how smart even the smallest living cells can be. (Shapiro 2007)

The idea of microbial cognition and intelligence has been gaining purchase in the life sciences (Shapiro 2007; Lyon 2015). Many of us have probably heard of the amazing learning and decision-making capabilities of the slime mold *Physarum polycephalum* (Tero et al. 2010; Tsuda 2009) or the collective social motility and pattern-forming of numerous bacterial species (Ben-Jacob 1997; Ben-Jacob, Cohen, and Levine 2000) (along with their sophisticated communication methods such as quorum sensing). More and more notions of cognition, collective intelligence, communication, and creativity in a number of "primitive species" are being more seriously considered, significantly challenging how these concepts are constructed (for example as requiring brains or neurons). This is a more complex adaptive systems of view of life and intelligence, with more of a focus on adaptive relationships with the environment. There is an emerging ontology here that is based upon a recognition of the agency of non-humans, expectations of complexity and ambiguity, self-organization and co-emergent interplay between all kinds of agents: human, non-human and computational. What I would like to focus on here is on how artists are building artifacts and systems that act out a kind of model of this unknowable and co-emergent world.

So in what ways are artists working with "primitive" living organisms to create experiences that suggest alternative ontological visions for understanding and acting in the more than human world. And in doing so, how are they laying the groundwork for a new kind of art-science? And what are the implications for Creative AI?

Many artworks in this emerging art-science field feature technological interfaces with non-human organisms, often involving some sort of interface with an electronic and/or AI-powered computational system. For example, the art collective Interspecifics have been constructing these types of organic-computational interfaces in various works that feature speculative attempts at bacterial interaction and communication as a source of sonic and visual variety. A recent work, *Speculative Communications* (2017–18) (Interspecifics 2017), features an AI-powered microscope that observes and learns from cultures of the bacteria *Bacillus circulans*. The tracking and analysis of the bacteria's growth and swarming behavior is then used a generative sound score. In *micro-rhythms* (2016) (Interspecifics 2016), another bacteria-focused work, the group utilized microbial fuel cell (MFC) technology (Logan 2008) along with machine learning to create a generative sound and light composition.

Here, voltages released from anaerobic bacteria in the MFCs are used to trigger lights which are then read by a computer vision and machine learning system that groups the light patterns into different categories or clusters. The appearance and repetition of these clusters are then used as modulation sources for a sound synthesis and spatialization system. The group's most recent project (currently in development at SETI), titled *Codex Virtualis* (Interspecifics 2021), proposes the use of deep learning generative models and cellular automata to create speculative hybrid bacterial-AI organisms.

Similarly, Kuai Shen Auson's *0 h!m1gas* (2010–13) (Auson 2012) explores the human-ant relationship. This piece uses computer vision to track the movement of leafcutter ants and piezo sensors to detect vibrations that the ants produce via their stridulation behavior. The collected data is then used to control the movement of a pair of turntables, and the sound of which is amplified. The movement and behaviors of the ant colony thus emerge as a soundscape of scratching effects, an obvious reference to the ants' stridulating behavior.

In Michael Sedbon's *CMD* (2019) (Sedbon 2019) (Fig. 1), bioreactors of photosynthetic bacteria can claim access to light thanks to credits earned for their oxygen production. They can sell it in a market, the rules of which are optimized through a genetic algorithm. The system tests different populations of financial systems on these two sets of cyanobacteria. Andie Gracie's *Autoinducer_Ph-1* (Gracie 2006) features bacterial cultures (*Anabaena azollae*) that interact with simulated bacteria powered by an AI model. Digitized stimuli produced by the real bacterial cultures are interpreted by the AI, which in turn dictates the supply of air, heat and light provided to the organic cultures. The piece borrows from a traditional Southeast Asian rice cultivation technique. The outcomes of this complex relationship (parasitic vs symbiotic) determine the behaviors of a robotic rice farming system that cares for and manages the rice growing.

Whether fanciful and provocative or strange and mysterious, these works provide a new lens to view conceptions of life, the environment and non-human "otherness" overall. While all very different, I believe they share at least three traits in common.

- First (and most obviously), they feature biocybernetics machine-organism interactions.
- Second, they all eschew the "look but don't touch" approach to environmentalism—trading in traditional conservation approaches (consume less, recycle, etc.) for active engagement and literal contact with living systems and the environment.
- Third, they explore speculative ontological and epistemological models with regard to the non-human world (or at least the "primitive" species such as bacteria and ants). In essence, they adopt a "Paskian" approach, interacting directly with organisms to learn from them and generate knowledge about them. In other words, adding an element of scientific subjectivity.

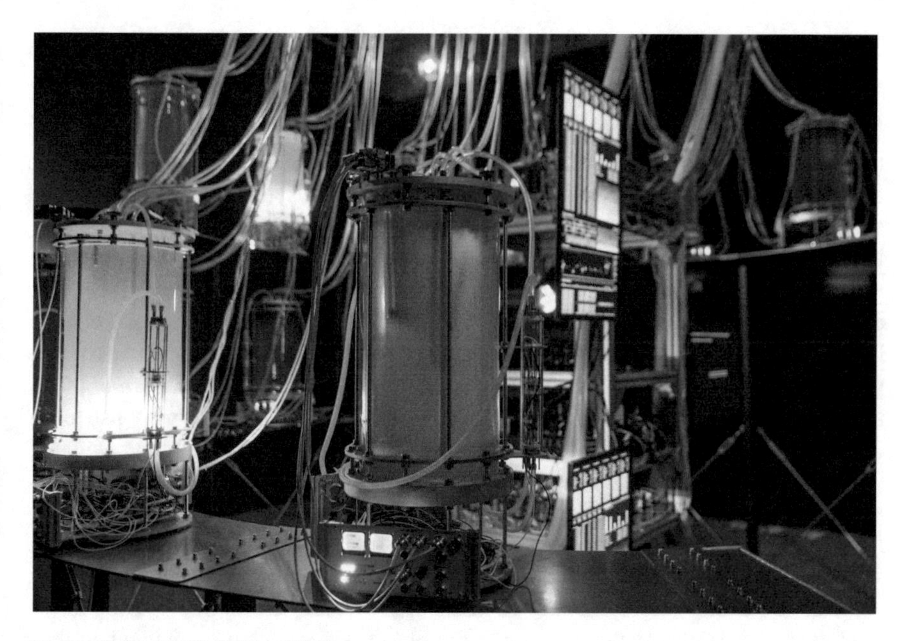

Fig. 1 CMD (2019), Michael Sedbon

3 PlantConnect and Biopoiesis

To supplement this theoretical discussion, I will also discuss two projects of my own:

- First *Biopoiesis* (Castellanos 2018; Castellanos and Barnes 2018), a cybernetic art project that explores the relationships between structure, matter and self-organization. Based upon Pask's experiments with the construction of electrochemical assemblages that were capable of growing their own sensors (a kind of analog AI), I will discuss the design and construction of the system and explore the relevance of Pask's electrochemical work to the arts. I also put forth the notion of a "philosophy of open-ended ambiguity" embedded within this work and discuss its resonance with the arts.
- Second, *PlantConnec*t (Castellanos 2020) (Fig. 3), an exploration of human-plant interaction via the human act of breathing, the bioelectrical and photosynthetic activity of plants and computational intelligence to bring the two together. The system measures the photosynthetic and bioelectrical activity from an array of plant microbial fuel cells (P-MFCs) and translates them into light and sound patterns using computer vision and machine learning. Part of larger investigations into alternative models for the creation of shared experiences and understanding with the natural world, the project explores complexity and emergent phenomena by harnessing the material agency of non-human organisms (plants and bacteria in this case) and the capacity of emerging technologies as mediums for information

transmission, communication and interconnectedness between the human and non-human. *PlantConnect* is a collaboration with Bello.

PlantConnect explores human-plant interaction via the human act of breathing; the bioelectrical and photosynthetic activity of plants and computational intelligence to bring the two together. The system measures the photosynthetic and bioelectrical activity from an array of plant microbial fuel cells (P-MFCs) and translates them into light and sound patterns using machine learning. The primary mode of participant interaction with the system is via breath. When a participant blows or whistles into a CO_2 sensor located within the array of plants, it triggers an array of 16 grow lights that are directed at the plants and thus contribute to their photosynthesis. The photosynthesis levels are obtained from small measuring chambers containing CO_2 sensors attached to each plant. In addition to an instantaneous audible response to the decreasing CO_2 levels caused by the increased photosynthesis, these photosynthesis levels are translated into interpolation parameters for the virtual sound instruments and spatialization module of the system. Meanwhile the voltage signals from the P-MFCs are amplified so they can be read by a standard microcontroller. These signals are then analyzed to find the minimum and maximum voltage values, which are used to generate a set of adaptive thresholds that are sent in binary code to the light array. These thresholds determine the on/off patterns of the lights when they are triggered by human breath/CO_2. Using a blob detection algorithm, the system detects the on/off state of the lights in the light array as well as the general shape produced by the lights, relative to the background. This data is then sent to a clustering algorithm (a form of unsupervised machine learning). This algorithm recognizes similarities and differences in the repeating light patterns and classifies them into groups or clusters. Essentially performing rudimentary pattern recognition. This data is then sent to a Max/MSP application via OSC/UDP messages that control a set of virtual instruments and a spatialization module within the Max/MSP environment. In this way, the machine learning algorithm—and by extension the plants—selects instruments and alters their amplitude, duration, frequency and spectral parameters. They also select a spatialization state (Fig. 2).

In *PlantConnect*, bioelectricity, light, sound, CO_2, photosynthesis and computational intelligence form a circuit that enhances informational linkages between human, plant, bacteria and the physical environment, enabling a mode of interaction that is experienced not just as a technologically enabled act of translation but as an embodied flow of information.

Biopoiesis (Fig. 3) is a series of experiments exploring the relationships between structure, matter and self-organization, in what might be described as a computational "primordial soup." This work builds on Gordon Pask's research into electrochemical control systems that could adapt to certain aspects of their environment (Pask 1960). A collaboration with neuroscientist Steven Barnes, these experiments explore the artistic potential of Paskian-like systems while also examining the interactive and computational possibilities of natural processes to serve as an alternative to the commonplace digital forms of computation, which might help (re)establish a dialog

Fig. 2 PlantConnect (2019)

between cybernetics, mainstream science and the arts. The piece entails the construction of several simple computational devices that are all based upon the process of electrochemical deposition: When electrical current is passed through a metallic ion solution (e.g., ferrous sulfate, stannous chloride), metal is deposited on the electrode that is the source of electrons (i.e., the cathode). In our experiments, information (in the form of an electrical current) is fed to a chamber filled with a solution of stannous chloride and ethanol via an array of electrodes (see Fig. 3, below). The resultant electrochemical reaction includes the growth and/or dissolution of metallic dendritic threads in the metallic ion solution, leading to a dynamic pattern of complex electrical and physical growth activity across the entire system. The dendrites are fluid and unstable, bifurcating and dissolving in seemingly unpredictable ways. Thread bifurcation and dissolution, in turn, lead to resistance changes that modify the flow of information (current) through the network. If a subset of electrodes in the electrochemical solution receives input from an environmental sensor (or via some other method), and the electrochemical output can affect that sensor (or otherwise influence the growth of threads), then the network may move toward a dynamic equilibrium with its environment. The dendritic network also carries a decremental memory trace of its previous activities: When the environment changes, the system is perturbed but not immediately reset. Thus, the prior activity and configuration of the system affect how it handles a change in its environment. It can thus learn from its interactions. Furthermore, the system can be trained by providing reinforcement for certain sorts

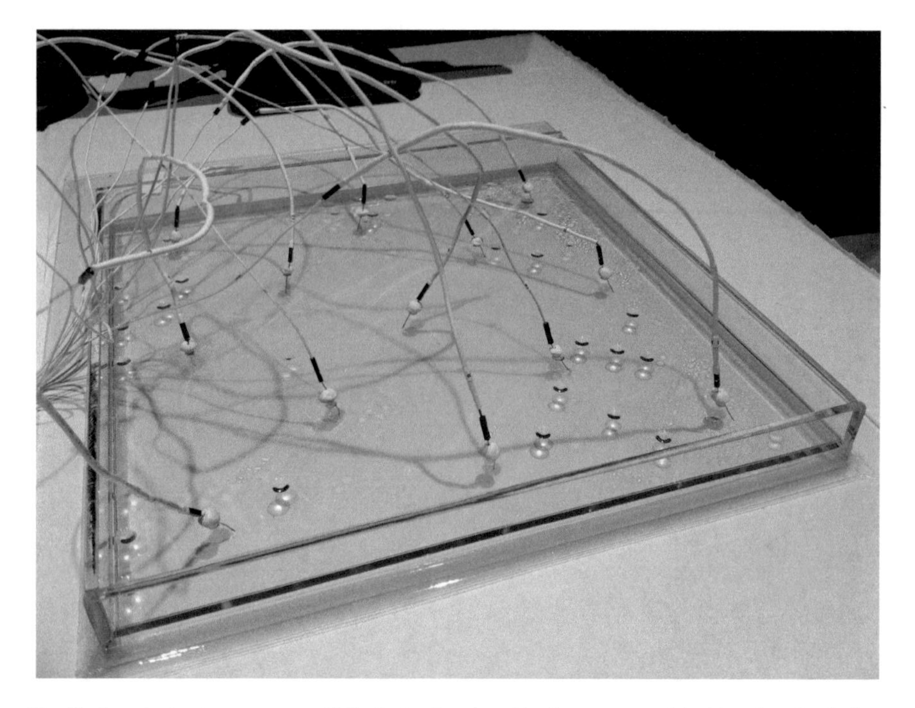

Fig. 3 A typical setup: an array of 13 electrodes placed in the stannous chloride-ethanol solution

of conductance changes that are produced in response to a particular environmental perturbation.[1]

Biopoiesis and Pask's electrochemical assemblages both serve to redirect our attention to the very material forms of the works and how they add a certain dimension of materiality and sensuous presence that is often lacking in digital and even robotic works. Some of these works display at least a hint of a certain kind of agency that can only come from these non-symbolic (i.e., non-digital), material forms grounded in processes of organic or quasi-organic growth. For many years now, artists have experimented with different mediums, techniques and locations without knowing exactly what the results would be. Thus, to an artist, Pask's approach might seem familiar and not that different from certain other artistic modes of experimentation. Pickering notes how in Pask work (and that of Stafford Beer, his sometime collaborator) there is a belief in the agency and variability of matter. He notes how rather than marshaling (or dominating) "inert lumps of matter" (as the building of computers and industrial machinery entails), there are attempts to couple this variability to human concerns

[1] Pask spoke of "rewarding" the system if it generated desired outputs. This is not reward in the Pavlovian sense or in the sense used in reinforcement learning applications in computer science where a response is rewarded so as to obtain more of that specific response in the future. Rather, it is reward in the sense Pask described: to give "permission" for the system to continue to develop its thread structures, but develop them in an unspecified way. A kind of adaptive steering. This reward usually entails sending more current to the solution. See (Pask 1959), Pask (1960).

(Pickering 2010, 236). A Paskian electrochemical system such as *Biopoiesis* encourages us to view the world as full of co-emergent, co-evolving systems too complex to be fully apprehended or objectively explained. A world that is in a perpetual state of becoming, characterized and brought forth via emergent relations of complexity that adumbrate an experience of the world that we characterize as open-endedly ambiguous. In other words, what we as artists who employ sophisticated technology in our work can learn from this Paskian philosophy is, in a sense what we already know.

4 Toward a Framework for Understanding Living and Machine Agencies in the Arts

These works are but a small sampling of a larger trend in the arts of increasingly hybridized practices and research agendas that necessarily require new ontological models for their analysis, critique and understanding. It is important to recognize that these works are created within a context infused with notions of art as research and art as experiment. This emerging paradigm—which can be said to foreground unpredictable emergence, rather than the rational clarity of a narrow and precise cause and effect—does not fit very well into the paradigm of most modern science, but it does fit quite well within the Paskian "cybernetic method" of maximizing interaction with an observed system to attain knowledge as a participant observer. This stands in stark contrast to the traditional scientific method, where interaction is minimized, as it is a potential source of confounding variables. I believe this model presents possibilities for new zones of negotiation and reciprocity between humans, non-humans and intelligent machines, as well as a reexamination of how art and science can come together. It offers a vision of the world as one that is filled with co-emergent, autonomous agents in reciprocal interplay—a world in a perpetual state of becoming, whose relations are too complex to fully apprehended expect through interactions and complex relations of alterity.

Beyond showcasing a new kind of art-science that embraces unknowable complexity, what more can we tease out of these works? What characteristics do they share that give rise to their performative ontologies? As no coherent analytical framework currently exists for understanding artworks that feature encounters between machine and non-human agencies, I present here a few key points of a provisional analytical model, which posits a concentration across four key areas (presented in pairs):

- ***Agency***, the ability to act in the world and exert influence, is innately bound up with ***Autonomy***, the ability to govern one's own interactions with the world. Autopoietic theory shows us that it is precisely the operational closure of the organism that gives it its autonomy and ensures the development of its own unique form of structural coupling to its environment (Maturana and Varela 1980). As Varela and Bourgine note, autonomy refers to a system's ability to assert itself and to "bring forth a world" for itself (Varela and Bourgine 1992). These works give us a view of intelligent machine-biological systems as a performance of agency, where autonomy is seen as arising from situated, contingent and perhaps most importantly (and a bit counterintuitively) collective interactions with the world (environment, other systems, etc.). These pieces emphasize the ontological nature of autonomous systems. Their capacity to simply be, "to assert their existence" and—through their interactions with their environment—"shape a world into significance" (Varela and Bourgine 1992). As such, they are good examples for how to think of autonomy as shaped through engagements with an "other(s)" (be they machines or other organisms).

- ***Adaptation*** is an adjustment of internal and external relations, and a change in internal structure to better perform in an environment. This is bound up with ***Emergence***. The latter is notoriously difficult concept to define. Indeed, its ambiguous and subjective nature is part of what gives the concept its appeal. Although there are highly formalized, computationally focused conceptions of emergence from fields such as computer science (Dessalles, Müller, and Phan 2007), in the context of art and Creative AI, neocybernetic models of emergence as observer-dependent and tied to our knowledge of the material/physical system and its internal/external adaptive relations (which is always incomplete) offer a more suitable definition that accounts for subjectivity—while not being overly reliant on rules and formalisms (Cariani 1992).

Analyzing these works through this lens may help us apprehend and leverage this incomplete knowledge by accepting the world as "exceedingly complex" and ultimately unknowable (Pickering 2010, 223). It thus allows us to understand our interactions as experiments or what Pickering calls "dance[s] of agency" (Pickering 2010). Or as Gordon Pask put it: "a participant observer decides upon a move which will modify the assemblage and, in general, will favor his interaction with it" (Pask 1958, 173). This approach of favoring alterity relations may open up window into how Creative AI can be leveraged to explore the "otherness" of other creatures in a way that a detached "view from nowhere" of traditional science (Nagel) may miss (Nagel 1989).

5 Conclusion

Understood within the theoretical context outlined above, and with the examples, we can begin to see that by creating these strange types of techno-ecological systems that can bridge heterogeneous lifeworlds, all kinds of heretofore unimagined possibilities for mutual understanding and influence emerge, which may give us new perspectives on AI-non-human alterities and may serve to question the anthropocentric divisions between humans, human technology and the more than human world, while also pointing toward a model of art-making where encounters between living organisms and intelligent machines can serve not only as vectors of novelty and unexpected variety, but also as a step toward developing a system of ideas focused on showcasing alternative possibilities of human–machine non-human relations.

Their works thematize and harness complexity, symbiosis and technological reciprocity, transformation and renewal between the human and non-human worlds. They disrupt nature/culture boundaries in ways a disciplined scientific method cannot, and in doing so, outline possible futures and alternative methods of relating our technologies to our environment.

By taking a non-anthropocentric approach, these art as experiments highlight the possibilities of creative partnerships between humans, computational systems and living organisms by reframing biological systems as active, intelligent agents capable of perception, cognition and sentience, rather than passive materials or design elements.

Furthermore, by developing intelligent computational environments that increase non-human participation and set-up states of constructive mutual influence between non-human organisms, humans and machines may make us better attuned to wider cultural shifts regarding the importance of non-human organisms and the larger environment, which may in turn lead to research that has important positive effects on our collective view of interspecies relationships. This kind of conceptual expansion is just one example of how perspectives from the experimental digital art field can contribute to Creative AI research by creating novel multispecies experiences that create meaning and encourage critical reflection.

While simultaneously drawing from established scientific tools and methods, artists working at this intersection of machine and biological agency are subverting radically—through the actual products they produce (i.e., the artworks)—the ontological premises of these very fields. The inherent tensions and challenges that artists must confront when working in this area (and art and science more broadly) also simultaneously offer opportunities for disruption and opening up of new perspectives. In so doing, they point to new avenues in Creative AI research and more broadly, the beginning elements of an alternative art-science formation that embraces complexity, ambiguity and unknowability.

References

Auson KS (2012) 0h!M1gas: a biomimetic stridulation environment. In: Ursyn A (ed) Biologically-inspired computing for the arts: scientific data through graphics. IGI Global, Hershey, PA, pp 59–80

Bateson G (2000) The roots of the ecological crisis. In: Steps to an ecology of mind. University Of Chicago Press, Chicago, pp 496–501

Ben-Jacob E (1997) From snowflake formation to growth of bacterial colonies II: cooperative formation of complex colonial patterns. Contemp Phys 38(3):205–41. https://doi.org/10.1080/001075197182405

Ben-Jacob E, Cohen I, Levine H (2000) Cooperative self-organization of microorganisms. Adv Phys 49(4):395–554. https://doi.org/10.1080/000187300405228

Cariani P (1992) Emergence and artificial life. In: Langton CG, Taylor C, Farmer JD, Rasmussen S (eds) Artificial life II, Santa Fe Institute Studies in the Sciences of Complexity, vol X. Addison-Wesley, Redwood City, CA, pp 775–97

Castellanos C (2018) Biopoiesis: electrochemical media. Leonardo 51(2):133–37

Castellanos C (2020) Plantconnect and microbial sonorities: exploring the intersection of plant, microbial and machine agencies. In: Proceedings of the 26th international symposium on electronic art. Printemps Numērique, Montreal, Quebec, Canada, pp 115–22

Castellanos C, Barnes S (2018) Biopoiesis: cybernetics, art and ambiguity. Leonardo Electronic Almanac 22(2). https://www.leoalmanac.org/cybernetics-revisited/

Dessalles JL, Müller JP, Phan D (2007) Emergence in multi-agent systems: conceptual and methodological issues. In: Amblard F, Phan D (eds) Multi-agent models and simulation for social and human sciences, 327–356. The Bardwell-Press, Oxford

Gracie A (2006) Autoinducer_Ph-1 (Cross Cultural Chemistry) [2006]. http://hostprods.net/projects/autoinducerph-1/.

Guattari F (2000) The three ecologies. Athlone Press, London

Interspecifics (2016) Micro-rhythms [2016]. http://interspecifics.cc/work/micro-ritmos-2016/

Interspecifics (2017) Speculative communications [2017]. http://interspecifics.cc/work/speculative-communications-2017/

Interspecifics (2021) Codex virtualis [2021]. http://interspecifics.cc/work/codex-virtualis-_/

Logan BE (2008) Microbial fuel cells, 1 edn. Wiley-Interscience, Hoboken, NJ

Lyon P (2015) The cognitive cell: bacterial behavior reconsidered. Front Microbiol 6. https://doi.org/10.3389/fmicb.2015.00264

Maturana HR, Varela FJ (1980) Autopoiesis and cognition: the realization of the living. D. Reidel, Dordrecht, Holland and Boston.

Nagel T (1989) The view from nowhere. Oxford University Press, USA

Pask G (1958) Organic control and the cybernetic method. Cursos Congr Univ Santiago De Compostela 1:155–173

Pask G (1959) Physical analogues to the growth of a concept. In: Mechanisation of thought processes: proceedings of a symposium held at the National Physical Laboratory on 24th, 25th, 26th and 27th November 1958, 877–928. H. M. Stationery Off, London

Pask G (1960) The natural history of networks. In: Yovits MC, Cameron S (eds) Self-organizing systems: proceedings of an international conference, 5 and 6 May. Pergamon Press, New York, pp 232–263.

Pickering A (2010) The cybernetic brain: sketches of another future. University of Chicago Press, Chicago

Pickering A (2016) Art, science and experiment. MaHKUscript J Fine Art Res 1(1):2. https://doi.org/10.5334/mjfar.2

Sedbon M (2019) CMD [2019]. https://michaelsedbon.com/CMD

Shapiro JA (2007) bacteria are small but not stupid: cognition, natural genetic engineering and socio-bacteriology. Stud Hist Philos Biol Biomed Sci 38(4):807–819. https://doi.org/10.1016/j.shpsc.2007.09.010

Tero A, Takagi S, Saigusa T, Ito K, Bebber DP, Fricker MD, Yumiki K, Kobayashi R, Nakagaki T (2010) Rules for biologically inspired adaptive network design. Science 327(5964):439–442. https://doi.org/10.1126/science.1177894

Tsuda S (2009) Robot with slime 'Brains.' Technoetic Arts J Speculative Res 7(2):133–140

Varela FJ, Bourgine P (1992) Towards a practice of autonomous systems. In: Varela FJ, Bourgine P (eds) Toward a practice of autonomous systems. MIT Press, Cambridge, MA, pp xi–xvii

https://ccastellanos.com/projects/plantconnect

Conformed Thoughts, Representational Systems, and Creative Procedures

Silvia Laurentiz

Abstract In our daily lives, we exercise and practice extensively logical–symbolic reasoning, which we call "conformed thoughts". These are self-controlled and deliberate and can be the result of concepts, models of knowledge of a culture or group. They are adaptable and can be changed by our habits, behaviors, beliefs, and actions. This, in turn, will cause changes in our way of thinking. Our main interest in this chapter is to present a representational model, promoted by information processing, that will be able to point out questions about computational algorithms—in particular, artificial intelligence—and present the contribution of art and creative procedures in this process.

Keywords Creativity · Thought · Logic · Language · Representation · Art

1 Introduction

The term 'conformed thoughts' (Laurentiz 2015, 2018, 2019) was created to describe how an internal set of codes, norms, algorithms, and standards are malleable by external factors. In everyday life, our actions are determined by habits and externalized factors. These are capable of modifying attitudes, behaviors, cultural practices that shape our thoughts. In creative AI, similar processes can occur: The visual forms generated by algorithms, say, (and in this context, we will be dealing with artificial intelligence/machine learning/deep learning), can be considered externalized thoughts. They use mathematical and statistical calculations for data analysis and model generation, and although invisible (generated inside a black box) they become externalized and therefore sensitive in a resulting image as they are fed back into the system. This leads us to consider that such 'conformed thoughts' do not have only formal characteristics—are not restricted to appearances—but it is an action that is determined by processes capable of modifying attitudes, behaviors, cultural practices that "in-form", "re-form", and "con-form" thoughts.

S. Laurentiz (✉)
University of Sao Paulo, Sao Paulo, Brazil
e-mail: laurentz@usp.br

C. Vear and F. Poltronieri (eds.), *The Language of Creative AI*, Springer Series on Cultural Computing, https://doi.org/10.1007/978-3-031-10960-7_10

From the principle of "conformed thoughts", we can make important considerations about representational aspects of creative AI. We can observe, for example, an encapsulation of the representational system starting from the "thing itself", which is perceived and made an "object" in order to be communicated and shared (Deely 2004), and that is then synthesized into a "modeled object", with the understanding that models are formed by objects, which in turn are "things themselves" that have been objectified.

In line with Flusser (2007), the passage from an "object" to a "model" carries levels of abstraction that would be important to note. In other words, the modeling and learning process, which we suggest is a meta-processing of/ for the generation of "conformed thoughts", shows the relationship of the "thing itself" in these representational nests; even if the "thing itself" distances itself from the model, it is preserved somehow as a motivator and trigger of these processes. All this happens in a context where systemic feedback occurs, with processes of evaluation, transformation, comparison with reference values, adaptation, regression, codification, following the principles of cyber systems.

The digital system based on these processes appropriates these experiences and is guided by models and patterns that in turn will guide the results obtained. In these procedures, there is a new tension between "feelings and conformed thoughts", and in an environment of mixtures of information and levels of abstractions, we find a potential to bring about new experiences, given this structural complexity. Faced with this scenario, artificial intelligence, machine learning, and deep learning are structures of/ for generating conformed thoughts.

This chapter intends to present some studies already conducted on this subject and to point out some consequences for language and thought. The main interest is to present a representational model promoted by information processing capable of discussing questions about computational algorithms—in particular, artificial intelligence—and the contribution of art in this process.

2 Justification and Previous Research

> Humans make their own brain, but they do not know that they make it. (Malabou 2008, p. 1)

Several mobile games were released in 2012. One of them, by the American studio Big Duck Games,[1] called *Flow Free*, is distributed for free to this day. It is classified as an electronic puzzle game, in which the player must solve a problem using logical reasoning and a simple engine.

---

[1] Big Duck Games has also released several expansions of the series, both paid and free. They have been planned for the Android, iOS, and Windows Phone platforms. [Available at https://www.bigduckgames.com/flowfree, accessed July 2021].

Flow Free is a game known as Numberlink,[2] which involves finding paths to connect numbers, dots, or colors in a grid. In this case, there is a grid of squares (organized in horizontal rows and vertical columns, as on a board) with colored dots occupying some of the squares. The goal is to connect the dots of the same color, creating a flow of "paths" between them so that the whole grid is occupied by such "connecting paths". The difficulty is determined mainly by the size of the grid, which can vary from 5×5 to 15×15 squares, and the variation in the number of colored dots, which will define how many empty squares to go through/fill. The challenge is to connect the colors to cover the whole grid with the number of steps set as ideal.

Let us understand how it works. Consider an $n \times n$ matrix of squares: Some of the squares are empty, and others are marked by colored circles (green, red, yellow, blue, orange…). Each color occupies exactly two different squares in the grid. The player's task is to connect the two occurrences of each color by a path of continuous lines made with horizontal and vertical movements; no two paths are allowed to cross. The game ends when all previously empty squares in the grid are filled by the created lines. Each path is measured by its difficulty in relation to the percentage of paths achieved. Each move is considered a step, and there is a best flow ratio (number of steps) to achieve. At first, there are valid paths that have already been predetermined so that they do not cross. In other words, initially, a solution to the problem has already been generated, and the player must find it. The idea is that we have a set of nodes related by reflexive and symmetrical equivalence, such that when "blue_1" is connected to "blue_2", it is the same as "blue_2" being connected to "blue_1", and the path between them will always be of equal length.

After a prolonged period of playing this puzzle daily, we had the hasty idea of comparing the results obtained as if they were visual enigmas deciphered. If we consider puzzles as problems to be solved by deduction, we are overestimating the process. It is about logical inference, certainly, as players perform heuristic research (in the computational enigmas sense),[3] and choose strategies based on previous experiences, have goals to achieve, and, for each heuristic, there is a pruning process that removes certain branches of the search tree if they cannot become a consistent option, deciding which branch to pursue. For example, "paths that cross" will be discarded, as well as "paths that do not fill all the grid square spaces". Bear in mind that, while some grids require more effort than others, there is no creative action involved, and it is a predetermined, finite grid.

We authored a paper in 2018 about this experiment (Laurentiz 2018), in which we eventually realized that, since decisions were driven by previously learned repetitions, we already had evidence of learning. Moreover, what can be successively reapplied to the structures resulting from its earlier application is the triggering principle of any language, that is, the property of recursion.[4] After some time of this training by repetition, the grid configurations were memorized, and the mind predicted the next moves almost automatically, by reflex, depending on the mechanical agility and

[2] *Numberlink* is a type of logic puzzle.

[3] A classic example of applying heuristic search is the "traveling salesman problem".

[4] It will be recovered in the end.

degree of difficulty of the grid. This means that, after a few levels, the player would learn trends, formats, and rules, thus getting better performance. From this point on, they start playing with the patterns and reacting from these configurations that have been memorized. For example, one would naturally first try to draw the outer lines with larger path lines and next solve the smaller central lines. Points that are far apart and resting on the edges of the grid are a foolproof clue to drawing a large line around the edge of the grid.

Our strategy was to find recognizable blocks—that we might call meaningful blocks—contours, paths, and repeated configurations that we had already learned. We would look first for easy-to-solve moves, i.e., those with obvious solutions (with the lowest percentage of moves index), for they were positioned in situations with no other options; then, we would try to complete the grid with the more problematic decisions. These were some strategies gotten by experience from repetition of previous moves. Tasks that require simple strategies and logical reasoning are exercised and repeated every turn. The point that interests us at this moment is the premise that repeating simple symbolic patterns from formal systems can generate traces of memories, trigger cognitive and sensory experiences, develop strategies, and promote changes in thinking (Laurentiz 2018).

This helps us, by analogy, to understand methods for generating training data, evaluating systems, modeling, and learning in AI algorithms. In other words, performing the same task to exhaustion can change our way of thinking. Recording all the experiences acquired also created opportunities for comparisons, references, identification, and recognition of blocks of meaning, and all this enhanced the process, considering that there is also a Web site for daily solutions of the game.[5] Now, multiply this by the countless logic games we currently have. We are training these "conformed thoughts" all the time. These self-controlled, self-contained, deliberate thoughts are everywhere. One only has to be somewhere public to notice that people are plugged into their cell phones, in many cases playing some of these puzzles. Add to that the current number of social network applications. What effect will this have on us?

We had already studied memory traces from Gestalt theory (Koffka 1975) in other articles (Laurentiz 2017a, 2018), but recent research speaks of "neuroplasticity", a term used by neuroscience for the brain's ability to change with experience and keep some of those changes. In this approach, researchers define the brain as a dynamic, adaptive, information-seeking system that is interconnected and networked (Vasconcelos et al. 2011; Eagleman 2020). David Eagleman explains that the brain needs repeated practice to learn an activity, which can be motor or cognitive. This practice expressively changes the brain configuration, so much so that "when medical students study for their final exams over the course of three months, the gray matter volume in their brains changes so much it can be seen on brain scans with the naked eye" (Eagleman 2020, p. 143). Therefore, since we are exhaustively practicing symbolic logics, especially in this period of 2019–2022, when our interactions and experiences are restricted almost exclusively to interfaces, screens, projections, and technical images. These logics are in themselves "conformed thoughts", self-controlled

[5] Available at https://flowfreesolutions.com/daily/, accessed July 2021.

and deliberate, the result of concepts, and knowledge models of a culture or group. Consequently, we are reorganizing ourselves, changing our habits, behaviors, cultural practices, and this will also cause changes in our way of thinking. Are we aware that this means that we are shaping our brains? Although it is always announced that machine learning networks learn from us, that we are the ones teaching them, is not it these "conformed thoughts" that are causing changes in our way of thinking? This means that both human and non-human systems are interconnected, and one interferes with the other.

There are several studies that relate cognitive activities to games (Laurentiz 2018; Baniqued et al. 2014; Oei and Patterson 2014; Nef et al. 2020), and although this subject still warrants further research to recognize a real cognitive pairing, "puzzle Numberlink games are promising as a tool to monitor the progression of motor impairment in neurodegenerative diseases" (Nef et al 2020, p.1), for example, and one can already suggest "future studies to create game-based adaptive computerized cognitive assessments." (Ibid., p. 3).

Here begins our investigation, which is focused on studies of logic and general laws of signs. Considering that:

1. all thinking is accomplished through signs: "logic is the theory of self-controlled, or deliberate, thought" (Peirce et al. 1994, CP 1.191);
2. we have a brain capacity that allows changes from experiences and can retain certain changes;
3. pattern recognition methods and "conformed thought" processes feed back into our everyday activities;

the goal is to present a representational model promoted by information processing capable of discussing questions about computational algorithms—in particular artificial intelligence—and the contribution of art in this process.

3 Defining Conformed Thought

Initially, it is necessary to define what we are naming "conformed thought". This concept was born from an attempt to relate different representational models based on Vilém Flusser's "escalation of abstraction". In particular, the author identifies a "new abstraction", referring to processes of technical image generation from photographic apparatuses to computational models; this abstraction results from a process of imagination different from the one that mentally and/or handcrafted happens from "something to the image of this thing" and from the process of conceptualization to generate a "concept of the image of this thing".

Something changes when a device starts to mediate this process, and later, when using computer models, argues Flusser, an imagination with a simulation character emerges. It is still called imagination, he explains, because the intention of generating images remains, but, in his understanding, it is a different kind of image. The images of the new imagination are projected by zero-dimensional[6] processes and are the result of calculations, numbers, and computation. Flusser also says that it is as if "imagination had become autonomous"[7] (Flusser 2007, p. 173).

At this point, we begin to delineate what we call "conformed thought" (Laurentiz 2015, 2018, 2019, 2021). "Conformed thought" is the result of a process of image generation by this new abstraction, starting from an idea of thinking that is configured through a numerical logic, and that ends up generating a mathematical and statistical reordering of its own code. As Norval Baitello explains in the preface of the book *Vilém Flusser: The Universe of Technical Images*, "the escalation of abstraction [...] is nothing more than an escalation of subtraction, it consists of the progressive removal of dimensions from objects, from three to two, to one and to zero dimensions"[8] (Baitello, in Flusser 2012, p. 10), in this path where the codes of representation go through processes of increasingly abstract symbolic systems. This abstractive retreat entails consequences, since "images are mediations between subject and objective world, and as such are susceptible to an internal dialectic: they imagine the objects they represent" (Flusser 2007, p. 166); with images projected by zero-dimensional calculations, one no longer confuses "what one imagines with what one has imagined"[9] (Flusser 2007, p. 172); aside from the fact that these "images do not hide their simulation character"[10] (Flusser 2007, p. 172). While images that mediate humans and their objective world are, in some way, copies of facts and circumstances, the images of this new abstraction do the mediation between calculations and their possible application in the surroundings, and this would indicate that these two imaginations walk in opposite directions (Flusser 2007, p. 173). Not only are they going in opposite directions, but computerized images can appear as if they were copies of circumstances, exactly like the images that used earlier processes, impersonating their predecessor representational models. Even so, computerized images will always carry a potential for situations belonging to a new field of possibilities. Flusser concludes by saying:

> only when images are made from calculations, and no longer from circumstances (even if these circumstances are quite 'abstract'), can 'pure aesthetics' (the pleasure in playing with 'pure forms') unfold; only thus can homo faber detach itself from *homo ludens*." (Flusser 2007, p. 175).[11]

---

[6] We will be calling as zero-dimensional processes those that have been translated as adimensional, null-dimensional, and even dimensionless among the different Flusser texts. We will not discuss such distinctions here.

[7] Translated from Portuguese: "[...] a imaginação tivesse se autonomizado".

[8] Translated from Portuguese: "[...] a escalada da abstração [...] nada mais é que uma escalada da subtração, consiste na retirada progressiva de dimensões dos objetos, de três para dois, para uma e para zero dimensões".

[9] Translated from Portuguese: "As imagens são mediações entre o sujeito e o mundo objetivo, e como tais, estão submetidas a uma dialética interna: elas imaginam os objetos que apresentam".

[10] Translated from Portuguese: "imagens não ocultam seu caráter de simulação".

[11] Translated from Portuguese: [...] somente quando as imagens são feitas a partir de cálculos, e não mais de circunstâncias (mesmo que estas circunstâncias sejam bem 'abstratas'), é que a 'estética

This leap to the zero-dimensional is a boldness that forces us to "renounce causal explanations in favor of the calculus of probabilities, and we must learn to renounce logical operations in favor of propositional calculus." (Flusser 2007, p. 176–177). It is noteworthy that he is referring to formal systems, which are calculations that represent formal objects, with the purpose of computing inferences, reaching goals, and that follow rules of an abstraction process from a set of a notation system that is also formal.

It is important to say that it is the representational aspects that concern us, and that our interest is exactly to try to understand this new abstraction and to insert it in a dynamic representational model. The main justification is that dealing with a zero-dimensional image would already indicate changes in cultural values, consequently possessing the ability to change habits and behaviors.

The term "conformed thought" was thus created to designate codes and set of codes, norms, algorithms, patterns, and interfaces. Hence, the forms generated by algorithms—and, in this context, we will be dealing with artificial intelligence/machine learning—are externalized thoughts, which use mathematical and statistical calculations (propositional logic) for data analysis and generation of models. These elements, although dematerialized, become sensitive in the resulting image when actualized. This leads us to consider that such conformed thoughts, whether internal (our mind also functions from "conformed thoughts") or external, act in a determined way and "form", "inform", and "conform" thoughts.

From the principle of "conformed thoughts", we can make important considerations regarding representational aspects. We can observe, for example, an encapsulation of the representational system starting from the "thing itself", which is perceived and made "object" in order to be communicated and shared (Deely 2004), and that is then synthesized in "modeled object", with the understanding that models are formed by objects, which in turn are "things themselves" that have been objectified (as we see in Fig. 1).

In our view, in line with Flusser (2007), the passage from an "object" to a "model" carries levels of abstraction that would be important to note. In other words, the modeling process, which is also "conformed thought", shows the relationship of the "thing itself" in these representational nests; even if the "thing itself" distances itself from the model, it is preserved somehow as a motivator and trigger of these processes. All this happens in a context where systemic feedback occurs, with processes of evaluation, transformation, comparison with reference values, adaptation, regression, codification, following the principles of cyber systems (in Fig. 2).

Moreover, since we are a part of this system and not just observers (Fig. 3), to say that conformed thoughts feed back into the system means that we share these processes in an ongoing relationship, even if we are not fully aware of it. Even though we do not have full control over these processes, they will interfere with our future decision-making in some way.

---

pura' (o prazer no jogo com 'formas puras') pode se desdobrar; somente assim é que o homo faber pode se desprender do homo ludens."

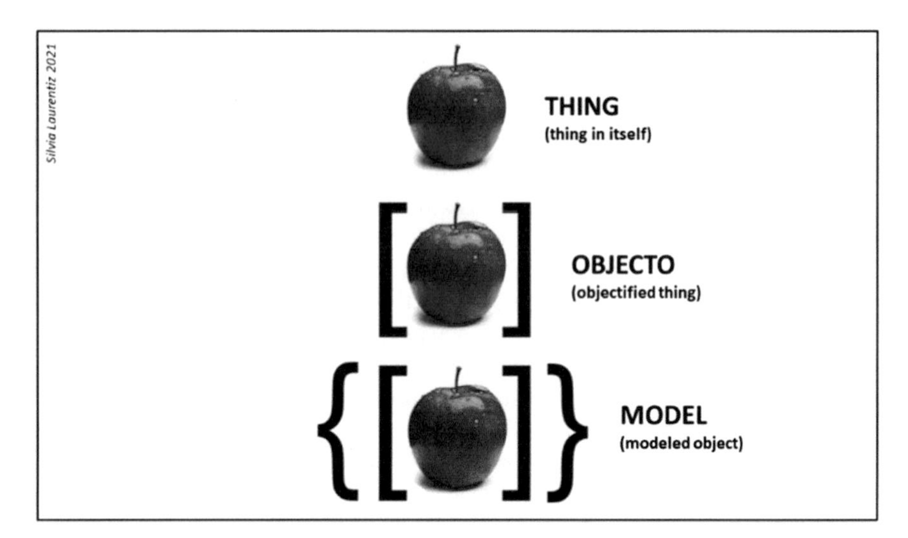

Fig. 1 Basic unit for the dynamic model. *Source* Author's personal collection

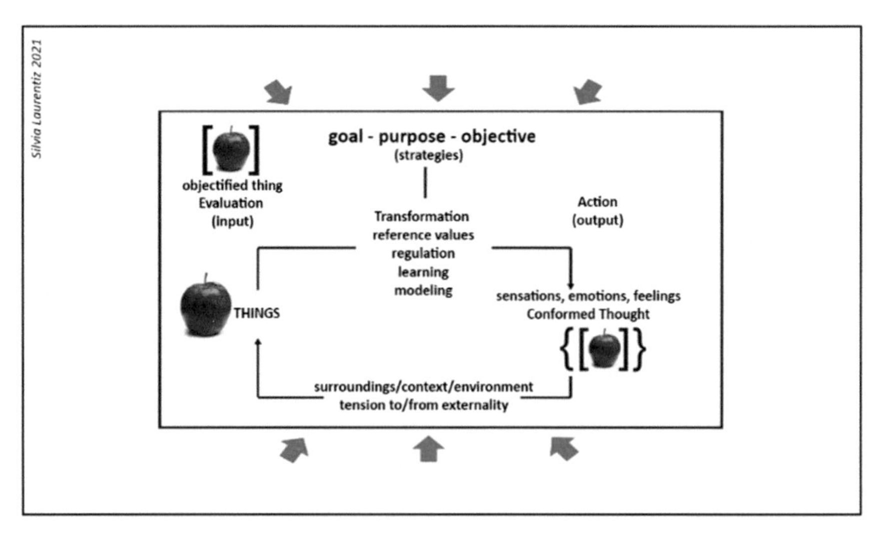

Fig. 2 Simulation from the cybernetic model for the basic unit of representation. *Source* Author's personal collection

Following this reasoning, models, which are formed by "objects that have been modeled", also objectify themselves so that they can be experienced by others, leading to the next level of abstraction, causing effects in these passages, and re-fueling the system again. These changes in levels of abstraction are significant because there are differences between the "things that are objectified" and the

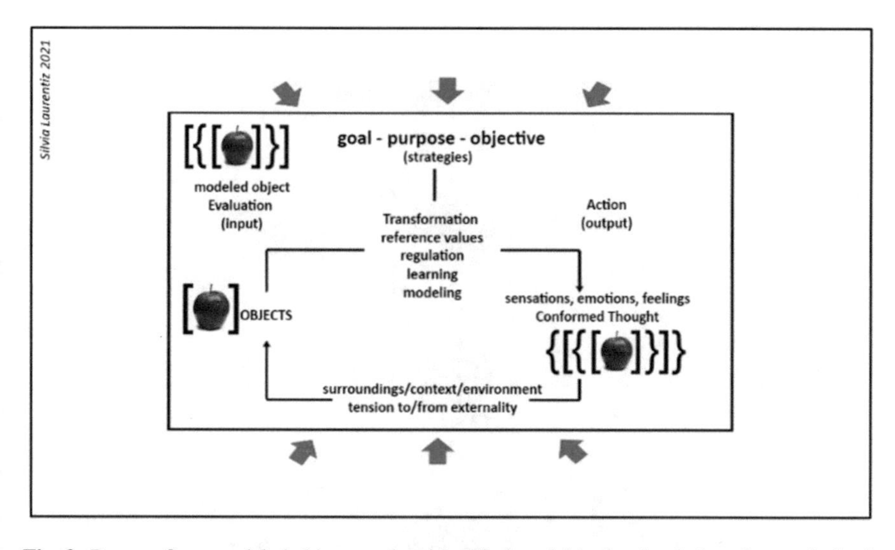

Fig. 3 Passage from modeled objects to the objectified model in the simulation. *Source* Author's personal collection

"models that are objectified". This is the point, although it is a speculative argument, from a proposition based on the hypothesis that we are training thought forms from "conformed thoughts", and this affects us even though it is not a response to direct stimuli from the "things" of the world (Fig. 4).

From what has been presented so far, we can already anticipate that we will be sharing two structural systems:

- the cybernetic model applied to a sign system—which would guarantee the intended dynamics between internal and external systemic actions;
- and Flusser's representational model of the "escalation of abstraction"—which would guarantee the perception of these passages between levels of abstraction.

"Conformed thoughts", therefore, are actualized forms of elaborated and deliberate knowledge. This condition already brings "conformed thought" closer to the very definition of sign; however, it is a special sign.

A first distinction is that it is an abstraction from the Flusser's technical image. This is an important fact, which already delimits a sign. Second, there is the approximation between "conformed thought" and Charles Sanders Peirce's concept of symbol. Despite this obvious relation, we point out that every "conformed thought" is in fact a symbol, but not every symbol will be a "conformed thought". For example, the word "house" represents a [house] symbolically, by convention of law; however, this is a process of conceptualization, of one dimension in Flusser's approach, and therefore, will not be in our scope.

It is also worth noting that thoughts that are 'Conformed' are not restricted only to forms, appearances, expressions of patterns which are recognized in images. Initially, because we will also be considering in this list the concept of model—which in

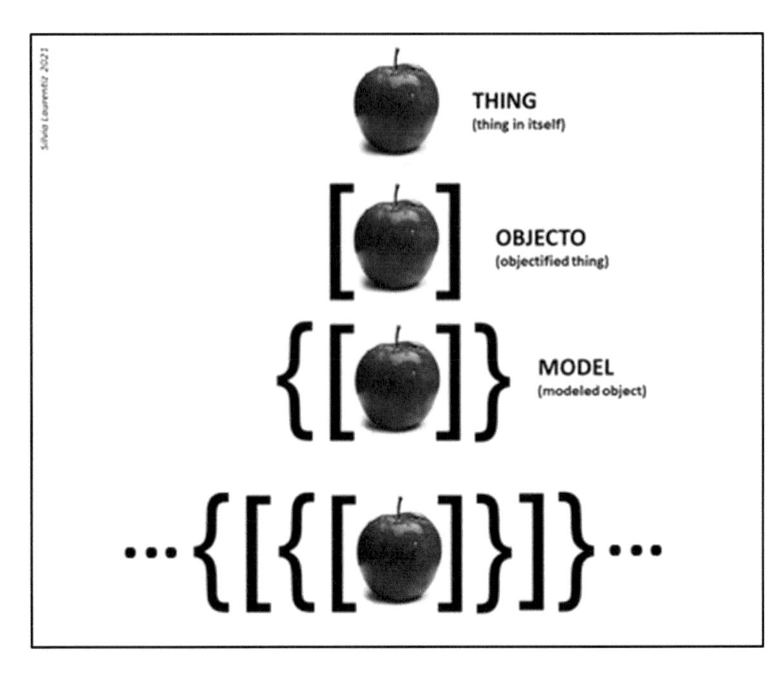

Fig. 4 Conformed thought's abstraction escalation. *Source* Author's personal collection

turn is formed by objects, and these are objectified things. Being a model already announces how "conformed thought" carries a conceptual and dematerialized charge. In addition, all thought is formed by distinct kinds of signs (cf. Peirce et al. 1994; Sebeok 2001). The type we are dealing with refers to the habit(s) acquired and formalized by a culture, which depends on the context in which it is embedded and is related to a notably computational technology.

We cannot fail to mention that there are various kinds of signs, which, although not recognized as "conformed thought'" also make up all thought (cf. Peirce et al. 1994). For example: vague formations of sensations, emotions, and feelings. This type is governed by our sensory system (including the sense of experience and observation) and drives our thinking in semiotic evolution, also promoting changes in habits and giving rise to new thoughts interdependent on each other, as demonstrated by Antonio Damasio, on the significant role of emotion in reason in the human brain (Damasio 1996). Thus, both the "conformed thoughts" and these other sign types of not so structured merge into an integral thought. More important than just recognizing this coexistence between signs of different nature is realizing that creative thinking will depend on this! Because the human mind acts from different processes and there are pre-interpretative states in "conformed thoughts".

To better explain how forms of knowledge conform thought, even before they establish a de facto interpretive action, we will return to Flusser's theory. For him, technical image is concept (Flusser 2011) and depends on a technological procedure of a period, within a context, determined by a society or group, and determines a way of seeing, from a point of view. We are not even evaluating the represented object (the diegetic object) of this technical image, nor the way it relates to the sign itself, which is presented in the image (a condition that occurs even in the face of an abstract form). Before any object in the image is recognized—or an abstract form is contemplated in its plastic and formal qualities—even before an interpretant is generated from this recognition/contemplation, we can already state that a technical image will bring a particular point of view, and this guides the new interpretations generated, in a vague and subjective way. That is, we are not only evaluating the sign's plastic characteristics (in its quali, sin and legisign conditions, according to the Peircean vision), nor are we recognizing a relation between sign and object (immediate and dynamic object in the Periclean vision), but the fundamental point is to understand how certain characteristics conform the thought beforehand, even before an interpretant is formalized. It is to perceive characteristics of the immediate interpretant through the recognition of the structuring principles of the sign itself. It is a quasi-interpretation capable of causing signifying effects and provoking changes in habits even if we do not always realize it. And, later, to understand how these principles feed back into the systems—sensory and cognitive—that, evolutionarily and by circularity, generate increasingly complex systems of interpretations. We thus begin to design a model that approximates the proposed cybernetic model.

Advancing further, we must still recognize that:

(I) Once conformed thought is actualized, in the very sense of taking shape in the world, it will have elements that will reflect its sensory aspects. Since things and signs, as well as experience and thought, are intertwined, conformed thoughts—codes, patterns, interfaces, etc., —are not only concepts, that is, abstractions of a certain degree, but also have sensible elements when instantiated in the world.

(II) Every abstract thought (whether of first, second or third degree) has the power to generate an interference of some kind in the way we perceive the world. Therefore, conformed thinking con-forms, in-forms, and forms.

(III) With the premise that there is a close relationship between signs and things in the world, we can extend this discussion by further recognizing that:

(IV) Our relationship to the world depends on relationship to our surroundings, an expanded Umwelt (von Uexküll 2007, cf. Jakob von Uexküll) formed by a complex network of interwoven interpretations of things, objects (objectified things), and models (formed by objects, which are objectified things). In this proposal, models will be considered "conformed thoughts" when they are a zero-dimensional result of abstraction—understanding, therefore, that not every model is a "conformed thought".

(V) As Umwelt acts as an interface that selects and filters information from the environment, and we internalize this information in codified form, any material used by living systems in knowledge construction is representational (even if vaguely or in the condition of quasi-representation), that is, it is formed by a myriad of "some things" that represent "external things", which are processed into a particular "kind of thing" of our Cognitive System (Peirce et al. 1994; Deely 1990). Moreover, this process feeds back into our sensory system (Albuquerque Vieira 2008).

(VI) Consequently, sensations and "conformed thoughts" depend on each other (Damásio 1996), in the same way that actions of a body (Innenwelt), and the environment (Umwelt) in which it is inserted are also associated (Sharov 2010, 2012). It is also worth noting that Alexei Sharov[12] introduces the principle of agent, to resolve the boundary between living and artificial systems in his "Functional Information" approach. Thus, living organisms are *agents* as much as non-living artificial devices. For the author, agents are broadly defined as "systems with goal-directed programmed behavior" (Sharov 2010, p. 1052). What would unite living and artificial agents "is their ability to perform functions for the purpose of reaching certain goals. Functions of agents are encoded and controlled by a set of signs which I call functional information" (Ibid.). And when we said that the "actions of a body" (Innenwelt) are associated with its environment (Umwelt), it should be understood as a body of an agent, as proposed by Sharov (2010, p. 1051).

(VII) Finally, one understands life and semiosis (sign action) as coextensive (Sebeok 2001).

In fact, Fig. 3 itself can be considered an agent, since it is already a system with goal-directed programmed behavior (or quasi-agent[13]), and that interacts with other agents, of distinct kinds (living or artificial systems), which would all act in this process (Fig. 5). Thus, the surroundings, the context, the environment, and the tensions to and from the externality are already being considered (see Fig. 3), and we also must consider that other agents act and interact dynamically with each other.

After this initial presentation, we can already begin to present the elements for the formation of our dynamic, generative, and time-evolving model. Diagrams 1, 2, and 3 illustrated the proposal in general and have already been presented in other publications (Laurentiz 2017b, 2019).

4 Proposed Model for Representational System Simulation

The main point so far is to realize how objects and models affect us and our emotions and conform the thought. And with the emergence of artificial intelligence and machine learning processes, which are conformed thoughts par excellence, we start to notice new triggers.

As we have seen, we go through differentiated processes of "objectified things", "modeled objects", and "objectified models". When we "think and feel" from

[12] In addressing a possible relationship between biosemiotics and cybernetics.

[13] The distinction between agents and quasi-agents will be more fully detailed in another article. At this point, we recall the reference by Sharov (2010) to introduce the agent principle. This author recognizes levels, and these include degrees of proto-icons, proto-indexes, and proto-symbols. This is just one of the aspects that led us to consider agents and quasi-agents.

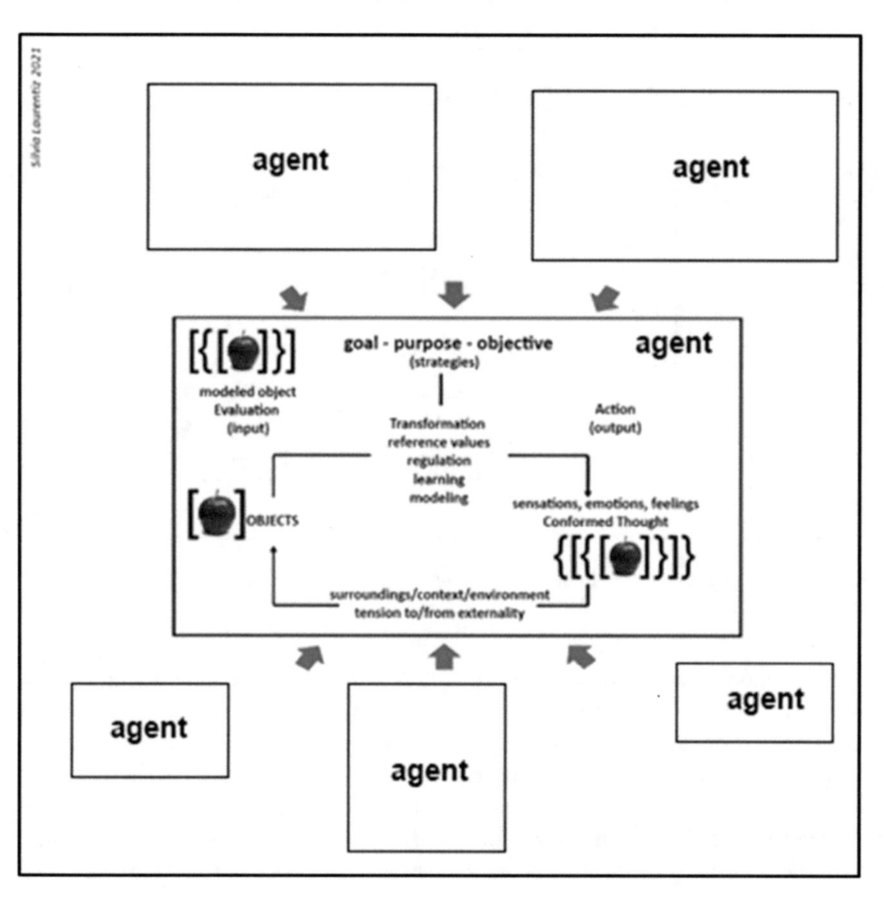

Fig. 5 Complex system among agents of different kinds. *Source* Author's personal collection

patterns, using "conformed thoughts", and react from these abilities obtained through repetition, we are affected by these memories and their interfaces. So far, the model would behave similarly to the case of the *Flow Free* game analysis; we did in the Justification and Previous Research section. It is important to remember that, at that moment, we asked if we were aware of how we are shaping our brains, and we took advantage of the systemic feedback between agents and their surroundings to question whether we are the ones teaching the machines or if they are shaping our way of thinking.

Following this argument, a digital system based on learning and modeling processes, will be guided by models and patterns, which in turn will guide the results obtained by a machine that will feed the system back again. In this scenario, artificial intelligence, machine learning, and deep learning suggest a meta-processing of/for the generation of "conformed thoughts" (Fig. 6). Therefore, in our reasoning from Fig. 5, since we would again have new goals and strategies, we could consider it other agent (or quasi-agent), acting on the entire system.

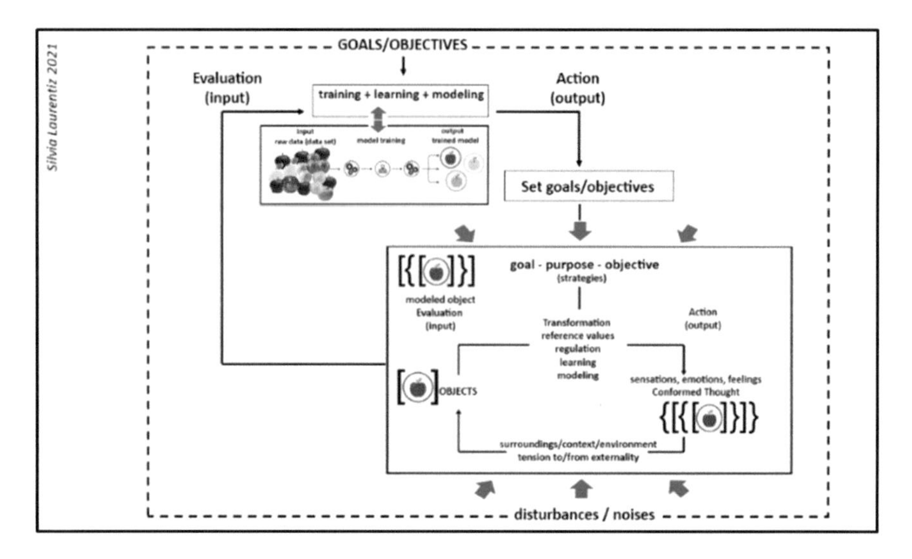

Fig. 6 Model (agent) inserted in meta-processing of/for the generation of "conformed thoughts". *Source* Author's personal collection

It is important at this point to pause to reflect on how computers perform cognitive tasks. For example, those that involve recognizing objects in images through computer vision, and how a computer program can name and detect objects in an image. To do this, a system must be based on visual concepts, with object descriptions, attributes, formal specifications, and structured relationships between the different elements detected in the image's regions (Kiros et al. 2014; Krishna et al. 2017), in addition to the analysis, classification, evaluation, regression, and training procedures that will guide the learning process.

It is worth noting that machine learning algorithm is a mathematical model that maps inputs to specific outputs and feeds the model with pairs of "input + expected output" to train it and will adjust its internal parameters from that training. Since the algorithm itself goes through a process of self-adjustment determined by what it has "learned" during its training phase, correcting, adjusting, and fitting results to its purpose, it leads us to think of a metacognitive quasi-process.[14] With training completed, the program can be used with new data inputs and even be easily adapted to new situations. It would be hasty to suggest that the system would be consolidating its own memory traces, but we should recognize that there is a significant difference between an algorithm that solves one problem and another that may solve different problems, as in the case of classical algorithms and machine learning algorithms (respectively).

It is also important to highlight that, in machine training procedures, we have different strategies involved. The classification technique, for example, widely used in machine learning, groups things that are similar by parameters that satisfy some

[14] On metacognition we recommend the article by Jou G and Sperb T (2006).

selection, norm, or law criteria. Discrimination processes, also commonly used, impose restrictions based on certain conditions and circumstances. Thus, we do not only have synthesis of objects transformed into models, but mathematical models that analyze, classify, and select models, which are formed by objects, which are things that have been objectified. In fact, there are classical algorithms within the machine learning algorithm, encapsulated algorithms, networks containing several coupled algorithms. This increases the complexity of these logical procedures and, in turn, the levels of representational abstraction. These new procedures should provoke a new tension between "sensations and conformed thought", in an environment of mixes of information and levels of abstractions. In any case, by the very nature of statistical computation, one works with frequency of values, averages, and systemic trends.

Returning to Fig. 6, the initial model is now considered encapsulated in the system of Fig. 5. Now, modeling and learning systems (as an a*gent*) will be able to adjust goals of the previous system. The main point is that the goal adjustments will be done by machine processes (zero-dimensional abstractions), the result of learning and modeling.

Despite this increase in complexity, the results we are getting with current machine learning systems show that we will have to revise and update our current bases and even generate new ontologies, from a more dense and complete set of descriptions about an image. Recall the ImageNet[15] case and Microsoft researcher Kate Crawford's criticism that we are injecting our own limitations into the algorithms. In her article *Excavating AI - The Politics of Images in Machine Learning Training Sets* (Crawford and Paglen 2019), she highlights that making a machine interpret images is much more of a social and political issue, rather than merely a technical one. ImageNet demonstrated how these processes can promote discrimination, misjudgments, biases, and that the technical process of categorizations and classifications proves to be a political act. Therefore, the production of images made by a machine carries social, political, and economic issues based on the context in which they are inserted, their surroundings, and because of their condition as an "allopoietic system" (Nöth 2001, p.66).

The entire basis of machine learning systems is built on training sets, and it is these that underlie how the AI will recognize and interpret data from the world. Joy Buolamwini, who identifies herself as the Poet of Code,[16] working with facial analysis software, realized that the software could not detect her face because the people who coded the algorithm had not taught it to identify a wide range of skin tones and facial structures. After that, she faced a mission to combat bias in machine learning, the result of a "Coded Gaze". Still according to Crawford, this is not an easy task since images are loaded with multiple senses and meanings. "Entire subfields of

[15] ImageNet (https://www.image-net.org/index.php) is an image database organized according to the WordNet hierarchy (currently only the nouns), in which each node of the hierarchy is depicted by hundreds and thousands of images. For further information about its functioning, we recommend: Krizhevsky et al. (2017).

[16] Available at https://www.poetofcode.com/, accessed July 2021.

philosophy, art history, and media theory are dedicated to teasing out all the nuances of the unstable relationship between images and meanings" (Crawford and Paglen 2019).

In this sense, it is also important to know the trajectory of the research that culminated with deep learning (Kurenkov 2020), so we can understand how the generation of models in the computer takes place and has a better understanding of what a simple image is. In fact, according to Lev Manovich (2018), the challenge is to try to go beyond the search for types, structures, and patterns from already existing and recognized ways of seeing the world (which we understand here as "already conformed thoughts").

In the presentation of the *Flow game*, we realized that if machine learning networks learn from us, what we teach them ends up causing changes in our way of thinking, which, again, denote those systems are formed by living and non-living agents that are interconnected, and one interferes with the other. Therefore, quasi-interpretive processes, sensations, and vague ideas must somehow participate in this system between agents, whether living or non-living systems.

5 The Contribution of Art

The dynamic model of representation proposed here is an interesting tool to understand which layers and procedures are being questioned and transgressed by artists. Let us look at the case of Shinseungback Kimyonghun, a Seoul-based artist duo consisting of Shin Seung Back and Kim Yong Hun, authors of the work *Cat or Human* (2013, at https://ssbkyh.com/works/cat_human/, accessed July 2021). The work is composed of two sets of one hundred photographs that use human face detection algorithm (OpenCV's) and detection for cat faces (KITTYDAR) in an inverted form. So, Flicker Photos were used, and the program recognized human faces by the cat face detection algorithm, and cat faces were recognized as human faces by a human face detection algorithm. When the artist uses tools and techniques in unconventional ways and uses them for a function for which they were not created, they explore this field of possibilities. Albuquerque Vieira says that the artist explores fields of possibilities of his/her surroundings (Umwelt) and ends up perceiving sophisticated articulations of reality that follow criteria of organization and coherence, which are associated with an "aesthetic root" (Albuquerque Vieira 2010).

After the machine is trained to identify one type of object (human faces), it is presented with another type of object (cat faces), and vice versa. The provocation of the work is that there is recognition of the objects despite having been trained for other models. We can see, evaluating the proposal through Fig. 7a (an update of Fig. 6), that in this case, the machine would be trained to recognize apples and would be placed to recognize oranges, and in some cases, it would identify them as apples (Fig. 7b is an update of Fig. 1). It is evident that the strength of the cat or human work is the unusual use of inverted models between animal and human. Furthermore, it signals how these resources are flawed, make mistakes and are incapable of perceiving subtleties between things/objects/models, which a human agent would easily perceive.

Since our present time is insistently faced with what we call "conformed thoughts", as we said in the beginning of this work with the example of games, social network applications, interfaces, and screens that we use to communicate in the confinement

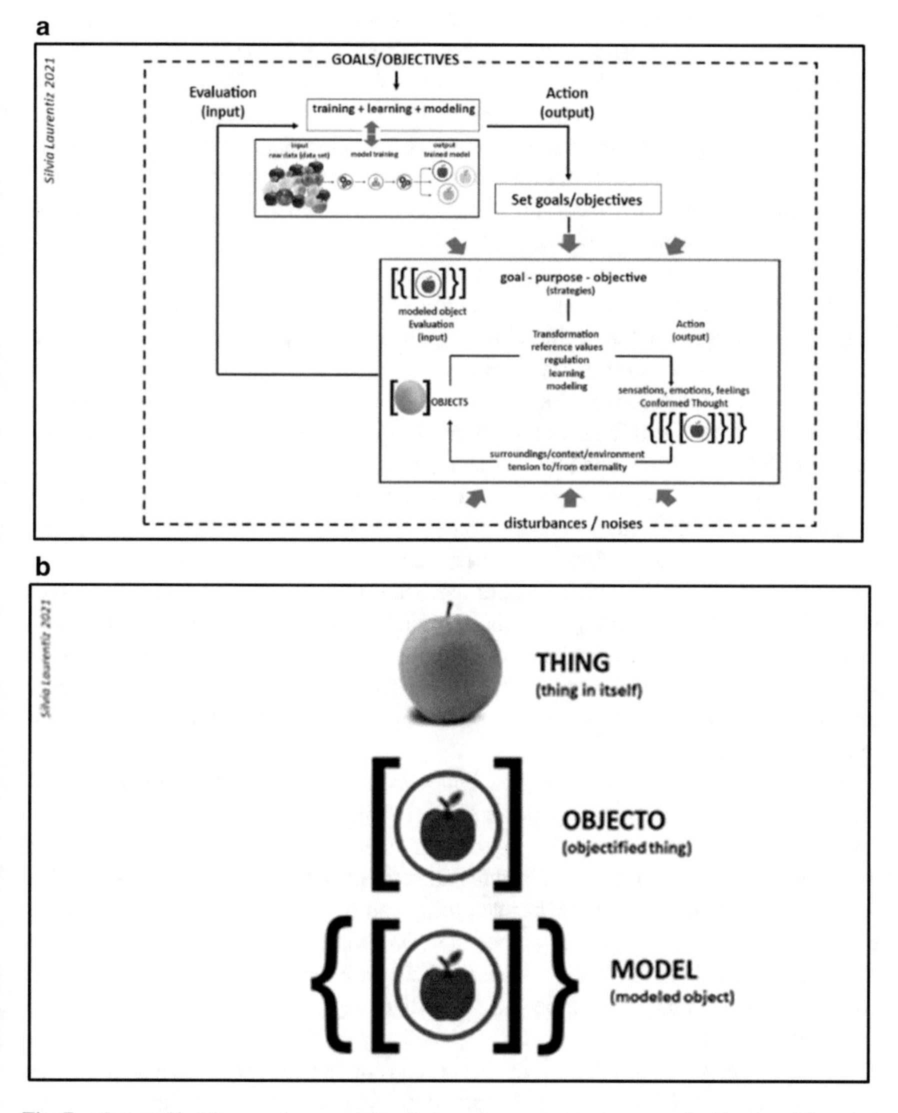

Fig. 7 **a** Inserted in Diagram 6 as an artistic strategy. *Source* Author's personal collection. **b** Inserted in Fig. 1 as an artistic strategy. *Source* Author's personal collection

we are living in, these must be somehow altering and reconfiguring our brain plasticity. Consequently, actions of a body (Innenwelt) and environment (Umwelt) that are always associated participate in this process.

Another artwork by this duo is the *Cloud Face*, which is a collection of images of clouds that were recognized by the face detection algorithm as human faces. The proposal follows a principle like that of the previous work, but here the machine error is compared by the authors to human imagination, which recognizes figures in the clouds of the sky. Abstract shapes always surprise the human mind by the degree of openness to possible interpretations they have. Even if it is a mistake, a machine, recognizing objects in shapeless masses is also revealing in some way.

Another strategy is, for example, for the artist to train a machine and make it "lose" on purpose, what it has learned, triggering the very human capacity in which remembering and forgetting are accomplices and not adversaries in the thought process. This is the case of the video work entitled *What I Saw Before Darkness* (AI Told Me 2019[17]) by an artist who simply goes by the name of "the girl who talk to AI". In this case, the artist makes an intervention in the training and modeling agent of Fig. 6. She programmed an artificial intelligence to generate a human-looking face, then shut off its own neurons one at a time—a process she recorded and shared in a time-lapse video. A normal-looking face slowly takes on strange glitches—lines shift, colors change, and features blur until the face is no longer a face at all, replaced instead by blobs of brown and white that eventually fade to black.

Also impressive is the work by the same artist *Grandmother of Man and Machine*,[18] where the neural network processes images from the concept of "grandmother". According to the Web site, "Approaches from neuroscience and computer vision helped to explore which traits of a grandmother the subject network grasped from all the millions of images it had previously seen". And a series of images featured on the site express what a grandmother is to AI. If recognizing figures in shapeless masses (as in "Cloud Face") by comparing formal similarities is something unexpected for a machine, but a machine recognizing shapes from concepts seems to be even more unusual.

The classic book *What computers can't do: a critique of Artificial reason* by Hubert Dreyfus (1972) already pointed out strong reasons for the difficulties of programming intelligent activities by the computer. An issue that he said should be taken into consideration was the discrete nature of all computer calculations. There is also the fact that the human mind has flexibility and is able to perform creative actions to solve problems, and a machine does not (Dreyfus 1972). Already at that time, a proposal was made to think about systems that would promote a symbiosis between computers and human beings, because together they could accomplish things that could not be done separately.

In this sense, we reinforce the idea that human and machines should work together. The artist Sougwen Chung can offer us an important contribution. The artist, with her

[17] Available at https://vimeo.com/337909277, accessed in July 2021.

[18] Available at https://aitold.me/portfolio/grandmother-of-man-and-machine/

emblematic work *On the collaborative space between humans and non-humans*,[19] presents human–robotic performances that generate drawings collaboratively and arise from the relationship of a living system with an artificial one. The performances utilize one or several robotic arms, which respond to a variety of data inputs that the artist has been developing for some time. Her idea is to think about ways in which humans connect to mechanical and artificial systems, and vice versa, and which would function as a "creative catalyst".

Very more relevant is how the artist says that she ended up adapting to the "inaccuracies of the machine", which led her to readjust her own gestures. It is a paradox, since a machine would not have, in its nature, principles of imprecision. It is imprecise compared to the complexity of the drawing performed by the artist's gesture, which the mathematically precise calculation and the flexibility of the robotic interface cannot achieve. It is important to say that the AI was trained on a second moment from the artist's numerous drawings. At first, it was only the real-time data that was captured and processed during the performance, but then the system was implemented by a learning process from a set of 20 years of data retrieved from the artist, which reflected a certain "trend of her style" as she adapted to the movements of the robot, in continuous circularity. One can clearly see the feedback process between the two, and that is described in the project by the collaborative involvement of the creative action itself.

Finally, Cesar & Lois created a system that feedbacks actions from agents of living systems and artificial systems, in a hybrid process capable of triggering improvisations, mishaps, accidents, and new stimuli during performances, including imperfections to the process, as the case of their artwork called *degenerative cultures*. The authors call it "A Post-Anthropocentric Intelligence" and explicitly are "Corrupting the Algorithms Of Modern Societies" (Cesar and Lois 2018[20]). These strategies are important escape valves for the pitfalls of "conformed thoughts". The aesthetic outcome of all these procedures will then be relearned, re-evaluated, and reconfigure experiences through circularity and systemic feedback.

6 Final Considerations

From the above, we can consider two main points about the contribution of art to representational systems:

1. In the words of Ivo Ibri, "[...] simply contemplating the world, in a disinterested experience because it has no practical purpose, allows us to demobilize the

[19] Available at https://sougwen.com/on-the-collaborative-space-between-humans-and-non-humans, accessed July 2021.

[20] Available at https://cesarandlois.org/digitalfungus/, accessed July 2021.

conceptual forms that mediate our acting in the world"[21] (IBRI 2020, p.6). This makes the artist and art fundamental pieces of this puzzle of conformed thoughts, representational systems, and creative cognitive procedures. In a dynamic representational system, the artist with his/her artwork has the role of fine-tuning the process, expanding our sensitivity, and exposing the fragility of representational models. In other words, the artist is the agent who will adjust the model, launching other perspectives, extrapolating, and testing rules and structures, subverting standards, taking the system to its extreme exaggeration, or denouncing its inconsistency. At the same time, he/she expands the model, suggesting deviations, causing noise, and promoting unforeseen degrees of opening. In our point of view, the artist is a calibrating agent of the system.

2. "Soft representation"[22] is the term that Paulo Laurentiz chose in 1991 to designate an artist's special attitude toward technology. This kind of representation implies that productive rules of technology are not being imposed on the world (Laurentiz 1991, p. 110). This means that, in terms of operative thinking, not to let there be internal interference of one language on the qualities of the other [...] (Laurentiz 1991, p. 113), of the systems involved. Nothing is more current than thinking about artificial intelligence algorithms in this way. I end with a quote from Paulo Laurentiz:

> The commitment of soft representation is not to mask or camouflage the information transmitted by the world, mediated by the organizing rules of signs. It discredits the authoritarian and arbitrary character of the sign, and at the same time, seeks to highlight another side of representation that despises the servile imitation of the sign of the real (Laurentiz 1991, p. 130).[23]

References

Albuquerque Vieira J (2008) Teoria do Conhecimento e Arte - Formas de conhecimento: Arte e Ciência – uma visão a partir da complexidade. Expressão Gráfica e Editora, Fortaleza. SBN-13:9788575631928 ISBN-10:8575631926, 2008, 135 p, Brasil

Albuquerque Vieira J (2010) Teoria do Conhecimento e Arte. Revista Música Hodie, 9(2). https://doi.org/10.5216/mh.v9i2.11088. Accessed on 27 July 2021

Baniqued P, Kranz M, Voss M, Lee H, Cosman J, Severson J, Kramer A (2014) Cognitive training with casual video games: points to consider. Front Psychol 4:1010. https://doi.org/10.3389/fpsyg.2013.01010. https://doi.org/10.3389/fpsyg.2013.01010. Accessed on 27 July 2021

---

[21] Translated from Portuguese: "[...] contemplarmos simplesmente o mundo, em uma experiência desinteressada porque sem propósito prático, permite desmobilizar as formas conceituais que medeiam nosso agir no mundo".

[22] Translated from Portuguese: "Representação branda".

[23] Translated from Portuguese: "O compromisso da representação branda é não mascarar ou camuflar as informações transmitidas pelo mundo, intermediadas pelas regras organizadoras dos signos. Ela desacredita no caráter autoritário e arbitrário do signo, e ao mesmo tempo, procura realçar um outro lado da representação que despreza a imitação servil do signo ao real" (Laurentiz 1991, p. 130).

Crawford K, Paglen T (2019) Excavating AI: the politics of training sets for machine learning (September 19, 2019). https://excavating.ai/. Accessed on 27 July 2021
Damásio A (1996) O Erro de Descartes: Emoção, Razão e o Cérebro Humano, São Paulo: Companhia das Letras, 1996. ISBN 85–7164–530–2, 336 páginas, Brasil
Deely J (1990) Basics of Semiotics. Indiana University Press, Bloomington
Deely J (2004) Semiotics and Jakob von Uexküll"s concept of Umwelt, Sign Systems Studies Journal (ISSN 1406–4243), 32 (1–2):11–33
Dreyfus HL (1972) What computers can't do: a critique of artificial reason. Harper & Row, Publishers, USA
Eagleman D (2020) LiveWires: the inside of the ever-changing Brain. Pantheon; Illustrated edition. August 25, 2020
Flusser V (2007) O mundo codificado. Ed. Cosac Naify, São Paulo, Brasil
Flusser V (2011) Filosofia da caixa preta: ensaios para uma futura filosofia da fotografia. Annablume, Brasil, São Paulo
Flusser V (2012) O Universo das Imagens Técnicas. Coedição: Annablume editora (Brasil) e Imprensa da Universidade de Coimbra, Portugal
Ibri I (2020) Semiotics and pragmatism—theoretical interfaces (vol 1). Editora Cultura Acadêmico, Marília, Brasil
Johnson J, Krishna R, Stark M, Li L-J, Shamma DA, Bernstein M et al (2015) Image retrieval using scene graphs. In IEEE conference on computer vision and pattern recognition (CVPR)
Jou G, Sperb T (2006) A metacognição como estatégia reguladora da aprendizagem [Metacognition as regulatory strategy of learning]. Psicol Reflex Crit 19(2). https://doi.org/10.1590/S0102-797 22006000200003. Accessed on 27 July 2021
Kiros R, Salakhutdinov R, Zemel R (2014) Multimodal neural language models. In: Proceedings of the 31st international conference on machine learning (ICML-14), pp 595–603. http://proceedings.mlr.press/v32/kiros14.pdf. Accessed on 27 July 2021
Koffka K (1975, [1935]). Princípios de Psicologia da Gestalt. Ed. Cultrix, São Paulo, Brasil
Krishna R, Zhu Y, Groth O et al (2017) Visual genome: connecting language and vision using crowdsourced dense image annotations. Int J Comput Vis 123:32–73. https://doi.org/10.1007/s11263-016-0981-7. Accessed on 27 July 2021
Krizhevsky A, Sutskever I, Hinton GE (2017) ImageNet classification with deep convolutional neural networks. Commun. ACM 60(6):84–90
Kurenkov A (2020) A brief history of neural nets and deep learning. Skynet Today, at https://www.skynettoday.com/overviews/neural-net-history. Accessed on 27 July 2021
Laurentiz P (1991) A Holarquia do Pensamento Artístico, Edunicamp, Campinas, Brasil
Laurentiz SRF. Sensoriality and Conformed Thought (2015). In: Antona M, Constantine S (Org) Universal access in human-computer interaction. In: Access to interaction 9th international conference, UAHCI 2015, Held as Part of HCI International 2015, Los Angeles, CA, USA, August 2–7, 2015, Proceedings, Part II. 1 edn. Springer International Publishing, New York, vol 9176, pp 217–225
Laurentiz S (2017a) Videogames e o desenvolvimento de habilidades cognitivas [Videogames and the development of cognitive abilities]. DAT Journal, 2(1), pp. 80–90. https://doi.org/10.29147/2526-1789.DAT.2017v2i1p79-89. Available in https://datjournal.anhembi.br/dat/article/view/45/37. Accessed on 27 July 2021
Laurentiz S (2017b). Notas sobre um pensamento conformado. In: Encontro da Associação Nacional de Pesquisadores em Artes Plásticas, 26o, 2017b, Campinas. Anais do 26o Encontro da Anpap. Campinas: Pontifícia Universidade Católica de Campinas, Brasil, pp 3603–3617
Laurentiz S (2018) Conformed thought: consolidating traces of memories, In: Marcus A, Wang W (eds) Design, user experience, and usability: users, contexts and case studies. DUXU 2018. Lecture notes in computer science, vol 10920. Publisher Springer, Cham
Laurentiz S (2019) Conformed thought and the art of algorithms. In: Proceedings of the 9th international conference on digital and interactive arts—ARTECH, 23–25 October 2019, Braga,

Portugal. ISBN/ISSN 9781450372503. Editora Universidade Católica Portuguesa, Portugal, pp 465–472

Laurentiz S (2021) Arte en el contexto de los procedimientos de lógica algorítmica. Arbor 197(800):a603. https://doi.org/10.3989/arbor.2021.800005. Accessed on 27 July 2021

Manovich L (2018) AI Aesthetics, Strelka Press, December 21, 2018, ASIN: B07M8ZJMYG. In http://manovich.net/index.php/projects/ai-aesthetics. Accessed on 27 July 2021

Malabou C (2008) What should we do with our brain? Fordham University Press, JSTOR. www.jstor.org/stable/j.ctt1c5chvm. Accessed on 27 July 2021

Nef T, Chesham A, Schütz N, Botros AA, Vanbellingen T, Burgunder J-M, Müllner J, Müri RM, Urwyler P (2020) Development and evaluation of maze-like puzzle games to assess cognitive and motor function in aging and neurodegenerative diseases. Front Aging Neurosci 12:87. https://doi.org/10.3389/fnagi.2020.00087. Accessed on 27 July 2021

Nöth W (2001) Máquinas semióticas, in Revista Galáxia: Revista do Programa de Pós-Graduação em Comunicação e Semiótica, n.1, 2001, Dossiê, pp 51–73

OEI AC, Patterson MD (2014) Are videogame training gains specific or general? Front Syst Neurosci 8:54. http://doi.org/https://doi.org/10.3389/fnsys.2014.00054. Accessed on 27 July 2021

Peirce CS, Hartshorne C, Weiss P, Burks AW (1994) The Collected Papers of Charles Sanders Peirce. Electronic edition. Charlottesville, Va, InteLex Corporation. (Quoted as CP)

Sebeok T (2001) A. Signs: an introduction to semiotics, 2nd edn. University of Toronto Press, Canada

Sharov AA (2010) "Functional information: towards synthesis of biosemiotics and cybernetics" entropy 12, no. 5: 1050–1070. https://doi.org/10.3390/e12051050. Accessed on 27 July 2021

Sharov AA (2012) A. The origin of mind. In: Maran T, Lindström K, Magnus R, Tønnensen M (eds) Semiotics in the wild. University of Tartu, Tartu, pp 63–69. http://alexei.nfshost.com/pdf/Sharov_origin_of_mind.pdf. Accessed on 27 July 2021

Vasconcelos N, Pantoja J, Belchior H, Caixeta FV, Faber J, Freire MAM, Cota VR, de Anibal Macedo E, Laplagne DA, Gomes AM, Ribeiro S (2011, September) Cross-modal responses in the primary visual cortex encode complex objects and correlate with tactile discrimination. In: Proceedings of the National Academy of Sciences, 108(37):15408–15413. https://doi.org/10.1073/pnas.1102780108. https://www.pnas.org/content/108/37/15408. Accessed on 27 July 2021

Von Uexküll T (2007) A teoria da Umwelt de Jakob von Uexküll. Galáxia. Revista do Programa de Pós-Graduação em Comunicação e Semiótica. ISSN 1982–2553, 0(7). Available at https://revistas.pucsp.br/index.php/galaxia/article/view/1369. Accessed on 27 July 2021

Practices

Creative AI, Embodiment, and Performance

Rob Saunders and Petra Gemeinboeck

Abstract In this chapter, we explore the relationship between creative AI, embodiment, and performance with reference to our artistic practice. The history of creative machines traces back to the automata of antiquity and featured at the dawn of computing. In the cognitive sciences, the study of human creativity has generally focused on creative thinking and the generation of novel and valuable ideas, rather than the role of embodiment in creative activity. The current renaissance of creative AI has seen remarkable advances in generative systems but has similarly shied away from the questions of embodiment. Our creative practice explores the possibility of creative AI in robotic systems and in turn, has highlighted the importance of embodiment in creative AI. As a result, we have shifted our focus from the development of computational systems as models of creative agents toward the realization of skillful performers able to facilitate the emergence of creative agency between humans and machines. The chapter outlines our inquiry through the lens of our arts-led research practice.

Keywords Robotics · Embodiment · Creativity · AI · Machine learning

1 Introduction

Human creativity is often defined as the ability to generate novel and valuable ideas, whether expressed as concepts, theories, literature, music, dance, sculpture, painting, or any other medium of expression (Boden 1998). But creativity does not occur in a vacuum, it is a situated activity embedded within cultural, social, personal, and physical contexts that determine the nature of novelty and value against which creativity is assessed (Lindqvist 2003; Csikszentmihalyi 1999). This chapter explores

R. Saunders (✉)
LIACS, Leiden University, Leiden, The Netherlands
e-mail: r.saunders@liacs.leidenuniv.nl

P. Gemeinboeck
CTMT, Swinburne University, Melbourne, VIC, Australia
e-mail: pgemeinboeck@swin.edu.au

C. Vear and F. Poltronieri (eds.), *The Language of Creative AI*, Springer Series on Cultural Computing, https://doi.org/10.1007/978-3-031-10960-7_11

the relationship between creative AI, embodiment, and performance through the lens of our creative practice.

The term 'creative AI' potentially covers a broad spectrum of activities from the use of AI as a medium of expression for human creativity (see Audry 2021) to the modeling of creative behavior in computational systems (see Bown 2021). At one end, creative practitioners including artists, musicians, writers, and designers work with AI as a medium, tool, or socio-cultural phenomenon to facilitate, amplify, or inspire their creative practices; as generations of creative practitioners have done in light of technological innovations (see Burnham 1968). At the other end of the spectrum, researchers attempt to construct computational models of creativity as a means of understanding the phenomenon and to engineer useful systems; for the purposes of this chapter, we consider this meaning of 'creative AI' to be synonymous with 'computational creativity' (Veale and Cardoso 2019).

The story of the development of automata from antiquity through to the modern era is littered with the imaginings of machines exhibiting creative abilities, from the self-playing instruments described by Aristotle and the Ingenious Devices of Al-Jazari to Jacques de Vaucanson's Flute Player and The Draftsman of Jaquet-Droz, for a detailed account see Husbands et al. (2008). This fascination with mechanical performers, coupled with shifting conceptions of creativity from a divine or mystical notion to a subject amenable to rational inquiry, naturally led to the discussion of machines exhibiting creative thought. Ada Augusta Countess of Lovelace in her translation of Luigi Manabrea's *Sketch of the Analytical Engine* commented on the possibility of the algorithmic composition of 'elaborate and scientific pieces of music' (Lovelace 1843) but made clear that any credit for producing creative works would be due to the engineer not the machine. A century later, at the dawn of the modern era of computing, Alan Turing reframed the countess' comment as *Lady Lovelace's Objection* to machine intelligence; that a machine would be incapable of taking its engineer by surprise (Turing 1950). Turing responded by noting that computers often surprised him, due to a faulty understanding on his part and the complex nature of the processes involved. He also explored the possibility of a machine that could organize itself as the result of its 'experiences'—a learning machine capable of distancing itself from its engineer.

The proposal for the *Dartmouth Summer Research Conference on Artificial Intelligence* (McCarthy et al. 1955) included the modeling of creative thinking as one of the 'grand challenges' facing the nascent field. Attendees went on to develop the first examples of creative AI; discovery systems capable of reproducing findings of eminent scientists (Langley et al. 1987). Such discovery systems were later criticized for their lack of autonomy because they required significant amounts of a priori knowledge and often relied on human supervision to determine when a creative result had been achieved (see Lenat and Brown 1984). As the field of AI matured, researchers focused their efforts on solving well-defined sub-problems of intelligence, e.g., classification, planning, or theorem proving (Colton et al. 2009). Spurred by developments in the cognitive sciences, however, a renewed enthusiasm for computationally modeling creativity saw the establishment of the field of computational creativity in the 1990s, which again seeks to uncover suitable algorithms and knowledge structures to support creative behavior (Veale et al.2019). As Guckelsberger et al. (2021) discuss, however, there has been a significant lack of studies concerning questions of embodiment in computational creativity.

Our creative practice sits along the spectrum of creative AI; we attempt to use models from computational creativity to produce robotic artworks. Dealing with the 'messiness' of the real world in robotics, however, has frequently raised questions about situatedness, embodiment, and the performance of creativity, i.e., the performative nature of the creative act of generative meaning-making in the human–machine encounter. Consequently, our creative practice has shifted our understanding of computational creativity away from the algorithmic perspective; de-emphasizing the development of generative systems capable of producing novel and valuable concepts and focusing our attention on the development of machine performers that skillfully participate in the enactment of creativity within, and with, a physical and social environment.

2 Background

Traditionally, cognitive science has considered cognition as computations over mental representations (e.g., Fodor 1975). This approach asserts cognition consists of abstract processes that mediate between internal representations of sensory inputs (perception) and motor outputs (action). In contrast, embodied cognition is a program of research centered around the key assumption that the body functions as an active *constituent* of cognitive processes rather than a passive perceiver and actor *serving* the mind (Shapiro 2007; Leitan and Chaffey 2014). Embodied cognition had a profound influence on computational sciences including robotics (e.g., Brooks 1990; Clancey 1997) but has yet to have a significant impact on the development of computational creativity (Guckelsberger et al. 2021). More broadly, Malinin (2016) argues that until recently research in human creativity has often overlooked the importance of embodiment. Glăveanu and Kaufman argue that the roles of the immediate physical and social environment have been 'blind spots' in creativity research (Glăveanu and Kaufman 2019), and it appears that the computational modeling of creativity has inherited these 'blind spots'.

Newen et al. (2018) identify different strands of embodied cognition research—*embodied*, *embedded*, *extended*, and *enactive* cognition—collectively known as '4E cognition'. A cognitive process is *embodied* if it relies on the body in a non-trivial way. Proponents of embodied cognition argue that much of human cognition encompasses both the mind and the body. A cognitive process is *embedded* if it relies on the physical, social, or cultural environment in non-trivial ways. Proponents of embedded cognition note that people often exploit features of their environment to increase their cognitive abilities. Proponents of *extended* cognition argue for a strong form of embedded cognition such as may be experienced by a person skilled in the use of a tool such that it becomes part of their cognitive apparatus. A cognitive process is *enactive* if it relies on an ability or disposition to act. Proponents of enactive cognition argue that a person's cognitive abilities are dependent on their interactions with the world.

Creative activity can be understood from this embodied, embedded, extended, and enactive perspective, as evidenced by Csikszentmihalyi's study of the daily habits of more than 100 exceptional creative people from diverse fields (Csikszentmihalyi 1996, p. 127). Ingold (2013) found that an artisan's craft is not simply a consequence of physical skill but emerges within a system of relationships and interactions situated in a material environment. Similarly, Glăveanu (2012) studied the creative process of perceiving, exploiting, and generating novel affordances as part of a socially and materially situated activity of traditional Easter egg decoration in rural Romania. These studies highlight the embedded nature of creative activities and foreground the complex dialog with materials that practitioners engage in. Malafouris (2008), for instance, observed that a potter makes ongoing, second-by-second decisions on how to respond to the *material agency* of the clay. Yokochi and Okada (2005) highlight the enactive processes unfolding through a traditional Chinese ink painter's hand movements as they explore and perceive, where, what and how to paint next.

Embodied cognition provides an empirically supported framework for understanding existing creative practices grounded in the body (Kirsch 2012), and a foundation for the development of new embodied practices, like our own, as we will explore later. Malinin (2019) identifies two emerging streams of empirical research in embodied creativity:

(1) studies of embodied metaphors associated with creativity, e.g., studies of free walking as enacting *thinking outside of the box* (Leung et al. 2012), and
(2) studies of creativity as an emergent phenomenon of dynamic systems, e.g., music as an emergent, dynamic interaction among musicians and instruments (Schiavio and van der Schyff 2018).

In general, however, Malinin (2019) argues that the traditional cognitivist separation of mind and body manifests in creativity research as disjoint views of creative cognition and creative action. This disconnect between the realms of ideas and actions is frequently mirrored in the computational modeling of creativity (Guckelsberger et al. 2021).

3 The Embodiment of Creative AI

Creative practitioners have always understood and experimented with the meaning-making potential of embodiment and movement in computational systems, as demonstrated by the history of cybernetic and robotic art (see Kac 1997; Penny 2012). For example, Gordon Pask's *The Colloquy of Mobiles* (1969) performed a dynamically evolving mating 'dance' between five 'mobiles'—three female soft fiberglass shapes and two male aluminum rectangles (Pickering 2010). Open to interference from visitors using mirrors and flashlights, the mobiles engaged in a dynamic performance as an endlessly emergent cycle of relations, meanings, and desires; a conversation across non-humans and humans (Fernandez 2008). Edward Ihnatowicz's cybernetic works anticipated the embodied approach of behavior-based robotics (Brooks 1990) by almost two decades. *SAM* (1969) and *The Senster* (1971) implemented a small set of simple behaviors that, when placed into the environment of a busy gallery, combined to produce complex social interactions (Zivanovic 2005).

3.1 Curious Whispers

Curious Whispers (2010–2011) was our first attempt to study embodiment in computational creativity (Saunders et al. 2013). *Curious Whispers* consists of three mobile robots—each equipped with a speaker, a microphone, and a movable plastic cover—and a three-button synthesizer, see Fig. 1. In operation, each robot performs simple melodies using three tones and listens for melodies performed nearby with the same three tones. When rehearsing melody variations, a robot makes use of its embodiment by closing its plastic cover allowing it to use the same hardware and software to evaluate melodies performed by itself and others.

The robot hardware consisted of a 3pi robot base, two microphones, a speaker, and a servo. The robot's software ran on the 3pi's AVR ATmega328 microcontroller. After a melody is heard, it is assessed against a small set of recently heard melodies, where each note is stored as a frequency and duration, to determine its novelty on a note-by-note basis. If a melody is determined to be novel, it is added to the set, replacing the oldest melody if it has reached its maximum size. Exposure to novel

Fig. 1 Curious Whispers (2010–2011)

melodies switches the mode of the robot from listening to rehearsing, if the melody came from another agent and from rehearsing to performing, if the melody was played by the robot. In rehearsal mode, a simple genetic algorithm is used to generate new melodies based on its set of previously novel melodies and played directly through the speaker to ensure that the internal representation reflects what is heard by the microphones.

Curious Whispers attempts to embed the robots within a wider social environment by providing a 'level playing field' between human and artificial agents using a three-button synthesizer and a simple interaction policy; if a robot considers a melody played using the synthesizer to be novel, it will be adopted. Closing its cover signals to human participants that the robot has adopted a melody and switched from listening to rehearsing. Using this simple interaction, human interactors introduce situated domain knowledge.

3.2 Zwischenräume and Accomplice

Our robotic installations *Zwischenräume* (2010–2012) and *Accomplice* (2013–2014) explore the affective-material potential of robotic systems that evolve based on the emergent cycles of relations that their material embeddedness gives rise to. *Zwischenräume* (2010–2012) was our first robotic installation; it embeds a pair of gantry robots into the architectural fabric of a gallery; sandwiched between a gallery wall and a temporary wall that resembles it. Each robot is equipped with a camera, a motorized tool, and a microphone. The control system for each robot combines machine vision and a model of intrinsic motivation (Gemeinboeck and Saunders 2013). Movements, shapes, sounds, and colors are processed using machine learning to allow each robot to develop expectations of their environment and the consequences of their actions. Multiple self-organizing maps (Kohonen 1995) are used to determine similarity between images taken by the camera based on shape, color, and optical flow. Dissimilar images provide a reward signal based on their degree of novelty, using a non-linear 'hedonic' function based on the Wundt Curve (Berlyne 1960), which maximizes reward for inputs with moderate levels of novelty, i.e., inputs that are similar-but-different to the previous inputs. Reinforcement learning (Watkins 1989) is used to develop strategies for moving about the wall and using the tool. Prediction errors between learned models of consequences and observed results are used as a measure of surprise.

Embodiment provides opportunities to expand the robots' behavioral range by taking advantage of properties of the physical environment, i.e., the wall. The robots are not equipped with an explicit representation of the wall's material, instead they learn to recognize salient features in the environment to support their learning. In this way, the computational model implemented in the robots is embedded in its immediate physical environment. The robot's capacity to act evolves over the course of the exhibition based on interactions with their surroundings, resulting in an enactive coupling of the sensorimotor loop, the physical environment, and the learning that drives behavior. The robots coordinate their movement, gaze, and tool activation to produce novel features in their environment to generate learning rewards. The result of their intrinsically motivated learning is a feedback process, which increases the complexity of the robot's environment relative to their perceptual abilities. Over time, sequences of movements and knocking actions develop into a behavioral repertoire of skills grounded in the robot's embodiment that can produce perceived changes in terms of color, shapes, and motion.

We explored different embodiments of *Zwischenräume* over multiple iterations between 2010 and 2012, most notably by equipping the robots with different motorized tools including a hammer, a chisel, and a punch. Using the same controller for each iteration, we could observe how the robots' behaviors were contingent on their embodiment as they learned to affect change in different ways. The coupling of the robots with their physical environment provided a simple model of enactive cognition; the robots learned to re-sculpt their physical environment through their actions and the perception of their consequences. Figure 2 is a collage of images taken by a single robot during an installation. Each image was taken when the robot discovered something 'interesting', i.e., when the hedonic function returned a reward higher than a predetermined threshold. Figure 2 illustrates how the evaluation of 'interesting' evolved over time and was affected by (a) positioning of the robot and camera, e.g., the discovery of lettering on the plasterboard wall; (b) use of the tool, e.g., the production of dents and holes; and finally (c) interaction of visitors. Visitors to the gallery were frequently a source of novelty for the robots, and the robots attended to them, as illustrated in the final row of Fig. 2, until the colors, shapes, and movement of visitors were learned and so ceased to be sufficiently novel to be interesting.

Accomplice (2013–2014) builds on *Zwischenräume*; like *Zwischenräume,* it encases robots into the walls of a gallery. Unlike *Zwischenräume*, the robots in Accomplice can move such that they have overlapping areas of operation. Sharing an immediate physical environment with other robots permits the indirect coordination of actions. Areas of common action become frequently visited by robots as repeated sources of learning rewards, modeling a form of physically embedded, social coordination through action. The robots communicate directly through knocking patterns of their motorized tools. Each robot evaluates the knocking pattern rhythms of other robots against its own and selects novel rhythms for its own actions. The collective result of sharing of rhythms is the use of similar knocking patterns across the gallery that change over time as the rhythms are performed, selected, and varied by each robot. The robots' embeddedness in a social space is materialized as a dynamic soundscape in the gallery.

In both *Zwischenräume* and *Accomplice,* the performance of the robots is shaped by their curious disposition, the drive to seek novelty, continuously expanding their behavioral envelope. Performance here is not about re-performing an existing script but rather emerges from the system's ongoing evolution, situated in and in-interaction-with its environment. Learning and adapting then are not goal driven but are based on what they discover and interpret as 'interesting'. The seemingly passive wall and its material capacities, resisting or accelerating the machines'

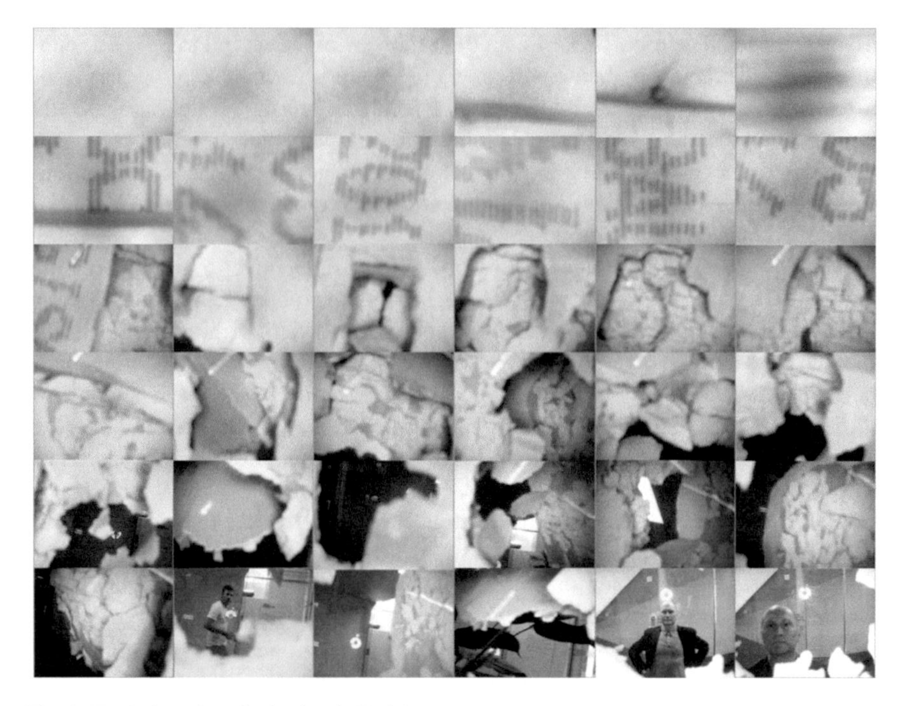

Fig. 2 Evolution of a robot's view in Zwichenräume

eager work, play an important role in the unpredictable evolution of this performance and the emergence of agency across computational, mechanical, and physical systems. The installation thus acts out a particular ecological niche—a dynamic co-mingling of processes, matter, and things, while foregrounding the affective potential of nonhuman, socially behaving, creative agents. In the next section, we take a closer look at this entanglement of embodiment and performance in our current practice.

4 The Performance of Creative AI

While *Zwischenräume* and *Accomplice* embodied and embedded machines that demonstrated an enactive cycle of learning through doing, our collaborative project, machine movement lab (MML), makes explicit our understanding of creativity as a distributed process, scaffolded by the performer's skills. The performance of creative AI here refers to both (1) the performativity of the creative act as distributed across the robot, other (e.g., human) agents, and the situation they are embedded in, and (2) the performance of the robot as a 'skillful participant', which scaffolds the performativity of the creative act.

Bringing together creative robotics, dance performance, and machine learning, MML is grounded in a performative framework to explore the enacted nature of creative agency and meaning-making. Questions of agency have been identified as central to the recognition of creativity in computational systems (Guckelsberger et al. 2017), and commonly, agency is understood as an attribute built into the system. Our experience of developing embodied systems, however, has shifted our focus from endowing robots with creative agency to that of developing skillful participants in the distributed enactment of agency. Rather than invested with agency, a machine as a 'skillful participant' engages in and facilitates the emergence of creative agency between machines, other agents, and their environment. From a performative viewpoint, agency is not a property that can be possessed but rather 'is a matter of intra-acting … an enactment' (Barad 2007, p. 178). MML builds on Barad's concept of intra-action to develop a performative approach to human–robot interaction through material, performance-based inquiries into the situated enactment of human–robot relations.

In contrast to many approaches to human–robot interaction that focus on relation-making with humanlike robotic agents, MML looks for differentiated starting points for the making of human–robot relationships by investigating the relational performative potential of abstract machinelike artifacts (Gemeinboeck 2021). Thereby, the performativity of the creative acts as part of the meaning-making in the interactive exchange has driven every aspect of our design process, including the robot's mechanical design in tandem with developing its behavioral language, learning, and improvisational capacities. Looking at design not as a method for developing an autonomous, creative agent but rather a socio-culturally situated, material process for scaffolding a robot's social and creative skills deeply integrates our understanding of embodied cognition with our design approach. Much of this integration has been driven by our collaboration with dancers and their bodily ways of knowing, which has allowed us to explore material, social, and cultural interrelations, and how they can mobilize enactments of creative meaning-making.

Our approach revolves around a novel performative body mapping (PBM) methodology, which involves dance performers wearing a robot costume to corporeally entangle with and 'feel into' a machine's different embodiment with its unique spatial-relational affordances and affective potential. Theater and performance have a history of using costumes to interfere with performers' bodies and their performance (see Suschke 2003, p. 205). Combining ideas from theatrical costume and demonstration learning, the PBM costume facilitates dancers' ability to bodily probe and kinesthetically extend into the robotic embodiment. The costume provides a material interface between human and robot bodies and their differing movement capacities, such that a performer can apply their embodied knowledge and their socio-culturally embedded understanding of movement. It allows the dancer to learn how to embody the machinic form and move with this unfamiliar embodiment (see Fig. 3) and for the robot to learn from the dancer-in-costume by imitating the recorded movements of the PBM costume.

MML explores how the relational, enactive potential of movement scaffolds a robot's ability to participate in the dynamic, creative processes of the social encounter. Movement in robotics is commonly a matter of safely navigating space, whereas human–robot interaction design also employs movement and its qualities to imbue

Fig. 3 Audrey Rochette in the PBM costume. Image copyright of the authors

robots with an expressive character or personality. Movement here serves as a medium for 'accurately expressing the robot's purpose, intent, state, mood, personality, attention, responsiveness, intelligence, and capabilities' (Hoffman and Ju 2012, p. 91). MML, in contrast, understands movement as a dynamic, relational phenomenon, unfolding through 'spatial, temporal, and energic qualities' (Sheets-Johnstone 2012, p. 49), whose generative potential can drive the relation-making dynamics of an encounter. Understanding movement, both performatively and from the perspective of phenomenology and embodied cognition, we bodily participate in the generation of meaning, 'often engaging in transformational and not merely informational interactions; [we] enact a world' (Di Paolo et al. 2010, p. 39). As we enact and experience meaning through movement, we also make sense of other bodies by resonating with them and their movements (see Fuchs 2016). In Fuchs and Koch's words, 'one is moved by movement [...] and moved to move' (Fuchs and Koch 2014, p. 1). Bodily feeling into the asymmetric relational potential of a robot's different embodiment enables the dancer to bodily resonate with it. This 'intra-bodily resonance' (Froese and Fuchs 2012, p. 212) then gives rise to a hybrid movement language, resulting from the dancer moving with or as part of the cube form without relying on expressions of inner states. Our performers frequently use mental imagery of nonhuman dynamics (e.g., that of a pressure cooker, melting, or heavy rain) to guide their search for new movement patterns and the body reconfigurations they require, together with attending to the costume's material affordances.

PBM harnesses the embodied expertise of dancers to inform every aspect of our process from the initial form-finding stage to the robot's movements and behavior. Instead of starting with a pre-defined form, PBM begins with an exploration of the agential potential of movement by collaborating with dancers to bodily investigate the performative potential of a wide range of materials and shapes. As dancers inhabit a variety of geometric forms made from different materials, form-finding unfolds along creative alliances between movement and materials and their emergent meanings, rather than a set of pre-defined social functions that reduce a robot's embodiment to a physical container (see Ziemke 2016). Views of machinelike robots lacking 'emotional displays' because they cannot 'express human facial expressions' (Hegel et al. 2009, p. 173) overlook the affective, agential capacities of situated movement. Entangling a dance performer with the unique spatial-material affordances of a becoming-robot allows for its dynamic, relational capacities to arise from a hybrid (human-nonhuman), interior perspective (Gemeinboeck and Saunders 2021).

The first robot prototype we realized using PBM has the shape of a cube. With its regular, omnidirectional geometry, a cube cannot be mistaken for a living 'thing' but instead offers a suitably blank canvas for dynamic relation-making. A dynamically or delicately moving cube, suddenly tilting up along one of its edges, gently swaying or rambunctiously thumping onto the ground, quickly loses its rootedness and transforms into something other than a familiar object. The mechanical design of the robot, referred to as cube performer, was derived from an analysis of the recorded motion patterns and their relational effects (Saunders and Gemeinboeck 2018). The PBM costume allows us to capture kinetic dynamics of a wide range of amplitudes; our goal for the machine learning as part of PBM is to utilize these dynamics to render the cube performer a highly skilled participant in the relational exchanges unfolding in a human–robot encounter without inscribing them directly onto the robot. Our approach builds on demonstration learning, also known as robot programming by demonstration (Billard et al. 2008), which involves a human demonstrator recording movements using motion capture that a robot learns to imitate from the captured data. A significant challenge when using this method is that it requires mapping between different embodiments, including different body shapes, sensorimotor capabilities, and movement repertoires (Dautenhahn et al. 2003), sometimes referred to as the 'correspondence problem' (Billard et al. 2008).

Our objective is for the robot to move according to its own abstract machine embodiment, while being 'seeded' with the movement qualities, textures, and nuances that support social sense-making. With respect to machine learning our challenge was thus to provide the necessary scaffolding for intra-bodily meaning-making informed by both the recordings from the PBM costume as well as the robot's own machinic embodiment. During the learning process, the robot engages in an embodied form of social learning, similar to what Kirsch describes as a 'sketch in dance' (Kirsch 2012). The term 'sketch' is used to highlight that imitated movements will inevitably be variations, due to differences in skill and the specifics of the embodiment. Recognizing and tapping into the differences of the machine's embodiment are at the core of the project. Hence, rather than looking at the robot's body as a mobile container, the machine learning approach has been developed in tandem with the robot's embodiment and capacity to move. The following outlines the three machine learning phases; *grounding*, *imitation*, and *improvization*.

In the *grounding* phase, the robot learns an initial movement repertoire, informed only by its unique physical embodiment in response to sensed environmental affordances. We deploy an 'illumination algorithm' (Mouret and Clune 2015), which allows a robot (in a simulated environment) to 'discover' its own body and possible kinesthetic relations in response to environmental affordances. The machine learner develops this repertoire by 'illuminating' a space of behaviors to find multiple possible ways of moving, rather than searching for a single optimal solution. We use a variation of the MAP-Elites (Mouret and Clune 2015) illumination algorithm, which combines the evolution of robot controllers with a dimensional reduction algorithm, e.g., an autoencoder, to define the behavior space being illuminated (Cully 2019). Through this active self-exploration, the robot begins to generate a movement repertoire unique to its physical form.

In the *imitation* phase, we bring together the repertoire of movements generated in the grounding phase with the repertoire captured from the dancers inhabiting the costume (PBM costume). Drawing from these two data sets, the grounded and the captured, the challenge for our machine learner is to create a new movement repertoire across these two differing data spaces. To facilitate this, we adapt the low-dimensional (latent) space of the autoencoder, which initially captures only the grounded movements, by training it on sequences drawn from the captured movements. The result of this additional training is a latent space where both data spaces are superposed and 'mingle' according to their similarities and differences, establishing niches and gaps. As the simulated robot learns to imitate the PBM costume's movements, the goal is for the robot to learn the constraints that produce the movement qualities and subtleties, which emerged from entangling dancer and robot costume.

Finally, in the *improvisation* phase, the robot learns to adapt previously learned patterns of movement to invent new movements. The hybrid, third data space, becomes the robot's learning ground for this phase, where the 'illumination algorithm' creates new movement repertoires by learning to fill in gaps. Our PBM methodology, including these three learning phases, thus allows us to reimagine robotic agents as skillful performers, capable of moving in uniquely machinelike ways while participating in creative human–robot meaning-making. Results from user studies involving audiences and experts vouch for the efficacy of this generative approach to produce robotic movement skills that are grounded in the physical embodiment of the robot while being embedded and informed by the social and cultural context of our interdisciplinary collaboration (Gemeinboeck and Saunders 2019).

5 Conclusion

The term 'creative AI' potentially covers a broad spectrum of activities from the use of AI as a medium of expression for human creativity to the modeling of creativity in computational systems. The computational modeling of creativity has seen remarkable advances but, like studies of human creativity, has typically focused on the generation of novel ideas, rather than the role of embodiment in creative activity. Our practice sits along the spectrum of creative AI; we attempt to use models from computational creativity to produce robotic artworks. Developing robots requires us to deal with the 'messiness' of the real world, highlighting for us the situatedness and embodiment of creativity, and the performative nature of the creative act. Consequently, our creative practice has shifted our understanding of creativity away from an algorithmic perspective and toward the development of skillful machine performers in the enactment of creativity. This conception of creativity as a form of intra-action has guided our current program of arts-led research. Designing from this performative viewpoint, where agency and meaning are no longer pre-defined, requires us to position ourselves in the middle of the encounter as part of the design process and attend to the ongoing dynamics, agencies, and meanings as they emerge.

Acknowledgements The research presented in this article has been partly supported by several funding bodies: the Australian Government through the Australian Research Council (DP160104706 and FT190100567); the Austrian Government through the Austrian Science Fund (FWF, AR545); and the EU Framework Program (FP7) European Research Area Chairs Scheme project (621403).

We would like to thank our collaborators for their contributions to this ongoing research: Roos van Berkel (Eindhoven University of Technology, NL), Maaike Bleeker (Utrecht University, NL), Katrina Brown (Falmouth University, UK), Rochelle Ha- ley (University of New South Wales, AU), Lesley van Hoek (Rotterdam, NL), Sarah Levinsky (Falmouth University, UK), Linda Luke (De Quincey Co., AU), Dillon McEwan (Sydney, AU), Kirsten Packham (Sydney, AU), Marie-Claude Poulin (University of Applied Arts Vienna, AT), Tess de Quincey (De Quincey Co., AU); Audrey Rochette (Montreal, Canada).

References

Audry S (2021) Art in the Age of machine learning. The MIT Press
Barad K (2007) Meeting the universe halfway: quantum physics and the entanglement of matter and meaning. Duke University Press, Durham, NC
Berlyne DE (1960) Conflict, arousal and curiosity. McGraw-Hill, New York, NY
Billard AG, Calinon S, Dillmann R (2008) Robot programming by demonstration, 1st edn. Springer, Berlin, Germany, pp 1371–1394
Boden MA (1998) Creativity and artificial intelligence. Artificial Intelligence (103):347–356
Bown O (2021) Beyond the creative species. The MIT Press
Brooks R (1990) Elephants don't play chess. Rob Auton Syst 6:3–15
Burnham J (1968) Beyond modern sculpture. George Braziller, New York, NY
Clancey WJ (1997) Situated cognition: on human knowledge and computer representations. Cambridge University Press, Cambridge, UK

Colton S, de Mantaras RL, Stock O (2009) Computational creativity: coming of age. AI Mag 30(3):11. https://doi.org/10.1609/aimag.v30i3.2257

Csikszentmihalyi M (1999) A systems perspective on creativity. In: Sternberg RJ (ed) Handbook of creativity. Cambridge University Press, Cambridge, UK, pp 313–338

Csikszentmihalyi M (1996) Creativity: flow and the psychology of discovery and invention. Harper Collins, New York, NY

Cully A (2019) Autonomous skill discovery with quality-diversity and unsupervised descriptors. In: Proceedings of the genetic and evolutionary computation conference, ACM. https://doi.org/10.1145/3321707.3321804

Dautenhahn K, Nehaniv CL, Alissandrakis A (2003) Learning by experience from others—social learning and imitation in animals and robots. In: Kühn R, Menzel R, Menzel W, Ratsch U, Richter M, Stamatescu I (eds) Adaptivity and learning: an Interdisciplinary debate. Springer, Berlin, Germany, pp 217–421

Di Paolo EA, Rohde M, De Jaegher H (2010) Horizons for the enactive mind: Values, social interaction, and play. MIT Press, Cambridge, MA, Chap 2:33–87

Fernandez M (2008) Gordon pask: cybernetic polymath. Leonardo 41(2):162–168

Fodor JA (1975) The language of thought. Harvard University Press

Froese T, Fuchs T (2012) The extended body: a case study in the neurophenomenology of social interaction. Phenomenol Cogn Sci 11(2):205–235

Fuchs T, Koch SC (2014) Embodied affectivity: on moving and being moved. Front Psychol 5(Article 508):1–12

Fuchs T (2016) Intercorporeality and interaffectivity. Oxford University Press, Oxford, UK, pp 3–24 (Chap. 1)

Gemeinboeck P, Saunders R (2013) Creative machine performance: Computational creativity and robotic art. In: Proceedings of the 4th international conference on computational creativity, pp 215–219

Gemeinboeck P, Saunders R (2019) Exploring social co-presence through movement in human-robot encounters. In: Proceedings of the AISB 2019 symposium on movement that shapes behaviour. AISB 2019 Convention, Falmouth University, UK

Gemeinboeck P, Saunders R (2021) Moving beyond the mirror: relational and performative meaning-making in human-robot communication. AI & Society

Gemeinboeck P (2021) The aesthetics of encounter: a relational-performative design approach to human-robot interaction. Front Rob AI 7

Glăveanu VP (2012) What can be done with an egg? Creativity, material objects, and the theory of affordances. J Creative Behavior 46(3):192–208

Glăveanu VP, Kaufman JC (2019) The creativity matrix: spotlights and blind spots in our understanding of the phenomenon. J Creative Behav 54(4):884–896

Guckelsberger C, Salge C, Colton S (2017) Addressing the "Why?" in computational creativity: a non-anthropocentric, minimal model of intentional creative agency. Georgia Institute of Technology, Association for Computational Creativity, Atlanta, GA, pp 128–135

Guckelsberger C, Kantosalo A, Negrete-Yankelevich S, Takala T (2021) Embodiment and computational creativity. In: Proceedings of 12th international conference on computational creativity (ICCC'21), pp 192–201

Hegel F, Muhl C, Wrede B, Hielscher-Fastabend M, Sagerer G (2009) Understanding social robots. In: Proceedings of the second international conferences on advances in computer-human interactions, IEEE, New York, NY, pp 169–174

Hoffman G, Ju W (2012) Designing robots with movement in mind. J Hum Rob Inter 1(1):78–95

Husbands P, Holland O, Wheeler M (2008) The mechanical mind in history. MIT Press, Cambridge, MA

Ingold T (2013) Making: anthropology, archaeology, art and architecture. Routledge, New York, NY

Kac E (1997) Origin and development of robotic art. Art J 56(3):60–67
Kirsch D (2012) Running it through the body. In: Proceedings of the 34th annual cognitive science society, pp 593–598
Kohonen T (1995) Self-organizing maps. Springer, Berlin, Germany
Langley P, Simon HA, Bradshaw GL, Zytkow JM (1987) Scientific discovery: computational explorations of the creative processes. MIT Press, Cambridge, MA
Leitan ND, Chaffey L (2014) Embodied cognition and its applications: a brief review. Sensoria J Mind Brain Culture 10(1):3–10
Lenat DB, Brown JS (1984) Why AM and EURISKO appear to work. Artif Intell 23(3):269–294
Leung AK, Kim S, Polman E, Ong LS, Qiu L, Goncalo JA, Sanchez-Burks J (2012) Embodied metaphors and creative "acts". Psychol Sci 23(5):502–509
Lindqvist G (2003) Vygotsky's theory of creativity. Creat Res J 15(2–3):245–251
Lovelace A (1843) Sketch of the analytical engine invented by Charles Babbage Esq. by L.F. Menabrea, of Turin, officer of the Military Engineers, with notes upon the memoir by the translator. Taylor's Sci Mem 3:666–731
Malafouris L (2008) At the potter's wheel: an argument for material agency. Springer, Boston, MA, pp 19–36
Malinin LH (2016) Creative practices embodied, embedded, and enacted in architectural settings: Toward an ecological model of creativity. Front Psychol 6:1978
Malinin LH (2019) How radical is embodied creativity? Implications of 4e approaches for creativity research and teaching. Frontiers in Psychology 10
McCarthy J, Minsky ML, Rochester N, Shannon CE (1955) A proposal for the dartmouth summer research project on artificial intelligence. AI Mag 27(4 (2006)):12–14
Mouret JB, Clune J (2015) Illuminating search spaces by mapping elites. arXiv preprint 1504.04909. Newen A, De Bruin L, Gallagher S (eds) (2018) The Oxford handbook of 4E cognition. Oxford University Press, Oxford, UK
Penny S (2012) Art and robotics: sixty years of situated machines. AI Soc 28(2):147–156
Pickering A (2010) The cybernetic brain. The University of Chicago Press, Chicago, IL
Saunders R, Gemeinboeck P (2018) Performative body mapping for designing expressive robots. In: Proceedings of the 9th international conference on computational creativity (ICCC 2018), pp 25–29
Saunders R, Chee E, Gemeinboeck P (2013) Evaluating human-robot interaction with embodied creative systems. In: Proceedings of the 4th international conference on computational creativity, pp 205–209
Schiavio A, van der Schyff D (2018) 4E music pedagogy and the principles of self-organization. Behav Sci 8(8):72
Shapiro L (2007) The embodied cognition research programme. Philos Compass 2(2):338–346
Sheets-Johnstone M (2012) From movement to dance. Phenomenol Cogn Sci 11:39–57
Suschke S (2003) Müller macht Theater: Zehn Inszenierungen und ein Epilog. Theater der Zeit, Berlin, Germany
Turing AM (1950) Computing machinery and intelligence. Mind 49:433–460
Veale T, Cardoso A (eds) (2019) Computational creativity: the philosophy and engineering of autonomously creative systems. Computational synthesis and creative systems, Springer International Publishing, Cham, Switzerland
Veale T, Cardoso A, Pérez y Pérez R (2019) Systematizing creativity: a computational view. In: Veale and Cardoso (2019)
Watkins CJCH (1989) Learning from delayed rewards. PhD thesis, Cambridge University, Cambridge, UK

Yokochi S, Okada T (2005) Creative cognitive process of art making: a field study of a traditional Chinese ink painter. Creat Res J 17(2):241–255

Ziemke T (2016) The body of knowledge: on the role of the living body in grounding embodied cognition. Biosystems 148:4–11

Zivanovic A (2005) SAM, The Senster and The Bandit: early cybernetic sculptures by Edward Ihnatowicz. In: Robotics, mechatronics and animatronics in the creative and entertainment industries and arts symposium, AISB 2005 Convention, Hatfield, UK

Musebots and I: Collaborating with Creative Systems

Arne Eigenfeldt

Abstract This chapter describes musebots and their specific use as collaborators within systems designed by an artist, rather than a computer scientist, as well as a brief personal history describing the trajectory leading to their use. The musebots are not proof of concept, but producers of genuine art: their work has been performed in a variety of concerts and festivals throughout the world. While AI systems have been developed that offer themselves as compositional assistants (Agres et al., Computers in Entertainment (CIE) 14:1–33, 2016), I view collaboration as an equal partner in the creative process and describe the unique relationship between composer and artificial agents in the creation of artworks that exist as artworks, rather than examples of computational creativity.

Keywords Robotics · Music · Creativity · AI · Human-AI collaboration

1 Introduction

Generative music offers the opportunity for the continual reinterpretation of a musical composition through the design and interaction of complex processes that can be rerun to produce new artworks. While generative art has a long history (Galanter 2003), the application of artificial intelligence, evolutionary algorithms and cognitive science has created a contemporary approach to generative art, known as metacreation (Whitelaw 2004); musical metacreation (MuMe) looks at all aspects of the creative process and their potential for systematic exploration through software (Pasquier et al. 2016).

One useful model borrowed from artificial intelligence is that of agents, specifically multi-agent systems. Agents have been defined as autonomous, social, reactive and proactive (Wooldridge 1995), similar attributes required of human performers in musical improvisation ensembles. Musebots (Bown et al. 2015) offer a structure

A. Eigenfeldt (✉)
Simon Fraser University, Vancouver, Canada
e-mail: arne_e@sfu.ca

C. Vear and F. Poltronieri (eds.), *The Language of Creative AI*, Springer Series on Cultural Computing, https://doi.org/10.1007/978-3-031-10960-7_12

for the design of musical agents, allowing for a communal compositional approach (Eigenfeldt et al. 2015a) as well as a unified model.

This chapter describes musebots and their specific use as collaborators within systems designed by an artist, rather than a computer scientist, as well as a brief personal history describing the trajectory leading to their use. The musebots are not proof of concept, but producers of genuine art: their work has been performed in a variety of concerts and festivals throughout the world. While AI systems have been developed that offer themselves as compositional assistants (Agres et al. 2016), I view collaboration as an equal partner in the creative process and describe the unique relationship between composer and artificial agents in the creation of artworks that exist as artworks, rather than examples of computational creativity.

1.1 An Artist's Approach

I am an artist, albeit an artist that has coded his own systems for over thirty years. All of my training is as a composer, and the focus of all of my software has been to create tools to produce music that I personally want to hear. Any validation of these tools has been through traditional artistic reflection: I, and I alone, decide if the system is producing artistically valid output.

My approach to the music production tools that I create is through the lens of a composer, as opposed to an improviser: the former organise sound in time, while the latter organise their sonic performances in time. The difference may seem subtle, particularly in my case, since almost all of my music has been concerned with live performance (as opposed to fixed media, such as acousmatic music) and has also included a great deal of improvisation. This paradox stems in part from my background as a jazz musician, one who revelled (and still revels) in mercurial emergences that may never be repeated again outside of a singular performance. I have come to realise that my own journey through generative systems has been centred around devising complex environments that could replicate the quicksilver of improvisation, but facilitating a level of control on my part that is impossible with human performers. As I will explain further, my arrival at the exploration of musebots—virtual musical agents—as collaborators allows for this blend of freedom and control.

2 Interactive Systems

My earliest systems were interactive, using those of Joel Chadabe (1984) and George Lewis (2000) as models. The balance between unexpectedness and predictability of such systems tended to reside in algorithms employing constrained random procedures (Winkler 1995)—the note-based music produced by these systems engendered certain requirements, such as harmonic and rhythmic consistency—thus random selection from preselected note pools and rhythmic sets was efficient and satisfying.

The complexity—or creativity, if one wishes to consider it as such—rested in the live performer, who tended to control the progression of time, as well as the flux of tension. However, as these reactive systems had predetermined expectations, unexpected actions or reactions from the live performer could stun the system into silence: providing a dumb algorithmic system with something it doesn't know what to do with is clearly problematic.

I personally solved this problem by restricting the interaction of the system to myself in performance, as I guided it based upon my interpretation of the live improvising musicians' actions. An interactive trinity was thus formed: I listened to the live performance as well as my system's output; the performer listened to the system, while the system only had to understand what it itself was doing (see Fig. 1). The benefit, and simplicity, of this design maintained the creativity and complexity of reaction with me.

This instrumental approach to interactive system design was successful insofar as the instrument remained "playable", which raised a twofold predicament: these systems were rarely stable, often being debugged and/or augmented right up until the performance and thus ripe to become unplayable at any time; and with the increase in computing power, the systems themselves became exponentially more complex, and interface design became an art in itself.

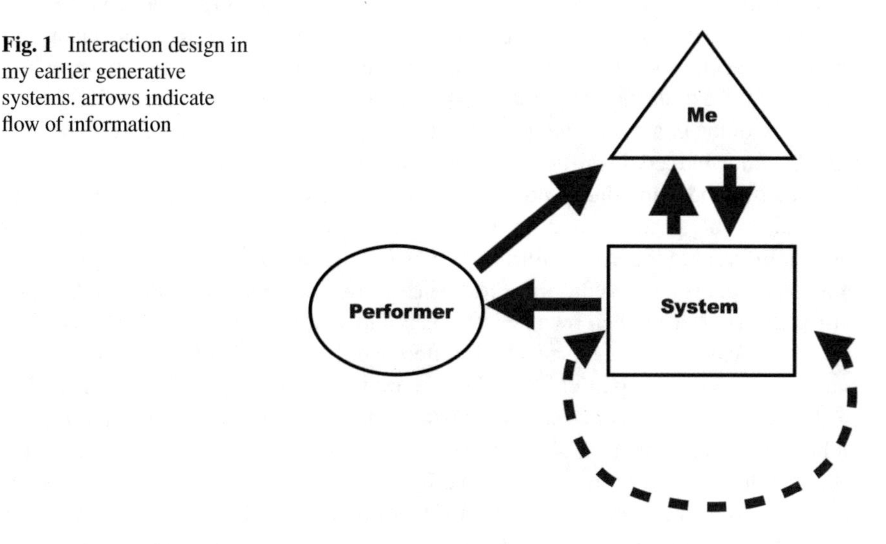

Fig. 1 Interaction design in my earlier generative systems. arrows indicate flow of information

2.1 Real-Time Composition

The concept of real-time composition (Eigenfeldt 2011a) was formulated as a way to assign more creative responsibility to the system. These systems' complexity now could be found in the interactions between the agents (Eigenfeldt 2011b), as I controlled higher level musical features, such as density (the number of active agents), timbre (from a menu of preselected samples), and structure (when to move to the next section) (see Coming Together: Notomoton 2010).

Because I remained determined to allow for the unfolding of a compositional structure during the performance itself—despite structure remaining an open problem in computational creativity (Eigenfeldt et al. 2016)—it was left to me to analyse those parameters during performance. This culminated in my final attempt at real-time composition, *An Unnatural Selection* (2014), in which the system used a genetic algorithm to generate musical material in performance, and pass this information as musical notation to eight live instrumentalists who sight-read the music from iPads (Eigenfeldt 2015). The system was designed to allow for live control over the genome, feeding it new corpora to direct it into new directions as the composition progressed. I found this impossible to do, as the task required concentrated listening to the material being generated and performed, and instantaneous decisions being made as to what material would be required one or two phrases in future. Not trusting myself in performance, I created templates which the system used as formal guides, predetermining an overall shape and direction (see *An Unnatural Selection* 2014).

While my desire to create a meaningful interaction between live musicians and intelligent agents has in no way waned, I have now removed myself from the process—at least during performance—as, demonstrated above, I seemed to be the weak link in the chain. In a move that might be considered as playing a composer's card, I acknowledged the difficulty of extracting meaningful information from a live performance and using that to not only produce creative responses, but also generate a logical overall structure, and instead designed a system where agents were potentially indifferent to the live performer. In *The Indifference Engine* (Eigenfeldt 2013), individual agents are given only a brief slice of live performance data, and use this information to inform their own beliefs, plans and actions, which are often in conflict with other agents' notions (see The Indifference Engine 2013). The musical result is a curious aloofness, particularly when the musician and agents attempt to end the work. During the premiere, the improvising musician had clearly reached a logical ending; however, one agent stubbornly continued on with its own plan for an extra four minutes. At any time, I could have forced an ending, but I felt that I owed it to the single agent to hear it out, as awkward as that might be. Later versions of the software, using musebots, set a predetermined duration, after which the agents would wind themselves down (see *The Indifference Engine vs. X* 2017).

2.2 *Towards Creative Compositional Systems*

As previously mentioned, one difficulty that I faced with interactive systems were the copious number of parameters that required control during performance. One solution was arrived at during a collaboration with a video artist and a sound artist in order to align our three separate generative systems: using only the parameters of valence (pleasantness) and arousal (eventfulness) as high-level controls (Eigenfeldt et al. 2015b), each systems responded independently to the human determined metadata ratings of a corpora. I have found that there is a great deal of possible musical mappings to these parameters, which make them general enough to be high level and thus controllable in performance if need be, but also useful enough to be applicable to musical details. Arousal is most easily maps to density of events and tempo: for example, high arousal can result in a high number of notes in a short amount of time. Valence is somewhat more flexible. It can be mapped to complexity—assuming more complex is less pleasant—and it can also be mapped to duration: longer compositions are usually less pleasing. In Russell's circumflex model (1980), low valence and high arousal create a stressed or tense reaction, a feeling I can attest to in listening to more than the occasional overly long contemporary music work.

This multiple mapping does present new issues: if both complexity and duration are mapped to valence, then this disallows short complex compositions, as all complex outputs will have long durations. This can (and has, to an extent) been alleviated by providing greater ranges from which agents can choose, or adding more randomness; however, this contradicts my original desire to replace randomness with musically intelligent and creative decision-making. In a position paper (Eigenfeldt et al. 2013), we posited that the final stage of musical metacreation would be volition: agents deciding before beginning a creation what that composition would entail, exactly how a human artist may approach such a work. No high-level controls would be in effect, and instead the agents may agree to create a short complex work that abandons the need for any parameter mappings. However, I know of no metacreative systems that have reached such a state.

3 Musebots

Since 2015, I have used musebots in a variety of artworks (Eigenfeldt 2016, 2017, 2018; Eigenfeldt and Ricketts 2019) and became their main evangelist. I had been working with agents for almost a decade, and the musebot protocol allowed for a consistency that kindled opportunities to adapt and reuse agents outside of their original creative work.

Musebots lack direct interactivity, as they are autonomous. As a result, my role during their performative actions is less as a performer or conductor, then as a critical listener. This is particularly the case during the long periods of fine-tuning every musebot ensemble, which entails a great deal of listening and note-taking in order to discern creative autonomy from buggy code.

The complexity of musebot interactions tends to produce interesting surface features; how these interactions evolve over time constitute the resulting music's structure and form, and as previously mentioned, these are aspects with which I as a composer am intrinsically obsessed. Like human improvisers, musebots are independent and autonomous; I can suggest and possibly provoke them, but, unlike chamber musicians reading fixed scores, I cannot force them to perform specific actions.

The musebot ensembles and their designed interactions result in a complex system. Like any such system, it is difficult to understand the intricate interactions that occur while it is operating, even as its designer. Parametric adjustments—such as raising an individual musebot's volume at a certain point—are not possible, since that parameter was calculated for that instant due to many underlying factors. Instead of interacting with the system directly to adjust the volume, it is necessary to understand why the musebot may be playing at a low volume: Does it's personality have too low a vitality attribute so as not to be able to sustain longer periods of higher volume? A higher vitality attribute may solve this, but it will also influence how the musebot behaves in other sections. Was the overall structural request for low volume being met by this agent, thus demonstrating a high compliance, while other agents were non-compliant and playing louder than requested? Matching all musebot compliance levels more closely may result in a more uniform volume level, but it will also disrupt the overall variety between them. One adage that has emerged in my work with musebots is to allow failure, because without this possibility, too much constraint will restrict true surprise in the results.

3.1 On Collaboration

Musebots can no longer be considered mere tools that I deploy in my creative work. With the advent of computing and its adoption into creative practices, a vast array of complex tools have become available to creative artists. These tools require acquired skill, or mastery, for them to be deployed successfully, and just as a violinist spends thousands of hours drawing a bow across the string of their violin in order to derive a precise and predictable relationship between their actions and the resulting sound, the same mastery is necessary with complex software and hardware. Musebots are not modelled on the *instrument* model and cannot be performed; instead, they are modelled upon the *musician* and the creative actions that they may make.

Earlier interactive and responsive systems that lacked the potential for creative actions and relied upon randomness for unpredictability were often intended (at least in my case) to produce an interesting environment for the improvising performer, and one that created musically meaningful responses to their actions.[1] If interactive systems produced musically meaningful reactions to a human performer—and many clearly did (see, e.g., performances by Michael Waisvisz (2004) and George Lewis (2020))—an interesting question can be posed as to the level of collaboration.

Referring again to Fig. 1 and the potential for collaboration and interaction between performer, system, and system designer, one needs to inquire as to the depth of collaboration between the coder and the system. Recursive system development is a standard way in which such systems are created, with the coder listening to the system output, and fine-tuning the responses. It may even be the case, and even desirable, that the responses are unexpected: Does this make the interaction collaborative, particularly if the responses rely on randomness? In my own case, I strongly believe that such systems remain a tool, executing actions and reactions that are scripted, and any unpredictability is the result of carefully constrained randomness within the algorithm.

As mentioned, my interaction with musebots has progressed beyond that of tool creation and use; instead, they are much closer to how would work with an ensemble of improvising musicians. Such a model should in itself also be noted: many live electroacoustic works that involve performer and live electronics reduce the performer to the level of score interpreter (see, e.g., Kaija Saariaho's stunning *NoaNoa*), continuing the traditional model established centuries ago of composer as artist and musician as hired labour. In no way belittling the contribution of such performers to the outcome of a fully scored work, I don't feel that one can consider their role as being in collaboration with either the system or its creator. On the other hand, when working with improvising musicians, as I have done in all of my live electronic works, assumes a significant contribution by the performer(s) to the final work: the role of a composer within such a work is to control its structure in a meaningful way. This is exactly how I have found my interaction with ensembles of musebots.

Let me further differentiate between what I consider to be interaction and collaboration. Collaboration between artists has existed for centuries, but has often been limited to what I would consider "old-school" collaboration: a bringing together of specialists, each of whom contributes to a final work their specific and unique expertise. Examples of such collaborations would be filmmaking, and the creation of operas. Such collaborations tend to involve a hierarchy, with one individual assuming the role of creative visionary, if only to avoid the messiness of conflicting artistic visions. Similarly, some disciplines involve the use of assistants who execute the visions of the main artist: major architectural designs, some visual art and some film music scoring houses come to mind. Whether a resulting creation is multidisciplinary is irrelevant here: to move beyond a hierarchical relationship in which individual artists have the potential to equally contribute to the final artwork is a level of collaboration which I am interested in emulating.

[1] I use the term musically meaningful rather than creative, as what constitutes the former seems clear to artists, while the latter term has become clouded. Artists have intuitively known what defines a creative work, but various academic definitions have been posited, the most basic being the production of something that is novel and useful (Mumford 2003).

Interaction is certainly a two-way process, and even if the interaction is between human and tool, it can lead to significant creative potential; however, the creativity will always remain on one side of the interaction. If, for example, I as a composer decided to create a work for the piano—itself a tool for musical creation—and I had never seen a piano previously, artistic conceptions my arise prior to my interaction with the tool, and these may drastically change once I directly interact with the instrument, but the relationship was never collaborative. Although the creative process may have relied upon direct interaction with the tool and may have been impossible without such interaction, once the work was complete, it could conceivably be performed on any other piano with negligible differences. If, on the other hand, I decided to create a work with a video artist, for a true collaboration to emerge, I would need to go into it with an open mind, allowing for equal creative input into the work from the other artist. And the resulting artwork would be completely different than if I had collaborated with another video artist.

4 Conclusion: Musebots and I

A final argument that I will make regarding my creative association with musebots being collaborative rather than as tools centres around a longer-term relationship: rather than the imaginary aforementioned collaboration with a visual artist, in which both artists can point to their specific creative input into a new work, my creative ideas themselves have begun to be shaped by musebots and their potential. Like any creative coder whose software has served more than one artwork, the musebots have appeared and reappeared in new situations, recombined with other, or new, musebots to form ensembles previously not considered. Early musebot ensembles (Eigenfeldt 2016) were described as being curated, and such a selective action is one I continue to follow; for example, musebots designed to create trap music (see Rattlestrap 2018) were adapted to work with a completely different corpus to produce ambient music (see Meryton 2022).

There have been numerous times that their creative output has surprised me, producing music that seemed to fulfil my original intention (i.e. trap, or ambient music), but done so in a way that I personally never would have created on my own. I want my generative systems to create music that I personally want to hear, yet to surprise me: musebots have certainly done that. But they have also suggested unintended paths that I've been unwilling to disregard.

More than once, musebots have produced "awkward successes", in that they produced music that was certainly surprising, but perhaps not appropriate: during an international festival in which two musebots controlled two Yamaha Disklaviers, they seemed to decide to play extremely sparsely and quietly, so much so that the audience could see keys on the piano being depressed, but little if any sound being produced. Such a conceptual work was not intended as a demonstration of the musical potential for musebots, and I could have stopped the performance and forced a restart (and reconceptualisation); however, I felt that it was important to allow them to produce their own work, staying just offstage, my face in my hands.

Sometimes I feel as if my relationship with musebots is more akin to that of a parent and set of precocious children: I can ask them to do something, something that they have previously executed without fanfare, but they are determined to follow their own path instead. And just like a parent, I adjust my expectations on a longer-term scale, allowing them to even lead me places I originally didn't expect to travel: my goals and artistic desires are evolving based on what I think the musebots are capable of. Maybe this will one day arrive at a point where the musebots evolve their own volition; I continue to code more and more intelligent musebots, and they may form ensembles on their own to produce works that they themselves want to hear.

References

Agres K, Forth J, Wiggins G (2016) Evaluation of Musical Creativity and Musical Metacreation Systems. Computers in Entertainment (CIE) 14(3):1–33

Bown O, Carey B, Eigenfeldt A (2015) Manifesto for a Musebot ensemble: a platform for live interactive performance between multiple autonomous musical agents. Proceedings of the international symposium of electronic art

Chadabe J (1984) Interactive composing: an overview. Comput Music J 8(1):22–27

Eigenfeldt A (2011a) Real-time composition as performance ecosystem. Organised Sound 16(2):145

Eigenfeldt A (2011b) Multi-agent modeling of complex rhythmic interactions in real-time performance. In: A-life for music. music and computer models of living systems. A-R Editions

Eigenfeldt A (2014) Generating structure—towards large-scale formal generation. In: Proceedings of the musical metacreation workshop, Raleigh

Eigenfeldt A (2015) Generative music for live musicians: an unnatural selection. In: International conference on computational creativity, Park City

Eigenfeldt A (2016) Musebots at one year: a review. In: Proceedings of the musical metacreation workshop, Paris

Eigenfeldt A (2017) Designing music with musebots. In: Proceedings of the fifth conference on computation, communication, Aesthetics and X, Lisbon

Eigenfeldt A (2018) Collaborative composition with creative systems. International symposium on electronic art, Durban

Eigenfeldt A, Bown O, Pasquier P, Martin A (2013) Towards a taxonomy of musical metacreation: reflections on the first musical metacreation weekend. Proc AAAI Conf Artific Intell Interactive Digital Entertain 9(1)

Eigenfeldt A, Bown O, y sey B (2015a) Collaborative composition with creative systems: reflections on the first musebot ensemble. International conference on computational creativity, Park City

Eigenfeldt A, Bizzocchi J, Thorogood M, Bizzocchi J (2015b) Applying valence and arousal values to a unified video, music, and sound generative multimedia work. Generative Art Conference, Venice

Eigenfeldt A, Bown O, Brown A, Gifford T (2016) Flexible generation of musical form: beyond mere generation. Proceedings of the 7th international conference on computational creativity, Paris

Eigenfeldt A, Bown O, Brown A, Gifford T (2017) Distributed musical decision-making in an ensemble of musebots: dramatic changes and endings. Proceedings of the 8th international conference on computational creativity, Atlanta

Eigenfeldt A, Ricketts K (2019) Unauthorized: collaborating with a performer collaborating with creative systems. Generative art conference, Rome
Galanter P (2003) What is generative art? complexity theory as a context for art theory. Generative art conference, Milan
Lewis G (2000) Too many notes: computers, complexity and culture in voyager. Leonardo Music J 12(10):33–39
Mumford M (2003) Where have we been, where are we going? taking stock in creativity research. Creat Res J 15(2–3):107–120
Pasquier P, Eigenfeldt A, Bown O, Dubnov S (2016) An introduction to musical metacreation. Comput Entertain (CIE) 14(2):1–14
Russell J (1980) A circumplex model of affect. J Pers Soc Psychol 39(6):1161–1178
Whitelaw M (2004) Metacreation: art and artificial life. MIT Press
Winkler T (1995) Strategies for interaction: computer music, performance, and multimedia. Proceedings of the 1995 connecticut college symposium on arts and technology
Wooldridge M, Jennings N (1995) Intelligent agents: theory and practice. Knowl Eng Rev 10(2):115–152

Musical Examples Cited in Text

A Walk to Meryton (2022) https://aeigenfeldt.wordpress.com/a-walk-to-meryton/
An Unnatural Selection (2014) https://aeigenfeldt.wordpress.com/an-unnatural-selection/
Coming Together: Notomoton (2010) https://aeigenfeldt.wordpress.com/works/music-by-agents/
George Lewis: Voyager (2020) https://youtu.be/o9UsLbsdA6s
Kaija Saaiaho—NoaNoa (1992) https://youtu.be/KbvbGkzMlg0
Michael Waisvisz (2004) https://youtu.be/U1L-mVGqug4
Rattlestrap (2018) https://aeigenfeldt.wordpress.com/works/rattlestrap/
The Indifference Engine (2013) https://aeigenfeldt.wordpress.com/works/indifference-engine-2/
The Indifference Engine versus X (2017) https://aeigenfeldt.wordpress.com/works/indifference-engine-musebots/

Composition with Computer Models of the Brain: An Alternative Approach to Music with Artificial Intelligence

Eduardo Reck Miranda

Abstract Artificial intelligence (AI) is aimed at endowing machines with some form of intelligence. Not surprisingly, AI scientists take much inspiration from the ways in which the brain—or the mind—works to build intelligent systems. This chapter proposes a different angle to harness the neurosciences for composition. Rather than building musical ANN to learn how to compose music, I shall introduce my forays into harnessing the behaviour of a type of neuronal model referred to as *spiking neuronal networks* to compose music. The discussion revolves around a piece for orchestra, choir and a solo mezzo-soprano entitled *Raster Plot*.

Keywords Creativity · Music-AI · Spike neurons · Orchestra · Performance

1 Introduction

Artificial intelligence (AI) is aimed at endowing machines with some form of intelligence. Not surprisingly, AI scientists take much inspiration from the ways in which the brain—or the mind—works to build intelligent systems. Hence, studies in philosophy, psychology, cognitive science and more recently, the neurosciences have been nourishing AI research since the field emerged in the 1950s, including, of course, AI for music (Miranda 2021).

The neurosciences have led to a deeper understanding of the behaviour of individual and large groups of biological neurones. We can now begin to apply biologically informed neuronal functional paradigms to problems of design and control, including applications pertaining to music technology and creativity (Magenta 2022). Artificial neuronal networks (ANN) technology owes much of its development to burgeoning neuroscientific insight.

However, this chapter proposes a different angle to harness the neurosciences for composition. Rather than building musical ANN to learn how to compose music, I shall introduce my forays into harnessing the behaviour of a type of neuronal model

E. R. Miranda (✉)
University of Plymouth, Plymouth, England
e-mail: eduardo.miranda@plymouth.ac.uk

C. Vear and F. Poltronieri (eds.), *The Language of Creative AI*, Springer Series on Cultural Computing, https://doi.org/10.1007/978-3-031-10960-7_13

referred to as *spiking neuronal networks* to compose music (Jang et al. 2019). The discussion revolves around a piece for orchestra, choir and a solo mezzo-soprano entitled *Raster Plot*.

2 Description of *Raster Plot*

Raster plot is a tribute to Plymouth-born explorer Robert Falcon Scott. It includes extracts from Scott's diary (Scott 2008) on the final moments of his expedition to the South Pole before he died in March of 1912; the extracts used in the piece are available in Appendix 1.

The mezzo-soprano sings the extracts using *sprechgesang*, a type of vocalization between singing and recitation: the voice sings the beginning of each note and then falls rapidly from the notated pitch, alluding to the endurance of Scott and his companions facing the imminent fatal ending of the expedition. A whispering choir echoes distressed thoughts amidst a plethora of jumbled mental activity represented by the sounds of the orchestra.

2.1 *New Models*

Inspired by the physiology of the human brain, I devised a method to represent the notion of mental activity musically. I used a computer simulation of a network of interconnected neurones that model the way in which information travels within the brain, to generate patterns that I subsequently turned into music. When the network is stimulated with an external signal (this will be clarified below), each neurone of the network produces sequences of bursts of activity, referred to as spikes, forming streams of rhythmic patterns. A *raster plot* is a graph plotting the spikes (Fig. 1): hence the title of the composition.

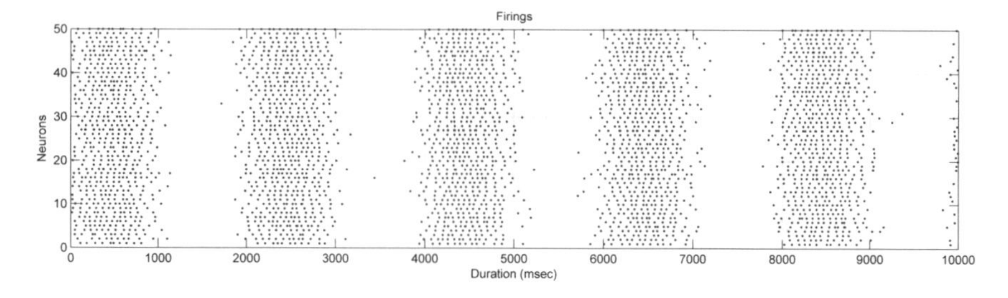

Fig. 1 A raster plot illustrating collective firing behaviour of a simulated network of spiking neurones. Neurone numbers are plotted (y-axis) against time (x-axis) for a simulation of 50 neurones over a period of ten seconds. Each dot represents a firing event

In a nutshell, I orchestrated raster plots by allocating each instrument of the orchestra to a different neurone of the network simulation. Each time a neurone produced a spike, its respective instrument was prompted to play a certain note. The notes were assigned based upon a series of chords, which served as frames to make simultaneous spikes sound in harmony.

The movement culminates with a transition from the orchestrated raster plots to a concluding passage bearing resemblance to a cathedral psalter chant. I wanted this to represent the moment Scott passed way; musically, it conveys a moment of *poiesis*: a moment of transition.

2.2 Music Neurotechnology

As briefly mentioned above, many recent advances in the neurosciences, especially in Computational Neuroscience, have led to a deeper understanding of the behaviour of individual neurones and their networks. I have coined the term *Music Neurotechnology* in a paper I co-authored for *Computer Music Journal* in 2009 (Miranda et al. 2009), to refer to a new research area that is emerging at the crossroads of Neurobiology, Engineering Sciences and Music. The compositional method described here is one of the outcomes of my continuing research in this field. Another important development in this area includes Brain–Computer Music Interfacing (BCMI) systems to enable persons with severe motor impairment to make music (Eaton et al. 2015).

The spiking neurones model that I used to compose the piece was originally developed by computational neuroscientist Eugene Izhikevich (2007). A biological neurone aggregates the electrical activity of its surroundings over time until it reaches a given threshold. At this point, it generates a sudden burst of electricity, referred to as an *action potential*. Izhikevich's model is interesting because it produces spiking behaviours that are identical to the spiking behaviour of neurones in real brains. Also, its equations are relatively easier to understand and program on a computer, compared to other, more complex models. Izhikevich's equations represent the electrical activity at the level of the membrane of neurones over time and can reproduce several properties of biological spiking neurones commonly observed in the brain.

The simulation contains two types of neurones, excitatory and inhibitory, which interact and influence the behaviour of the whole network. Each action potential produced by a neurone is registered and transmitted to other neurons, producing waves of activation, which spread over the entire network. A raster plot showing an example of such collective firing behaviour, taken from a simulation of a network of neurones, is shown in Fig. 1. Here, the spikes result from a simulation of the activity of a network of 50 artificial neurones over a period of ten seconds: the neurones are numbered on the y-axis (with neurone number 1 at the bottom, and neurone number 50 at the top) and time, which runs from zero to 10,000 ms, is on the x-axis. Every time one such neurone fires, a dot is placed on the graph at the respective time.

Figure 1 shows periods of intense collective spiking activity separated by quieter moments. These moments of relative quietness in the network are due to both the action of the inhibitory neurones and the refractory period during which a neurone that has spiked remains silent as its electrical potential decays back to a baseline value.

The network model needs to be stimulated to produce these patterns of activation. For the composition of *Raster Plot*, I stimulated the network with a sinusoidal signal that was input to all neurones of the network simultaneously. Generally speaking, the amplitude of this signal controlled the overall intensity of firing through the network. For instance, the bottom of Fig. 2 shows a raster plot generated by a network of spiking neurones stimulated by the sinusoid shown at the top of the figure. As the undulating line rises, the spiking activity is intensified. Conversely, as the undulating line falls, the spiking activity becomes quieter. As a gross generalization, if one thinks of the spiking neuronal network model as the brain of some sort of organism, then the stimulating sinusoid would represent perceived sensory information. Albeit simplistic, I find this model inspiring in the sense that it captures the essence of how our brain responds to sensorial information. Of course, a more complex signal could replace the sinusoid; for instance, a sampled sound could be used to simulate the network. In this case, the raster plots would look considerably more complex than the ones I am presenting in this chapter.

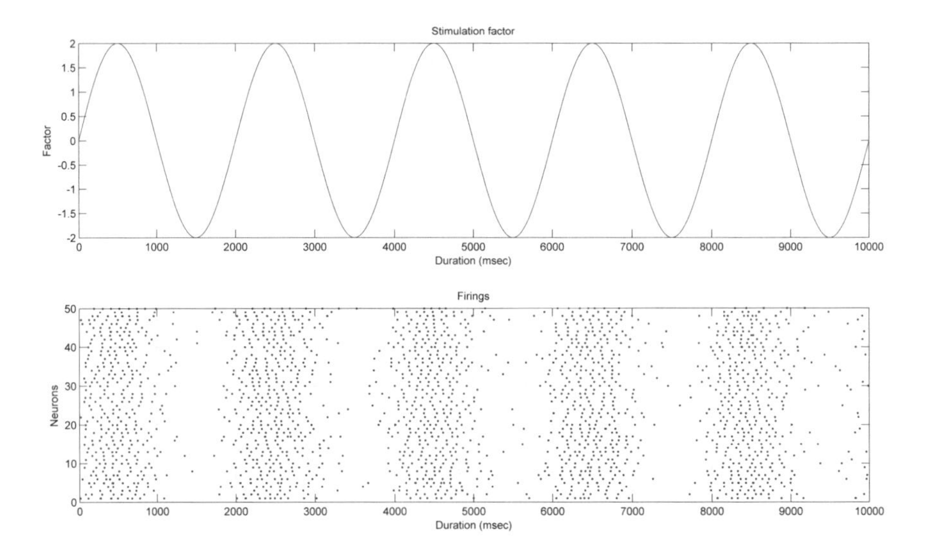

Fig. 2 At the top is a sinusoid signal that stimulated the network that produced the spiking activity represented by the raster plot at the bottom

3 Compositional Process

To compose the piece, I set up a network with 50 neurones and ran the simulation 12 times, lasting for 10 s each. For all runs of the simulation, I set the stimulating sinusoid to a frequency of 0.0005 Hz, which means that each cycle of the wave lasted for 2 s. Therefore, each simulation took five cycles of the wave, which can be seen at the top of Figs. 2, 3 and 4, respectively.

Compositionally, the top of Figs. 2, 3 and 4 suggests musical form to me, whereas the bottom suggests musical content. Hopefully, this will become clearer below as I unpack the process by which I composed this piece.

For each run, I varied the amplitude of the sinusoid, that is, the power of the stimulating signal, and the sensitivity of the neurones to fire. The power of the stimulating signal could be varied from 0.0 (no power at all) to 5.0 (maximum power) and the sensitivity of the neurons could be varied from 0.0 (no sensitivity at all; would never fire) to 5.0 (very sensitive). For instance, for the first run of the simulation, I set the power of the signal to 1.10 and the sensitivity of the neurons to 2.0 (Fig. 3), whereas in the tenth run I set these to 2.0 and 4.4, respectively (Fig. 1). One can see that the higher the power of the stimuli and the higher the sensitivity, the more likely the neurons are to fire and therefore the more spikes the network produces overall. One can observe a considerable increase in spiking activity in Fig. 4, which corresponds to the fourth run. And in Fig. 2, which corresponds to the tenth run, there is a substantial increase in the intensity of spiking activity. Table 1 shows the values for the 12 runs. I had envisaged at this stage a composition where the music

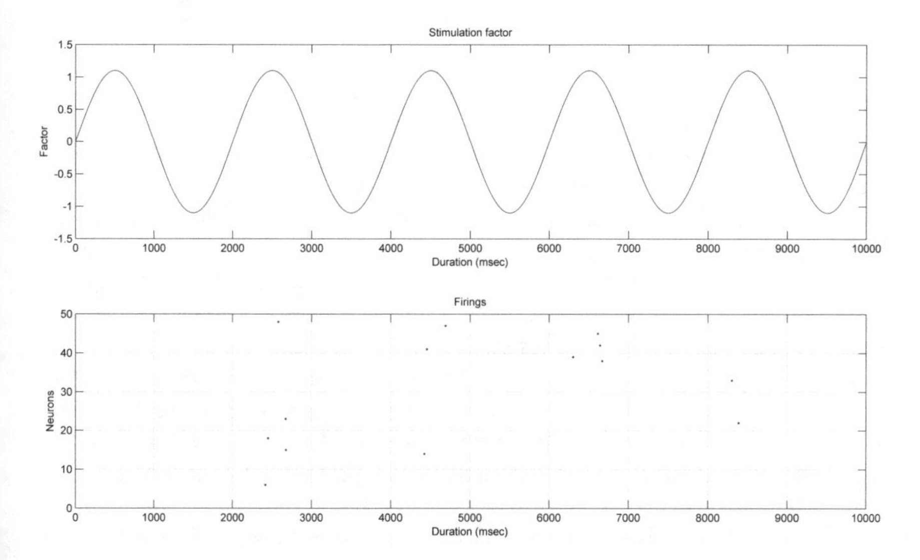

Fig. 3 First run of the simulation produced sparse spiking activity because the amplitude of the sinewave and the sensitivity of the neurons were set relatively low

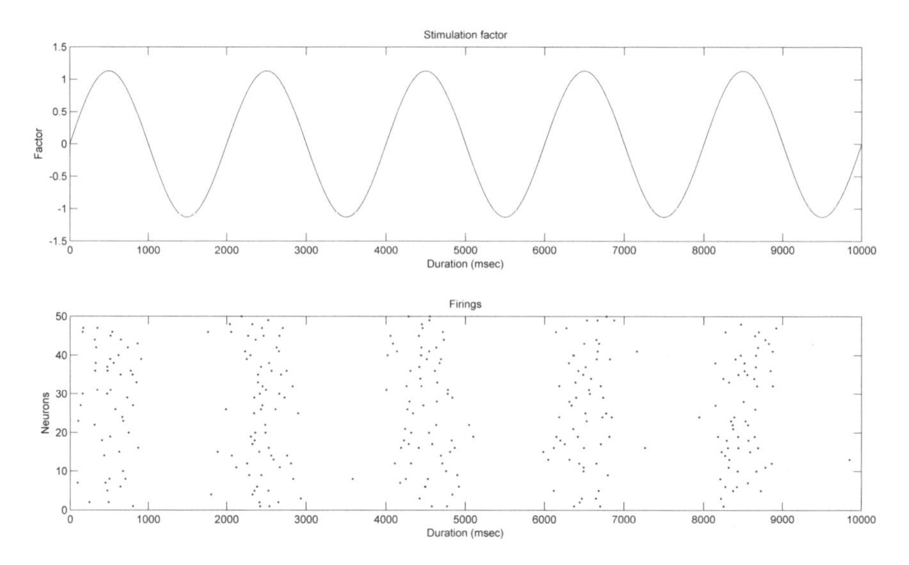

Fig. 4 Sensitivity of the neurons to fire was increased slightly in the fourth run of the simulations, resulting in more spiking activity than in previous runs

Table 1 Parameters for the 12 runs of the spiking neurones network

Run	1	2	3	4	5	6	7	8	9	10	11	12
Power	1.10	1.11	1.12	1.13	1.14	1.2	1.21	1.22	1.3	2.0	2.2	3.0
Sensitivity	2.0	2.3	2.6	2.9	3.2	3.5	3.8	4.0	4.2	4.4	4.8	5.0

would become increasingly complex and tense, culminating with the transition to the psalter-like chant I mentioned earlier.

I established that each cycle of the stimulating sinusoid would produce spiking data for three measures of music, with the following time signatures: 4/4, 3/4 and 4/4, respectively. Therefore, each run of the simulation would produce spiking data for fifteen measures of music. Twelve runs resulted in a total of 180 measures, but as we shall see below, I finished the spiking section at measure number 160. I felt that the resulting music was beginning to linger and loose interest at about this measure. Thus, the time was ripe for the transition to the psalter-like chant.

With the settings shown in Table 1, I noticed that the neurones did not produce more than 44 spikes in one cycle of the stimulating sinusoid. This meant that if I turned each spike into a musical note, then each cycle of the sinusoid would produce up to 44 notes. In order to transcribe the spikes as musical notes, I decided to quantize[1] them to fit a metric of semiquavers, where the first and the last of the three measures could hold up to 16 spikes each, and the second measure could hold up to 12. Next,

[1] To quantize means to restrict a variable quantity to discrete values. For example, an ordinary clock normally quantizes time to seconds; each tick of the clock corresponds to a second.

I associated each instrument of the orchestra, excepting the choir and the mezzo-soprano parts, to a neurone or group of neurones. This is shown in Table 2. From the 50 neurones of the network, I ended up using only the first 40, counting from the bottom of the raster plots upwards. Polyphonic instruments, such as the organ, were associated with a group of neurons because they can play more than one stream of notes simultaneously.

The compositional process progressed through three major steps:

(a) the establishment of a rhythmic template,
(b) the assignment of pitches to the template and
(c) the articulation of the musical material.

In order to establish the rhythmic template, firstly I transcribed the spikes as semi-quavers onto the score. Figure 5 shows an excerpt of the result of this transcription for a section of the strings.

Although I could have written a piece of software to transcribe the spikes, I ended up transcribing the spikes manually. I printed the raster plots for each cycle of the stimulating signal (Fig. 6). Then, I used a template drawn on an acetate sheet to

Table 2 Instruments are associated with neurones

Neurones	Instruments	Neurones	Instruments
1	Contrabass 2	17	1st Violin 1
2	Contrabass 2	18, 19, 20, 21	Organ
3	Cello 3	22, 23, 24, 25, 26	Celesta
4	Cello 2	27, 28, 29	Vibraphone, Timpani
5	Cello 1	30	Snare drum, Cymbal, Tam-tam
6	Viola 3	31	Tuba
7	Viola 2	32	Trombone 3
8	Viola 1	33	Trombone 2
9	2nd Violin 4	34	Trombone 1
10	2nd Violin 3	35	Trumpet 2
11	2nd Violin 2	36	Trumpet 1
12	2nd Violin 1	37	Horn 3
13	1st Violin 5	38	Horn 2
14	1st Violin 4	39	Horn 1
15	1st Violin 3	40	Clarinet bass clarinet
16	1st Violin 2		

Each instrument plays the spikes produced by its respective neurone or group of neurones

Fig. 5 Transcribing spikes from a raster plot as semiquavers on a score

establish the positions of the spikes and transcribe the information into the score (Fig. 7).

To forge rhythmic patterns that would be recognized as such by performing musicians, I altered the duration of the notes and rests, while preserving the original spiking pattern as much as I could. Figure 8 shows the new version of the score

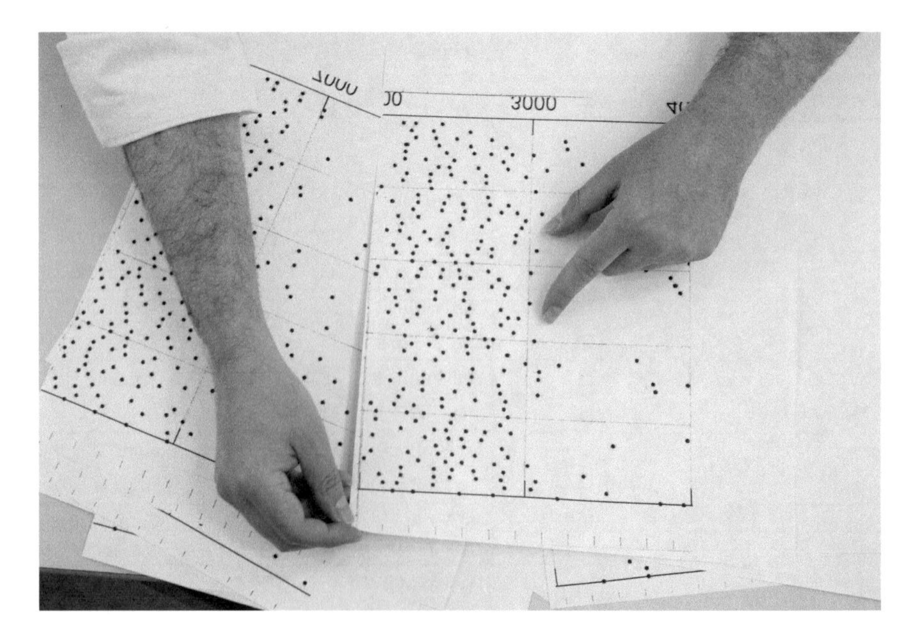

Fig. 6 A raster plots for each cycle of the stimulating signal produce spiking data for three measures of music

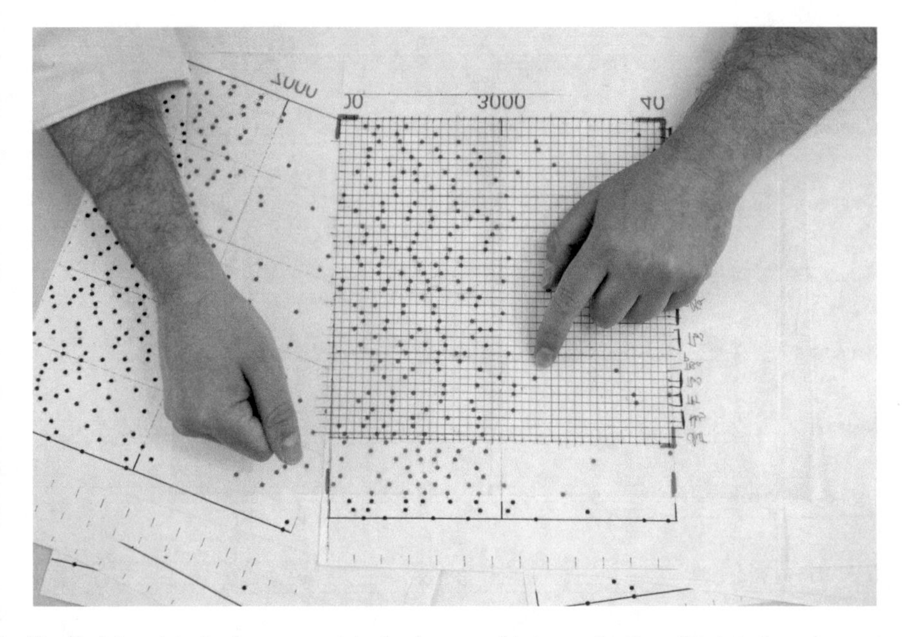

Fig. 7 A template drawn on an acetate sheet was used to transcribe the spikes into the score

shown in Fig. 5 after this process. Figure 9 shows the result of the compositional process, with pitches and articulation.

I would say that in many ways the compositional method that I developed for raster plot draws on Pierre Boulez serialism (Griffiths 1979). In order to assign pitches to the rhythmic template, I defined a series of 36 chords of 12 notes each, as shown in Fig. 10. These chords sprang on the back of the napkin after a conversation I had with composer Peter Nelson on a rainy afternoon in a café in Edinburgh. I had mentioned to him that I was struggling to find a decent way to assign pitches to the spiking rhythms. Peter suggested using matrices representing harmonic topologies. It was a eureka moment.

I started by creating 12 chords based on the harmonic series. Then, I established additional 24 chords firstly by inverting only a portion of those 12 chords (e.g., only the notes on the G clef) and then by inverting chords entirely. I do not remember the exact rationale for the different key signatures; most probably, I defined them in haste in order to avoid having to write all accidents next to the respective notes on the score.

To begin with, I used the first chord of the series to furnish pitches for the first 21 measures of music. As the spiking activity up to this point was not so intense, I decided to use only this chord to begin with. Then, from measure 22 onwards I used each subsequent chord of the series to furnish pitches for every three measures, and so on. Once I had furnished the pitches for measures 124–126 with the 36th chord, I subsequently selected chords unsystematically to continue the process until measure number 160.

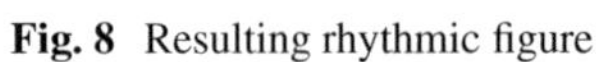
Fig. 8 Resulting rhythmic figure

Fig. 9 Resulting music

Fig. 10 Series of chords for the harmonic structure of *raster plot*

The actual allocation of pitches of the chords to notes of the rhythmic figures was arbitrary. I did this differently as the movement progressed. In general, those figures to be played by instruments of lower tessitura were assigned the lower pitches of the chords and those to be played by instruments of higher tessitura were assigned the higher pitches, and so on. An example is shown in Fig. 11, which shows the allocation of pitches from the G clef portion of chord number 22 to the rhythmic figures for

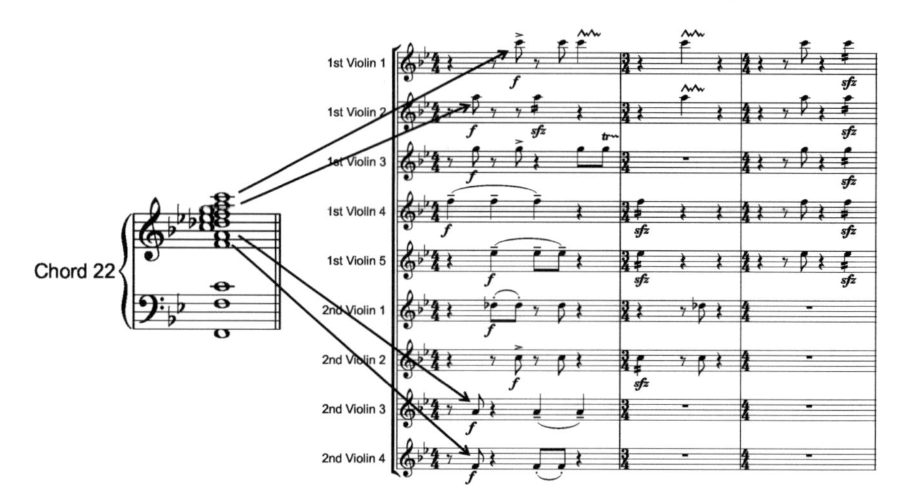

Fig. 11 An excerpt from *raster plot* illustrating the assignment of pitches to rhythmic figures

the violins in measures 82–84. There were occasions where I decided to transpose pitches one octave upwards or downwards in order to best fit specific contexts or technical constraints of the respective instrument. Other adjustments also occurred during the process of articulating the musical materials.

3.1 Limitations

A caveat of my method to turn raster plots into music is that it limits my ability to compose with the parameters composers would normally expect to work with, that is, duration and pitch. In a way, this limitation forced me to work with other musical parameters, such as articulation and timbre, to fashion the materials. To this end, I employed several non-standard playing techniques to forge new musical gestures.

The process of articulating the musical material is a difficult one to explain objectively because it was much less systematic than the processes described thus far. The vocal part was composed at the same time as I worked on the articulations. But it was not directly constrained by the spiking neurones method. The mezzo-soprano, which sings in *sprechgesang mode*, appears in measures corresponding to periods of rarefactive spiking activity. Musically, I wanted to create an effect akin to responsorial singing. Metaphorically, I wanted to allude to an imaginary process, whereby the neurones were sending commands to control the muscles of the vocal mechanisms of a hypothetic singer, but not so efficiently. Hence, the undefined effect of hearing neither clear singing nor clear speaking. The bass clarinet often doubles the mezzo-soprano, representing the hypothetic singer's mind's ear; it plays the melodic lines she intends to sing. Technically, this aids the singer to find the right pitch to enter passages that are difficult to ascertain the pitch unaided.

4 Reflection on Process

By way of introspection, I often find myself confronting the following dichotomy whenever I attempt to articulate my compositional practice. On the one hand, I think of music as the intuitive expression of ineffable thoughts, highly personal impressions of the world around me, and the irrational manifestation of emotions. On the other hand, I am keen to maintain that music should be logical, systematic, and follow guiding rules. In general, I think that rationality does play an important role in music composition, especially classical music. Hence, formalisms, rules, schemes, methods, number crunching, computing, and so on, are of foremost importance for my *métier*: But I also think that music that is totally generated automatically by a machine is rather meaningless. Music needs to be embedded in cultural and emotionally meaningful contexts, which composers express in subtle, often ineffable ways. A computer would not be capable of composing a piece such as Beethoven's *Symphony No. 9*. Its backstory, myriad of references, drama, and so on, are aspects of musicianship that computers, as we know them today, cannot grasp. The composition of *Raster Plot* is a good example of this dichotomy.

All the same, one of the reasons I find it exciting working with artificial intelligence, and computers in general, is because they can generate musical materials that I would not have produced on my own manually. This mindset is akin to John Cage's thinking when he preferred to set up the conditions for music to happen rather than composing music set in stone. Cage liked being surprised by the outcomes of such happenings (Cage 1994). By the same token, I enjoy being surprised by the outcomes of a computer. But I am not willing to just leave these materials intact I am afraid.

A recording of the premiere of *raster plot* by Ten Tors Orchestra, under the baton of the late Simon Ible, is published by Da Vinci and a free version is available on YouTube.[2] A short excerpt of the score is shown in Appendix 2.

Appendix 1 The Lyrics for Raster Plot

For God's sake,

look after our people.

Had we lived,

I should have had a tail to tell,

of the hardihood endurance and courage

---

[2] *Raster Plot* on YouTube: https://www.youtube.com/watch?v=xEywlAbP8Vs

of my companions,

which would have stirred

the heart of every Englishman.

These rough notes

and our dead bodies

must tell the tale.

We shall stick it out

to the end,

but we are getting weaker.

Of course

and the end cannot be far.

It seems a pity,

but I do not think

I can write more.

For God's sake,

look after our people.

Appendix 2 Excerpts from the Score

See Figs. 12 and 13.

Fig. 12 Page 17 from the score

Fig. 13 Page 20 from the score

References

Cage J (1994) Silence: lectures and writings. London, Marion Boyars. ISBN 978-0714510439

Eaton J, Williams D, Miranda E (2015) The space between us: evaluating a multi-user affective brain–computer music interface. Brain Comput Interfaces 2:103–116. https://doi.org/10.1080/2326263X.2015.1101922

Griffiths P (1979) Boulez. Oxford University Press. ISBN: 978–0193154421.

Izhikevich ER (2007) Dynamical systems in neuroscience. Cambridge (MA), The MIT Press. ISBN 978-0262090438

Jang H, Simeone O, Gardner B, Grüning A (2019) An introduction to spiking neural networks: probabilistic models, learning rules, and applications. https://arxiv.org/abs/1910.01059

Magenta Studio. https://magenta.tensorflow.org/studio-announce

Miranda ER (ed) (2021) Handbook of artificial intelligence for music. Springer. ISBN: 978–3–030–72116–9

Miranda ER, Bull L, Gueguen F, Uroukov IS (2009) Computer music meets unconventional computing: towards sound synthesis with in vitro neural networks. Comput Music J 33(1):9–18

Scott RF (2008), in: Jones M (ed) Journals captain Scott's last expedition. Oxford University Press. ISBN: 978–0199536801

Tuning Topological Morphologies: Creative Processes of Natural and Artificial Cognitive Systems

Johnny DiBlasi

Abstract Artificial intelligence is at the least an enhancement of computational systems, technologies, and processes, and at the most represents a stepping-stone on the road to developing an extension of human intelligence. In this chapter the AI model used serves as a model or framework for thinking about the aesthetics and structures of creative processes. The aim is to discuss factors which affect approaches to the creative process in general and how these influence the relationships between creators, technologies, and the resulting works. This chapter is an inquiry into how AI has altered our theoretical framework in the arts as well as to explore the properties or the language of creative AI. In this context, this chapter will ask the question what is the language of artificial intelligence (AI) in the artist's own creative practice? The author uses theories of embodied cognition and nonconscious cognitive systems to provide a foundation for creative practice as the creation of enacted, embodied meaning or aesthetic experience through numbers as exemplified in the performance piece *432Hz*. Through this piece, the author's practice, and the making of aesthetic experiences, numbers are expressed through sound frequencies and are then 'tuned' by the machine (AI) over time by way of playing or performing the machine.

Keywords Creativity · AI · Practice · Relationships · Aesthetics

1 Introduction

> With numbers nothing is impossible. Modulation, transformation, synchronization; delay, memory, transposition; scrambling, scanning, mapping - a total connection of all media on a digital base erases the notion of the medium itself. Instead of hooking up technologies to people, absolute knowledge can run as an endless loop. (Kittler 2012: 31–33)

Computation and computational media have a rich history in the meandering march of the arts and artworld. Computational technologies have long been ubiquitous within our daily experience, and they are now deeply embedded within all avenues of culture.

J. DiBlasi (✉)
Iowa State University, Ames, IA, USA
e-mail: jdiblasi@iastate.edu

C. Vear and F. Poltronieri (eds.), *The Language of Creative AI*, Springer Series on Cultural Computing, https://doi.org/10.1007/978-3-031-10960-7_14

They have become tools of our intelligence and at the same time, computational technologies, through extension, have increased the powers of our intelligence. Like all technologies, they have increased our abilities to manipulate our environment and understand it simultaneously. One has only to consider quantum mechanics or the Voyager 1 probe to get a glimpse at an example of how this particular tool has expanded our capacities to touch, smell, taste, see, and hear our reality(ies).

We are profoundly affected by this new extension of ourselves. Therefore, due to the nature of the work of an artist, even if an artist only works with analogue or non-computational/ traditional media, the artwork is informed by and influenced heavily by computational media. I am focused now more broadly on computational media because it is the foundation on which artificial intelligence rests. AI is at least an enhancement of computational systems, technologies, and processes, and at the most represents the goal, of which the development of computational media is the stepping-stone, on the road to developing an extension of human intelligence.

In this chapter the AI model used in creative practice is my own stepping-stone to think about the aesthetics and structures of creative processes in working on the *432Hz* project and others, as well as how these projects highlight a framework for a foundational language of creative AI which affects these approaches to the creative process and the relationships between creators, technologies, and the resulting works. This book, and this chapter, is an inquiry into how AI has altered our theoretical framework in the arts as well as to explore the properties or the language of creative AI.

1.1 AI and Artistic Experimentation

In this section I will explore the language of creative AI through artistic experimentation and processes of building and interfacing with artificial neural networks and generative deep learning models. Namely, an audio-visual performance piece titled *432Hz* that is an experiment in building deep artificial neural networks to calculate, train, and tune numerical expressions of computer-generated sound waves. This project represents an iterative process that explores one way in which artificial intelligence (AI) can be embedded in creative practice. This project and others in the field offer a look at unique aspects to creative practice where AI is embedded as a medium for making or where the AI system becomes the art object. First, we will cover some background on the machine learning (ML) developments as they relate to creative fields and then explore ideas relevant to cognitive systems more broadly before returning to the questions raised by the confluence of AI and creative practice.

1.1.1 Topologies

Topologies in mathematics and in networking are the description of a structure or object in space made from points that have an underlying logic. However, this logic and structure are highly variable and contingent. When exploring this field, and looking at these topological objects and spaces, it becomes apparent how these spaces are a construct that can be easily morphed, stretched, pulled, enlarged, squashed, and expanded. In other words, these objects and spaces are tested and highly contingent, and to a certain extent, they represent non-unique, modular possibilities much like in contemporary architecture and the built spaces we occupy, where "cookie cutter" homes and apartments serve as our spaces for existing. For example, consider building projects like the Hudson Yards project with modular, climbable sculpture *The Vessel* in Manhattan. If we apply these concepts of topological properties of constructed space to examples in visual arts, we can use the example of Sol LeWitt's artworks such as his wall drawings or sculptures.

Through my practice, I am interested in these morphologies and contingencies. In my own art making where I utilize data as a media, I have begun to think a lot about how nature and our artificial extensions of nature are expressed and transformed through numerical data and mathematical expressions. At its core, an artificial neural network or deep learning model is nothing more than a huge array of numbers being computed repeatedly. It is this "digital base" which becomes a starting point of which I propose as a lens in considering what are the properties of computational media and subsequently, creative AI.

1.1.2 Convergence as AI

The convergence of technical media, that is evident in the introductory quote by Kittler above stating the "total connection of all media on a digital base," comes to fruition through the translation of computational systems to simulate our own cognitive systems—this, I believe, is artificial intelligence. This scaffolding is realized at its core through the application of mathematics as means for the control of nature (and natural processes) as exhibited primarily through the harnessing of the electromagnetic spectrum and electrical current to compute bits in the form of electrical pulses in a CPU or across a circuit (i.e., through computation or which is down to its core frequencies of electrical pulses). Put more simply, in the context of this text, *what is the language of artificial intelligence (AI) in my creative practice?*

Finally, in pondering this question, I will use theories of embodied cognition and nonconscious cognitive systems to provide a model for creative practice as the creation of enacted, embodied meaning or aesthetic experience through numbers as exemplified in the performance piece *432Hz*. Through this piece, my practice, and the making of aesthetic experience, numbers are expressed through sound frequencies and are then "tuned" by the machine (AI) over time by way of playing or performing the machine (as instrument or medium). I will use *432Hz* as a model for a contingent and embodied experience of feedback loop between artist and machine, and through this case, numbers (or AI) become the medium for an aesthetic object or experience. Through the work, the artist's playing of the AI instrument is an exchange that generates an aesthetic experience, an exchange between artist and audience, and thus the work arises out of two cognitive systems exchanging sensorimotor feedback and operations.

1.2 Converging at Artificial Intelligence: The Successor to Computation

As I previously mentioned, the long history of the development of media is a history of the development of the expansion or the extension of our capacities for seeing and hearing (as well as our other three main senses) through technical means. This history or development does include not only the advancement of technical media, but also includes, and is represented by, the development of how we produce knowledge (and our perception of our own brains, bodies, and selves) more broadly.

As Siegfried Zielinski states in his comprehensive archeological unearthing of technical development, the history of our technical media (and production of knowledge) can be compared to geological deep time (Zielinski 2006: xx). Against Zielinski's urgings, I like to compare this history to a root system of a tree (sort of a flipped family tree of media) where the farther back we look, the more diversity in technology we see. Zielinski uses the history of geology and the evolution of the Earth as a starting point to begin to think about a concept of deep time for technical media:

> Earth's history could be explained exactly and scientifically from the actual state of the 'natural bodies' at any given moment in time, which became known as the doctrine of uniformitarianism. Further Hutton did not describe the Earth's evolution as a linear and irreversible process but as a dynamic cycle of erosion, deposition, consolidation, and uplifting before erosion starts the cycle anew. (Zielinski 2006: xx)

Using this example of "geological deep time," Zielinski applies these concepts more broadly to think about the history of the human species and its progress through technology. He urges us to draw a different picture of progress, and that from a paleontological perspective, a picture or metaphor of progress represented with models of "simple to complex" or tree structures should be rescinded (Zielinski 2006: 5–6). Rather, from this deep paleontological position, Zielinski reminds us of the branching diversity found as we look back on nature's and our own technical progress:

> From this deep perspective, looking back over the time that nature has taken to evolve on Earth, even at our current level of knowledge we can recognize past events where a considerable reduction in diversity occurred. Now, if we make a horizontal cut across such events when represented as a tree structure, for example, branching diversity will be far greater below the cut—that is, in the Earth's more distant past—than above. (Zielinski 2006: 5–6)

Against Zielinski's better judgment, I will adopt the tree metaphor to visualize this concept of the "deep time." What Zielinski is describing here could be thought of as a root structure where when you look into the deep past or below a marker in time, you see much more branching or diversity below or before that point. Or perhaps a forest metaphor would be more apt with its branching roots interwoven together and connected by even more intricate networks of fungal—mycorrhizal networks (perhaps computation would be the eldest tree at the center—the parent of the forest—the conductor orchestrating the sharing of resources and warning messages). As Zielinski has done with his crucial research in *Deep Time of the Media*, we could make a cut at any point on the timeline of history and find many diverse examples of the various tendrils of thought working out the concept of computation.

Now if we pick up one of these (or several) tendrils, we can examine its inherent properties which exhibit the humans' capabilities to understand nature through computation or numerical translation, driven by the exercise of pattern recognition. This foundation of mathematical understanding and numerical representation becomes the current running through the syntax of technical media. One of these examples that Zielinski explores is Athanasius Kircher's personal desktop device called the *cassetta matematica*. He gives an in-depth description of the "box," now residing in the Institute and Museum of the History of Science in Florence, and which sits nicely on a desk and has:

> a menu of nine different branches and applications of mathematics: arithmetic, geometry, fortificatoria (dealing with calculations for military fortifications), chronologia (measuring time by regular divisions, in his case, the cycles of the moon and movements of the planets), horologia (science of constructing sundials), astronomy, astrology, steganography, and music. Assigned to each of these headings are twenty-four wooden slats, one behind the other, which according to each of the nine mathematical areas, are of different colors and marked with the letters of the alphabet A through I. (Zielinski 2006: 141–143)

Kircher's apparatus serves as a kind of early calculator or algorithmic/mathematics database. Each of the slats has spaces which contain operations from the different fields and which can be arranged with other slats to recombine or arrange into different components (Zielinski 2006: 141–143). This provides another example of the digitization—the translation to a numerical code of knowledge—of media. As I will discuss later, and which is a theme that runs throughout this chapter, this numerical, mathematical translation of nature seems to be a key aspect of any sort of "language" of this media.

Evidence of the intertwined relationship between histories in art and human history and history of civilization more broadly are apparent whenever one considers the so-called "revolutions" in the history of progress, technologies, and collective thinking. Much like the industrial revolution triggered the scene for post-impressionism and then modernism, let's consider how computation affords the convergence of media and leads us toward the "post" eras in art and culture. In a *Fresh Air* interview, historian Walter Isaacson discusses his book *The Code Breaker* which dives into the history and development of the CRISPR[1] technology. In the interview Isaacson gives his version of the three revolutions in modern times—revolutions of culture founded on basic particles or kernels:

---

[1] CRISPR is a technology that can be used to edit genes.

1. The revolution in physics with a foundational kernel of the atom leads to "atom bomb, space travel, GPS, and semiconductors"
2. The revolution in computing with a kernel of the bit, "And it meant that all information could be coded in zeros and ones and binary digits"
3. The revolution in gene sequencing, with the kernel of the gene, "a fundamental particle of our existence...And in the beginning of this century, in 2000 or so, we sequenced the entire human genome. And now with [the invention of CRISPR] we found ways to rewrite the genome. And so this part of the twenty-first century…will be a biotech revolution, a life sciences revolution" (Isaacson 2021).

Now these revolutions can be debated—and have been—over the course of history. I'm sure some historians would take issue with the idea of leaving out the industrial revolution and other milestones in the history of science–culture–technology. Though here I want to focus on Isaacson's second revolution: the revolution of the bit. Although rather than focusing on the bit or 1's and 0's, I would like to go deeper—and refer to the bit as the harnessing of electrical current or its physical manifestation. The bit comes from the "ons" and "offs" of electrical current pulses—and from that we get computation. We have this convergence of mathematics, physics, electrical engineering, and so on to bring us to computation. Computation also represents a convergence of media—or a convergence of media by which we hear and see through a technical means. Again the introductory quote by Kittler rings true. However, because it is a convergence of this media, computation is also heterogenous in the sense it represents all of our media or mediums. Like the invention of photography, the development of computation was the next inevitable or determined step in the convergence of our technologies and thus a post-medium condition in the arts. The evolution of the arts into its post-medium condition was informed by a computational perspective in culture and assisted by the adoption of the technology itself.

1.3 The Language of the Post-Medium Condition

In this section I outline several key concepts that help us understand convergence of our technologies and thus a post-medium condition in the arts. In their book *Rethinking Curating*, Beryl Graham and Sarah Cook grapple with the various ways technologies or computational media have changed the endeavor of artists, curators, and the audience in relation to experiencing art (or having an aesthetic experience of a work of art). They define and explore all the different ways "New" media art have affected artists' and curators' processes, practices and the properties of art and aesthetic experiences in this new age of digital technology, information, and electronic data networks. The authors highlight some properties of new media art such as variable and hybrid materiality, systematized, time-based, and networked, and they remind us of the definitions of interaction and participation. The curators state:

> *Interaction: 'acting upon each other.'* Interaction might occur between people, between people and machines, between machines, or between artwork and audience. However, examples of human and machines or humans and artworks truly acting upon each other are relatively rare. What is popularly termed interaction in these cases is often a simpler 'reaction'—a human. (Graham and Cook 2010: 112–113)

They go on to discuss and define other aspects of artworks as systems of change between work and audience and define other areas on the spectrum of artistic agency. In the same way that artworks that utilize computation as a medium can be interactive at times, they can also be participatory as well as collaborative in their often evolving realizations. Graham and Cook highlight these various natures of this medium (or post-medium) of artistic practice:

> *Participation: 'to have a share in or take part in.* Participation implies that the participant can have some kind of input that is recorded…that is, not just getting reactions, but also changing the artwork's content…*Collaboration: 'working jointly with'*. Unlike *interaction* and *participation*, the term *collaboration* implies the production of something with a degree of equality between the participants. (2010: 113–114)

Later in the chapter I will discuss the collaborative nature between artist and machine (AI) within the piece *432Hz*. One could argue that all artists "collaborate" in some way with their media, and indeed many artists create works that are participatory or collaborative. Usually this is in reference between artwork (made through computational media or software) and audience. However, I will argue here that there is a certain level of equality—because this equality is afforded to the machine when we consider it to be its own working cognitive system working in tandem with the artist—another cognitive system—to produce the final work. For now, I want to consider the machine—which I use to describe this AI cognitive system—as the medium through which the artist works. What does it mean to consider computation as a medium, and what are the aspects of this medium's language? Luckily, this "language" and its theory of artistic practice have been around for many decades, and therefore, there are already very solid measuring sticks in place.

In her text *Voyage on the North Sea*, Rosalind Krauss posits the task of Modernism coming to its conclusion as the implosion of specificity and the heterogenous artworld of comprised of distinct media. She eloquently guides us through a narrative of the modernist endeavor which culminates and then is followed by conceptual art to push the post-medium condition past the finish line (Krauss 1999: 10–20). Rather, I think of this as the beginning or start of something for the artworld and art theory. This collapse of boundaries of or the concept of a medium speaks to the new terrain we find ourselves in as a culture as well as art practitioners. This is where I believe the concept of *The Deep Time of the Media* and the Kittler quote become very useful tools in framing the idea of a medium in art—and the idea of a non-medium framework (or post-medium) for the artworld and art practitioners.

We can also look to the recent history of the language of contemporary art within the digital and then post-digital context. We can look to, or even borrow from, the previous syntactical systems that undergirded the arts going back to the advent of the adoption of video as a medium for practitioners. Of course, as we've seen with Zielinski's research, we could go back further to find instances of a "language" of the arts that is fueled by concepts driven by digital systems or at least conceptually digital systems. Though I think a good place to start is period directly following the Second World War. This is the time when fields were all newly developing based on theories of cybernetics and computation—a time when computational media began to influence, and in some instances aid in the convergence of, the fields of cybernetics/cognitive sciences (science), technology/engineering (technology–military), and the avant-garde (art).

Krauss also adopts this time frame as she positions the post-medium condition in arts practices. She marks the advent of the post-medium age with the artist Marcel Broodthaers and the beginning of Conceptual art in the 60s:

> Twenty-five years later, all over the world, in every biennial and at every art fair, the eagle principle functions as the new Academy. Whether it calls itself installation art or institutional critique, the international spread of the mixed-media installation has become ubiquitous. Triumphantly declaring that we now inhabit a post-medium age, the post-medium condition of this form traces its lineage, of course, not so much to Joseph Kosuth as to Marcel Broodthaers. (Krauss 1999: 20)

With this new art movement, the concept of the Modernist quest to find a medium's essential essence—or its specificity—was abolished and medium specificity was upended. Starting in the post-war era, art was adopting a more heterogenous application and concept of the medium. As we'll see later in this section, this time period in art ran parallel to the adoption of heterogenous technologies such as video and computation to make art.

In *The Language of New Media*, Lev Manovich completes an exhaustive synopsis of new media as an medium for creative practice as well as a form for media more broadly. This "new" medium largely encompasses digital or computational media, but extends and overlaps other media used to create aesthetic objects (such as video or cinema, photography, mixed-media, performance, etc.) Manovich outlines five "Principles of New Media." These five principles include Numerical Representation, Modularity, Automation, Variability, and Transcoding (Manovich 2001: 27–48). Of course those principles were only the beginning of Manovich's description of the language of new media. Though these principles set a good foundation for the properties of electronic and computational media to establish its language. Two decades later, there are many, well-established theories and languages of the media.

In their exhaustive and effective survey of the field(s), Casey Reas and Chandler McWilliams outline formal design principles of computational media or, in this instance, of "code." Many of these principles overlap with Manovich's while also elaborating on them. These principles included repetition, transformation, parameterization, visualization, and simulation (Reas and McWilliams 2010). These are the various principles or outcomes when computational processes are used to manipulate form (2D, 3D, 4D), and, thus inherently, content. We could also analyze (or include) the principles that come about just from including the inherent properties of the resulting forms brought about by computational media's physical outputs. That is, we could also include more principles that are unique to say (RGB) light from a projector, 3D printed plastics, and printed pigmented inks. We could, for example, say that light from a projector is described as additive numerical representation of color and light which is transcoded by the machine. It is apparent though that this is not necessary as we already have consensus on the various principles of form-concept resulting from computational media. Even by looking at the example I've given of the projector's light, we already see various principles at work: numbers are parameters that are transcoded into simulated light.

These principles of computational media are already well-established, and so there is no need to elaborate on them here. We can use these principles or these previously mapped aspects of computational media's language as building blocks—a starting point—for the "language" of creative artificial intelligence. We will return to these aspects of this newly updated language later in the chapter. In addition, I want to focus on what these principles tell us—or how they can shine a light on why creativity and art have morphed into the place we find it in now. I want to utilize these principles in the event they can elucidate the various ways computational media and technology were the catalysts for this "post" era—these eras of post-digital and post-medium.

By Krauss's measure, it was the introduction of the complex system of the Portapack (video) as a medium of art, which shattered the Modernist dream—like a Benjaminesque moment, where we crossed a threshold. The main part of the Modernist dream or endeavor that was ended by the medium of video was the endeavor of medium specificity. Because like film or the cinematic apparatus, there was no indivisible essence or quality that this media could be broken down into. Video was film's electronic update in the 60s—adding telepresence, broadcast to film's industrial qualities of repetitive reproduction and time (among other qualities). The language, or Krauss's "essence," of film and then video was too complex, too heterogenous to be reduced down to a homogenous specificity (Krauss 1999: 24–26). One could also argue other points in time as well as point to other technological developments to reference this crossing of a threshold—or ending of medium specificity. What about the artists beginning to work with computers for the very first time?—which by the way, was happening simultaneously with the adoption of the more popular video apparatus as artistic medium. Perhaps it was our embedded experience of our environment which was utilizing our new sensorimotor systems of telepresence, computation, was driving a new perception of ourselves and the world which was more variable and heterogenous. And these new perceptions, like anything in our history, work its way into every avenue of culture—including art which has always been a communicative mirror reflecting our culture's current epistemological state.

2 The Language of Embodied Cognition

The current state of the field of cognitive science (referred by some as Post-Cognitive era) puts forth new ideas about how cognitive systems, consciousness, and the mind works through the theory of enaction or embodied cognition. In the seminal text by Francisco Varela, Evan Thompson, and Eleanor Rosch, after a survey of the past theories of mind that guided the field, the authors define and present a kind of "none" but "all of the above" theory through their idea of the mind as experiencing reality through a process of enaction.

What is key to these ideas of enaction, is that there is no such thing as a separation between the two entities of mind and body, but actually the mind–body is part of one cognitive system that experiences and takes actions in the world. It is worth explaining this idea as we consider the authors' concept of perceptually guided action as it relates to the different approaches to cognitive sciences as they draw out differences in the opposing theories. For example, the authors' state of perceptually guided action that: "We have already seen that for the representationist the point of departure for understanding perception is the information-processing problem of recovering pregiven properties of the world." (Varela et al. 1991: 173) The authors then go onto speak about how the theory of enaction is different as it is not based on a concept of a pregiven, independent world, but that "the point of departure for the enactive approach is the study of how the perceiver can guide his actions in his local situation" (Varela et al. 1991: 173). In addition, Varela and his co-authors remind us that these "local situations" are constantly in flux and change, and that some of these changes are a result of the perceiver's activity. Therefore, "the reference point for understanding perception is no longer a pregiven, perceiver-independent world but rather the sensorimotor structure of the perceiver (the way in which the nervous system links sensory and motor surfaces)" (Varela et al. 1991: 173).

I want to highlight this approach or this concept or this theory of perception as it relates to experience and the link between the mind, the body, and the experience of the world. As the author's state:

> The structure—the manner in which the perceiver is embodied—rather than some pregiven world determines how the perceiver can act and be modulated by environmental events. Thus the overall concern of an enactive approach to perception is not to determine how some perceiver-independent world is to recovered; it is, rather, to determine the common principles or lawful linkages between sensory and motor systems that explain how action can be perceptually guided in a perceiver-dependent world (Varela et al. 1991: 173).

So we see that the mind and sensorimotor system that is our body is actually a part of one cognitive system that experiences the world through a process of enaction where there is a constant feedback loop between this cognitive system and its environment through its sensorimotor functions it takes actions in the environment through a complex back and forth of tuning the environment and tuning its own reaction to the environment as it gathers information and takes subsequent action.

This model of the brain or cognition (and consciousness) is built on top of the previous connectionist strategy to model cognition/brains. The connectionist model is based on principles of emergence and self-organization that result from interconnected ensembles of neurons. Varela et al. summarize discussions taking place as far back as the Macy Conferences (the "formative" years of cybernetics):

> Rather, brains can be seen to operate on the basis of massive interconnections in a distributed form, for that the actual connections among ensembles of neurons change as a result of experience. In brief, these ensembles present us with a self-organizing capacity that is nowhere to be found in the paradigm for symbol manipulation. (Varela et al 1991: 85–86) (aka the computationist's model which was connnectionism's preceding theory of cognition)

The authors go on to zero in on the theory's foundational, singular explanation by citing 'Hebb's Rule' which:

> suggested that learning could be based in changes in the brain that stem from the degree of correlated activity between neurons: if two neurons tend to be active together, their connection is strengthened; otherwise it is diminished. Therefore, the system's connectivity becomes inseparable from its history of transformation and related to the kind of task defined for the system. (Varela et al. 1991: 87)

So thinking is a process of "learning" which is the transformation of complex connections between neurons that fire together in certain ways for certain events or thoughts or conceptualizations. In this way, the network of neurons or these connections self-organize over time as the brain experiences the world through the body–mind's sensors inputting physical touch, light/sight, eardrums, taste buds, and the olfactory system.

Let us consider further their trimmed-down example of this process. Take a total number of neurons and reciprocally connect them together. Connect some of the nodes to an input mechanism—say the retina. Then present the retina with a succession of patterns (images made of reflected light bouncing off objects). After each presentation of these patterns to the system, the system reorganizes itself by rearranging its connections to send signals in a very specific way. This rearranging is a process where the system is "increasing the links between those neurons that happen to be active together" during the time when the item is presented to the retinal inputs. This presentation of a whole collection of these patterns makes up the system's learning phase. Finally, after this learning phase, when the system is presented again with one of the patterns, the system recognizes it because the system "falls into a unique global state or internal configuration that is said to represent the learned item" (ibid.: 87–88).

I am taking the trouble to explain these concepts that elucidate the connectionist model because it helps understand the model of enaction as well as my own use of this type of system in the artworks presented later in this chapter (*432Hz*). In a very real sense, this is what is happening during my performance while I am 'training' the system. I give the system a set of inputs. It feeds forward these inputs (in his case a sequence of numbers that represent soundwave frequencies) through the fully connected network of nodes (again which are only placeholders in the computer's memory). Over time, connections are built up between specific nodes in the network. So that when I present the system with a set of frequencies, it recognizes the pattern through its subsequent connected firing of certain nodes, and the system responds by "answering" me with a new pattern of soundwaves. I will return to this process of training and my collaboration with the machine later in the chapter when I discuss the performance *432Hz* in more depth.

This brings me to the discussion and exploration of the current state of cognitive science through its latest theory of cognitive systems or embodied cognition/experience—that to enaction. In this model of cognition, the mind–body or the entire cognitive system and its environment arise together through enaction within this embodied experience. Varela, Thompson, and Rosch explain their model of cognitive science by defining their theory of "embodied action." The authors do this by focusing on explaining what "embodied" means in relation to cognition, and they highlight the first point "that cognition depends upon the kinds of experience that come from having a body with various sensorimotor capacities" (Varela et al. 1991: 173). Secondly, the authors point out that they use the term "embodied" because "these individual sensorimotor capacities are themselves embedded in a more encompassing biological, psychological, and cultural context" (Varela et al. 1991: 173). Furthermore, they define the term "action" and their intentions for using the term "action" in order to "emphasize once again that sensory and motor processes, perception and action, are fundamentally inseparable in lived cognition," and they also emphasize that "the two are not merely contingently linked in individuals; they have also evolved together." Finally Varela, Thompson, and Rosch define the concept of "enaction" as a model for cognitive systems:

> We can now give a preliminary formulation of what we mean by *enaction*. In a nutshell, the enactive approach consists of two points: (1) perception consists in perceptually guided action and (2) cognitive structures emerge from the recurrent sensorimotor patterns that enable action to be perceptually guided. (Varela et al. 1991: 173)

In this view of cognition and experience, the perceiving actor is embedded within its environment, and as it perceives its environment, through its actions, it alters itself and the environment and thus further alters its perceived experience of it. The two things are entangled together, or what Varela et al. refer to as "structural coupling." These concepts of embodied cognition can be found in studies in linguistics as well. Varela, Thompson, and Rosch use studies in linguistics by Mark Johnson (who I will look at in more detail later in this chapter) to exemplify these processes of cognition through enaction. Johnson explains how even during our basic categorization process cognitive structures are created based on our bodily experience. Varela et al. explain that these "…image schemas emerge from certain basic forms of sensorimotor activities and interactions and so provide a preconceptual structure to our experience…These concepts have a basic logic, which imparts structure to the cognitive domains into which they are imaginatively projected. Finally, these projections are not arbitrary but are accomplished through metaphorical and metonymical mapping procedures that are themselves motivated by the structures of bodily experience" (Varela et al. 1991: 177–178).

We will look more in depth at Johnson's studies regarding metaphor and aesthetics later. Though I bring up this study here as it is a good example of the relationship between cognition and embedded, bodily experience. In conclusion, our experience through our sensorimotor actions within our environment dictates the recurrent, neuronal mappings that represent both our thoughts (brain's activity, conceptualization of the world) and our experience of reality/the world and our feelings—which is to say this is all a sort of embedded feedback loop where we affect change in our sensorimotor experience as the wider physical world, society, culture is affecting changes to our cognitive structures/system. When we understand that this is how our cognitive system(s) works, we can begin to conceptualize different types of cognitive systems that exist in nature or in our constructed environments—different types of cognitive systems that are linked to ours but are not necessarily human. This also leads us to think about Hayles's concept of nonconscious cognition.

As we chip away at our world, trying to grasp it and generate new knowledge, over the past twenty years, our species' dive into this unknown has brought up some amazing theories and unearthed so many awe-inspiring discoveries of mysteries that still confound us—those mysteries involving things such as our brains (the interior) and our universe (the exterior). All the while, our ability(s) to hear and see through technical means—to borrow Zielinski's phrase—has become increasingly sophisticated (continuing to hit milestones along the evolutionary trajectory of our species). This of course has changed the way we see and understand ourselves and our world—within our specie's own embodied and enacted feedback loop at this moment of its evolutionary history where we find ourselves exactly where we should be given our actions and perceptions within our embedded environment. If we consider again these processes or systems of embodied cognition as a model for our own cognition, we see how these enacted patterns represent our cognition as interconnected or embedded within our environments and experience through our sensorimotor body–mind.

This leads us to think about cognitive systems—what they are and how they work—in a different light. In her text *Unthought*, N. Katherine Hayles effectively proposes a more encompassing perspective of cognition. The vehicle she uses for this model is a concept of a type of cognitive system she refers to as nonconscious cognition. In order to introduce and define this cognitive system, she first draws from various fields such as neuroscience and cognitive psychology to delineate definitions of cognition, consciousness, and higher consciousness among others. To start, consciousness is the cognitive function that comprises a core position in our thinking stemming from an awareness of self and others (found in humans, many mammals, and some aquatic species). Extended (or secondary) consciousness is associated with abstract thought, conceptualizing meaning, symbolic reasoning, verbal language, mathematics, and so on. Higher consciousness is the "autobiographical self" which is augmented by our inner monologue playing in our heads all day, and this then induces the "emergence of a self-aware of itself as a self" (Hayles 2017: 9–10). She then contrasts these self-aware cognitive processes with the cognitive system she calls "nonconscious cognition" which "operates at a level of neuronal processing inaccessible to the modes of awareness but nevertheless performing functions essential to consciousness" (Hayles 2017: 10).

Again she draws on the past few decades of neuroscientific research to detail some of these functions which include translating somatic markers into coherent body representations and discerning patterns too complex and subtle for consciousness to process (Hayles 2017: 10). Hayles generates a more inclusive definition of cognition as "a process that interprets information within contexts that connect it with meaning" (Hayles 2017: 22). She goes on to unpack the framework for cognition and provides various examples and applications at various levels for this model of cognition. In her parsing she dives into the definition and writes about each part of the definition and provides additional context. About the first part of the definition that states "*cognition is a process*," she writes that "this implies that cognition is not an attribute, such as intelligence is sometimes considered to be, but rather a dynamic unfolding within an environment in which its activity makes a difference," and she goes on to provide an example of a computer algorithm written on paper which is not cognitive until it is deployed to a platform capable of understanding the instructions and carrying out the process.

The next part of the definition "that interprets information," Hayles points out that this interpretation implies there is more than one option for which a choice is made, and for example, a computational choice would be between true or false or 1 or 0. The connection to the generation of meaning becomes a key part of the definition, and Hayles emphasizes how meaning comes from contingent contexts. Regarding the final portion "*In contexts that connect it with meaning*," Hayles writes that "the implication is that meaning is non an absolute but evolves in relation to specific contexts in which interpretations performed by the cognitive processes lead to outcomes relevant to the situation at that moment. Note that context *includes* embodiment" (Hayles 2017: 25–26, authors emphasis). I want to emphasize how these contexts are pointed out as being contexts of embodiment which serve as contexts for biological cognitive systems. As we discussed previously, or cognitive system is a sensorimotor system connected to its environment. Hayles also wants emphasize this point and states:

> let me emphasize that technical systems have completely different instantiations than biological life-forms, which are not only embodied but also embedded within milieus quite different from those of technical systems. These differences notwithstanding, both technical and biological systems engage in meaning-making within their relevant instantiated/embodied/embedded contexts. (Hayles 2017: 25–26, authors emphasis)

When considering these thoughtful definitions and examples of cognitive systems, we see their relation to concepts espoused in embodied cognition. Cognitive systems are enactors that are embodied or embedded within a milieu or context where they are constantly receiving information coming into a sensorimotor system and make a conscious/unconscious/nonconscious enaction (choice) and/or feeding forward new meaning. Following these examples, Hayles outlines a "tripartite framework" specific to human cognition but also used as a way to conceptualize how these various levels interact and also how these ecologies or systems can include biological systems and technical systems. Specifically referring to human or self-aware cognitive systems, she developed:

> A tripartite framework that may be envisioned as a pyramid with three distinct layers. At the top are consciousness and unconsciousness, grouped together as modes of awareness […] The second part of the [framework] is nonconscious cognition […] The even broader bottom layer comprises material processes. Although these processes are not in themselves cognitive, they are the dynamic actions through which all cognitive activities emerge. (Hayles 2017: 27–28)

This tripartite framework highlights the inner workings of the various aspects of the interwoven cognitive assemblages, and we can see how other nonconscious cognitive systems (biological or technical) are embedded within the environment and exact changes within these assemblages with "material processes."

Hayles urges us to expand/broaden our anthropocentrically derived definition of cognition and cognitive systems. Everywhere we look, at various levels of magnification, we see interconnected systems that are cognitive systems at the core, in biological systems and in our complex technical systems as well as our own systems of thought and consciousness. When defining her framework for technical cognitive systems, Hayles, rightly I think, gives a description of this purview, both macro and micro, of technical systems characterized through nonconscious cognition. As we saw above, she highlights the importance of cognitive systems' workings within a *context.* Furthermore, she elaborates on how interpretation embedded in contexts as it applies to nonconscious cognitive systems of technical devices such as "Medical diagnostic systems, automated satellite imagery identification, ship navigation systems, weather prediction programs, and a host of other nonconscious cognitive devices interpret ambiguous or conflicting information to arrive at conclusions that rarely if ever are completely certain." Hayles uses this example to point out how this ambiguous process is exemplified in human cognitive nonconscious by stating "Integrating multiple somatic markers, it too must synthesize conflicting and/or ambiguous information to arrive at interpretations that may feed forward into consciousness, emerging as emotions, feelings, and other kinds of awareness upon which further interpretive activities take place."

I want to highlight the parallels Hayles is making between these complex cognitive systems, namely that these systems involve feedback loops between input, interpretation, and decisions that feed forward into actions taken in the world. Hayles paints this picture of a complex technological cognitive system when she states that "In automated technical systems, nonconscious cognitions are increasingly embedded in complex systems in which low-level interpretive processes are connected to a wide variety of sensors, and these processes in turn are integrated with higher-level systems that use recursive loops to perform more sophisticated cognitive activities such as drawing inferences, developing proclivities, and making decisions that feed forward into actuators, which perform actions in the world." And key to her argument she writes how these systems and their architecture work in the same way:

> In an important sense, *these multi-level systems represent externalizations of human cognitive processes.* Although the material bases for their operations differ significantly from the analogue chemical/electrical signaling in biological bodies, the kinds of processes have similar informational architectures. (Hayles 2017: 24–25)

I also want to highlight what the author is emphasizing here that these multi-level cognitive systems represent a system that is similar to the system we saw in the embodied cognitive systems in previous paragraphs. If we consider various interconnected computational or technical systems we've employed around and above the earth alongside various micro- and macro-biome systems occurring in nature, we can see how all these systems are cognitive systems that work in a similar fashion as our own embodied cognitive system. I propose to take Varela's, Hayles's, and others' model of embodied cognition and enaction (and nonconscious cognitive assemblages), and take the entity of the embedded mind–body subject and apply this system-body to other assemblages or networks or systems. These concepts will hold up as a model for how these systems (technological and biological) are embedded or intertwined within a cybernetic, feedback loop such as the feedback system I will discuss later exemplified by the performance piece *432Hz*. Furthermore, this is an integral aspect of "creative AI" as well as the artist's "creative intelligence"—that is the recursive, iterative feedback between an actor and what she outputs.

2.1 Embodied Aesthetic Experience

What does embodied cognition or embodied experience mean for the language and models of aesthetic experience? In *The Aesthetics of Meaning and Thought: The Bodily Roots of Philosophy, Science, Morality, and Art*, Mark Johnson puts forth a case that brings aesthetics and aesthetic experience into a central role of cognition and therefore our conceptualization processes, knowledge production systems, and onto our cultural assemblages/practices. Throughout this collection of essays, Johnson lays out his argument where he contends that.

> we need to transcend this overly narrow, fragmenting, and reductionist view in order to recognize that aesthetics is not merely a matter of constructing theories of something called aesthetic experience, but instead extends broadly too encompass all the processes by which we enact meaning through perception, bodily movement, feeling, and imagination. In other words, all meaningful experience is aesthetic experience (Johnson 2018: 2).

Here Johnson posits his central argument that all experience is intertwined with the generation of meaning which by nature makes it aesthetic experience. He outlines his task to "... construct an argument for expanding the scope of aesthetics to recognize the central role of body-based meaning in how we understand, reason, and communicate" (Johnson 2018: 2).

Johnson goes onto speak to using the arts as a model for how we generate meaning through this merged (or recombined) lens of "body-based" perception or imagination-emotion-sensorimotor experience. Thus, the arts are instances of deep and rich "enactments of meaning," and therefore, the arts and their subsequent enactments of meaning, Johnson argues, "give us profound insight into our general processes of meaning-making that underlie our conceptual systems and our cultural institutions and practices" (Johnson 2018: 2). He then goes on to punctuate this 'embodied cognition' line of thinking about all experience as inherently aesthetic experience because we are constantly using our experience and perception to derive meaning from our environment. Johnson states:

> From this embodied cognition perspective, it becomes. possible to see the aesthetic aspects of experience as giving rise to mind, meaning, and thought. The view of meaning that emerges highlights the body-based, affective, and imaginative dimensions of our interactions with

our environments as they shape the ways we make sense of, and reason about, our world. (Johnson 2018: 2)

Here again we have embodied cognition where the cognitive system is enacted through a sensorimotor system within the environment. Furthermore, here the embodied cognitive system arises not only with experience, but with experience that is aesthetic in its nature. As we've seen and discussed with the cognitive systems of embodied cognition and then nonconscious cognition, the mind–body, through its sensorimotor enaction perception, is folded into experience of the environment. Johnson points out that this is inherently a creative process as this is a system where cognition is ultimately enacting metaphor and meaning. While referencing Varela, Thompson, and Rosch's process of enaction, he states that with this view of experience, "it is not correct so say that the mind is merely the brain, since experience encompasses the entire arc of organism–environment engagement, which is an enactive process. Sometimes neuroscientists are criticized (and rightly so) for claiming that all experience and thought take place within the brain. What they should say, according to a pragmatist nondualist ontology and according to good cognitive science, is that thought takes place via structures and processes operating at many levels: in neurons, in a cortex, in a brain, in chemicals in the blood, in an active body, in bodily interactions with one's surroundings, in social interactions, within cultural institutions, and thus in a multidimensional environment. In other words, any satisfactory account of cognition will have to include the whole creative process of organism–environment engagement" (Johnson 2018: 40). These ideas regarding the functioning of cognitive systems are in line with Hayles's concept of nonconscious cognitive systems. This seems to expand upon (or confirm) the concept of nonconscious cognition as Johnson points out is a multidimensional, intertwined collection of systems—or assemblages.

2.2 *Artificial Intelligence Through Machine Learning: Cognitive, Creative Systems*

As I stated previously, artificial intelligence (AI) has developed alongside computation and could even be seen as end by which computation is the means. In this context, it makes sense that AI was mostly theoretical up until only the recent past few decades—computation had to get ironed out first. Jürgen Schmidhuber's survey article on neural networks goes over the long (although he admits that he may not have caught everything in this complex and rich field) history of supervised learning, unsupervised learning, reinforcement learning, deep learning, and evolutionary computation. All of these techniques, algorithms, machine learning (ML) systems, and models amount to the developments, and the field, of AI. We can look back to Weiner's cybernetics and the Macy Conferences or Frank Rosenblatt's trailblazing Perceptron in 1958. Though things didn't really pick up until the 90s when artificial neural networks became more and more sophisticated, and after that, computer scientists were off to the races solving all kinds of problems and developing very sophisticated unsupervised, generative, and deep learning techniques. Schmidhuber writes:

> In the decade around 2000, many practical and commercial pattern recognition applications were dominated by non-neural machine learning methods such as Support Vector Machines (SVMs). Nevertheless, at least in certain domains, NNs outperformed other techniques [...] Important for many present competition-winning pattern recognizers were developments in the [Convolutional Neural Network] CNN department [...] Good image interpretation results were achieved with rather deep NNs trained by BP variant R-prop; here feedback through recurrent connections helped to improve image interpretation [...] Deep [Long Short-Term Memory Recurrent Neural Networks] LSTM RNNs started to obtain certain first speech recognition results comparable to those of HMM-based systems. (Schmidhuber 2015: 96)

These machines are becoming more and more effective and successful which in turn results in a renewed focus on their development in the twenty-first century. Computer scientists in both industry (Google) and academia are now using the ML models in place of normal computation in order to recognize and analyze patterns in datasets. Now of course, they are being put to use in order to learn and predict commodity consumption and subsequently target fine-tuned or personalized ads to the consumer, among other uses (robotics, etc.). This leads to the next generation of computational machine—deep learning generative machines.

To see an example of this development, around seven years ago the engineers at Google wanted to learn more about how the hidden layers in the machine operate which resulted in a new, more powerful machine learning system (Mordvinstev et al. 2015). As I discussed on the ISEA 2020 panel on AI in creative practice, these engineers realized through their investigation, that these models have the means to be able to generate novel images (DiBlasi et al. 2020). Other AI machines were developed and followed this DeepDream model's footsteps, but now being applied to text and music generation. As I said at the time, I was less interested in this latest iteration of the longstanding debate over who is the artist or author, but rather wanted to focus on what this creative act by machine can reveal about agency and cognitive systems.

As we've seen in the previous outlined history of AI and ML, in the last six years or so there have been rapid advances in this machine learning branch of artificial intelligence. As a result of these advances in deep learning and deep generative modeling, these machines are now able to generate novel, creative output such as a musical score, an image, or a piece of text. There are countless examples of artists using this technology in all sorts of interesting ways. Although I am not going to get into too much details regarding these examples. Rather, I want to consider a specific type of use of AI in my creative practice—through performing AI. In this usage of the machine, the AI is one cognitive system, and myself, the artist, is another.

3 Performing AI: 432Hz

The project *432Hz*, seen in Fig. 1, is an experiment in building artificial neural networks to calculate, train, and tune numerical expressions that are transcoded into computer-generated sound waves. *432Hz* is a live, generative soundscape performance that utilizes the act of training neural networks to generate various soundwaves that evolve over time and fluctuate between the harmonic and the discordant. The piece explores the aesthetics of sound and movement expressed as data in order to create an experience of this information into generative imagery and computer-generated sound waves.

In the past, tuning pitches tended to vary widely before tuning was standardized and based on the 440 Hz frequency. Before this standardization, this pitch was expressed in lower frequencies, and for a time, composers promoted a scientific pitch based on 256 Hz or *432Hz*. *432Hz* is an exploration of these tuning frequencies and how sound is expressed through these numerical relationships. The multimedia performance consists of generative imagery that evolves over time and mapped to computer-generated sound waves. Various soundwaves or oscillators expressed by the computer through assignment of these numerical values are layered and altered throughout the performance by a custom digital synthesizer created by the artist. The synthesizer is also a custom-built neural network that the performer trains throughout the performance to learn to generate a combination of various sine wave frequencies.

For the Festival Internacional de la Imagen *Inter/Species* festival I performed the work within the category of Soundscapes through the live production of *432Hz*, and

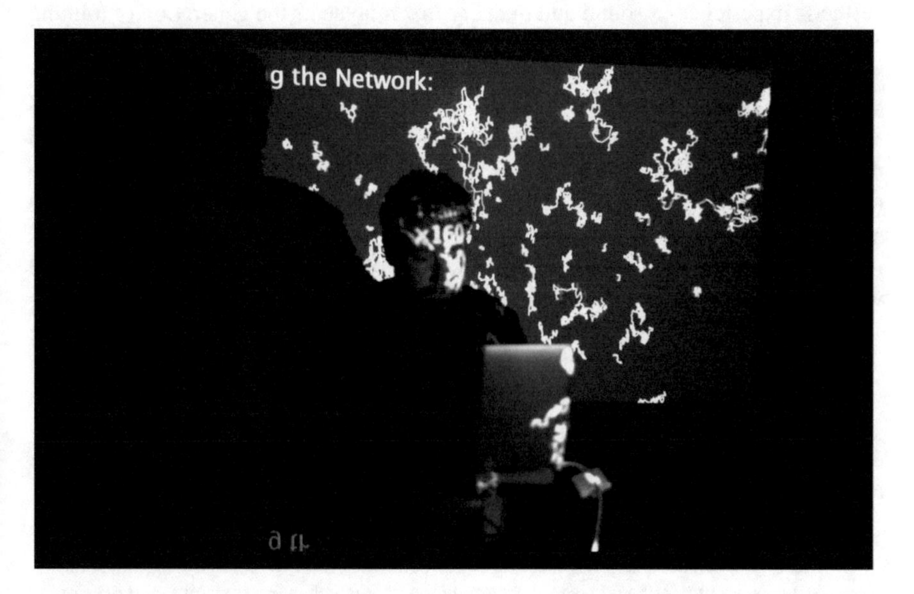

Fig. 1 Still from performance of *432Hz*. Custom Neural Network, multi-channel audio, single-channel video. 2021

the piece was previously performed (with an intimate, spaced audience due to the pandemic) at the Museums Quartier in Vienna. Inspired by the emergent relationships between naturally occurring and artificially generated oscillations, and the evolving relationship over time between the audience and machine (AI) agent and experience of the auditory output, *432Hz* involves a performance of a generative audio-visual experience. The development and the performance of the artwork take the form of a live computer-generated set of evolving projection and sound.

The performance *432Hz* is an exploration of these tuning frequencies and how sound is expressed through these numerical relationships. The multimedia performance consists of generative imagery that evolves over time and mapped to computer-generated sound waves. Various soundwaves or oscillators generated by the computer through assignment of these numerical values are layered and altered through performance and a custom digital synthesizer created by the artist. The synthesizer is also a custom-built neural network that the performer trains throughout the performance to learn to generate a combination of various sine wave frequencies. The machine "learns" and tries different emerging patterns of combined oscillators. So with this project, I explore AI and the generative neural network as itself the media for artistic output as well as the resulting art object. So rather than having the AI create something for the artist—or program the AI to generate the novel aesthetic object (i.e., to make something under the guidance of the artist)—the performance becomes a conversation between the performer and the AI as it is being trained. Through the performance of this system, *432Hz* explores the idea of the performer as simultaneously the builder and trainer of artificial intelligence through the construction of a neural network as itself the media of production. This media outputs an evolving aesthetic experience of sound and imagery that represents the generation of training over time but can also reveal the state of the learning AI at any moment in time.

I proposed as a model to elucidate a series of properties or principles for the use of AI for creative means. First of all, the neural network, or AI itself, becomes the created object—the aesthetic object to experience, rather than the AI's generated output. Secondly, throughout the work, the AI represents a cognitive system, or technical nonconscious cognitive system, with which the artist, another cognitive system, engages in a conversation or dialogue with the AI system through the process of tuning—or training of the AI. Lastly, I want to consider a certain model postulated within the fields of architecture and experience design. In Richard Coyne's text *The Tuning of Place*, he proposes what he calls a metaphor of "tuning" when constructing a theory of how we construct and manage experience within our places and spaces which we can think of in the context of nonconscious cognitive systems. Therefore, our places are cognitive systems that are made up of physical space as well as embedded, integrated, and pervasive digital media, and we tune these systems as we experience and interact with them. He writes that his use of tuning

> is intended to embrace tuning-in and attunement, opening up an examination of the micro-practices by which designers and users engage with the materiality of pervasive digital media and devices, including the inexorable accumulation of small changes, divisions, and ticks of such devices. So tuning provides a richer metaphor for the interconnected digital age than Mumford's trope of synchronization. (Coyne 2010: xv)

This is what happens throughout my performance of the piece: my cognitive system tunes or trains the AI's cognitive system over time. This is also an integral and unique aspect of the AI system. AI is trained over time where the connections between nodes in the network are tuned to be stronger or weaker based on the relationship between the inputs and the desired outcome. The piece and the experience are contingent as the two systems tune and morph over time based on different sensorimotor actions taken in response to the machine's generated light waves and sound oscillations.

Through projects such as *432Hz*, I want to explore the idea of artificial intelligence—and its evolution—as a medium for creative expression. As a medium for aesthetic experience in itself—the act of training is an act of tuning simulated "neurons"—which at its core are data expressed as a number occupying a space of memory within the larger interconnected network. Using the new research in the field of cognitive science—that of embodied cognition or enaction—as a lens to understand the relationship between myself, as an artist, in the act of creating, but also as the interplay between myself—a cognitive system—interacting or exchanging with another cognitive system. But wouldn't that make the two parts simply one cognitive system? And what of the audience who is also connected to the work through their own aesthetic experience of the piece which generates various levels of meaning reflected in the work of art or aesthetic experience?

In these various ways, variable and hybrid nonconscious (and conscious) cognitive assemblages are generated and enacted in an embedded aesthetic experience. This idea has always been at the core of my interest in the landscape as an artist and my exploration of concepts surrounding the landscape in my work. How we move through our environment which is changing, as we alter it with technologies, etc., and we change to adjust to new alterations to the surroundings. I'm interested in this feedback loop between sensorimotor data, our navigation through the landscape's infrastructure, learning its features, and then designing alterations to the constructed and experienced landscape. It's truly inspiring how I'm engaged in a feedback loop between all of these biological, technological, and cultural systems that make up the environment and that make up myself as an embodied mind–body system.

4 Conclusion

In conclusion, I aim to highlight and propose a model for thinking about AI in creative practice by generating properties or the so-called syntax of the language of creative AI. As the chapter title suggests, I wanted to explore the formal and structural relationship between overlapping, contingent, and fluctuating cognitive systems that collaborate, or more aptly, tune each other and bring about changed states in each system.

In this current moment at the culmination of the interwoven histories of computation, AI and art, we seek to define the properties and structures of the language of creative AI which [I argue] can be seen as a culmination of a variety of languages rooted in aesthetics, artistic practice, and cognitive science. The framework created here is elucidated by a dialogue between various cognitive systems which use this language to create aesthetic experiences and which represent a collaboration between various creative actors and agents involved in this conversation. We investigated the histories of computation and AI and how these technologies have affected the language of the arts as both areas of culture developed and grew.

Through using the lens and theories of embodied cognition and enaction, I propose a collaboration or generative feedback loop that arises between various cognitive systems or assemblages. Finally, I used the concepts of enaction, aesthetic experience, and the sensorimotor cognitive system (or cognitive assemblages) to describe the relationship between various levels of aesthetic experience and artistic production. By creating a custom AI agent (or building the algorithms and mathematical system of artificial neural network) as the art object in itself, and then through performing and 'tuning' and training this AI, creative agency and aesthetic experience take shape as a collaboration between these two cognitive systems: the AI and the artist. Which in turn is experienced by an audience which then makes up a collection of other cognitive assemblages or systems.

As we ponder the convergence of mind, body, and experience into a cybernetic feedback loop, I propose to think about how we constantly tune and adjust to our experience, our mind–body systems of the environment which are applied to aesthetic experience and the artist's research and production of aesthetic experiences and objects. The aesthetic experience (or the object of creative production) becomes a collaboration or a dialogue between various cognitive systems that are enmeshed together: the artist, the AI agent, and the audience. The language used in this dialogue exhibits the topology of embodied, aesthetic experiences that fold into one another and this, in turn, generates a possible model of the highly contingent morphology of these creative cognitive systems.

References

Coyne R (2010) The tuning of place: socialable spaces and pervasive digital media. MIT Press, Cambridge

DiBlasi J et al, Panelists (2020) Agency & autonomy: intersections of artificial intelligence and creative practice. In: Why sentience, 26th international symposium on electronic art. 13–18 Oct. Printemps Numērique, Montreal

Graham B, Cook S (2010) Rethinking curating: art after new media. MIT Press, Cambridge

Hayles K (2017) Unthought: the power of the cognitive nonconscious. University of Chicago Press, Chicago

Isaacson W (2021) Interview with Terry Gross. CRISPR scientist's biography explores ethics of rewriting the code of life. Fresh Air

Johnson M (2018) The aesthetics of meaning and thought: the bodily roots of philosophy, science, morality, and art. University of Chicago Press, Chicago

Kittler F (2012) Gramophone, film, typewriter. In: Johnston J (ed) Literature media: information systems. Routledge, New York

Krauss R (1999) A voyage on the north sea: art in the age of the post medium condition. Thames & Hudson, New York
Manovich L (2001) The language of new media. MIT Press, Cambridge
Mordvinstev A et al (2015) Inceptionism: going deeper into neural networks. Google AI Blog. Google. Accessed 2 July 2021
Reas C, McWilliams C, LUST (2010) Form + code in design, art, and architecture. Princeton Architectural Press, New York
Schmidhuber J (2015) Deep learning in neural networks. Neural Netw 61:85–117
Varela FJ, Thompson E, Rosch E (1991) The embodied mind: cognitive science and human experience. MIT Press, Cambridge
Zielinski S (2006) Deep time of the media: towards an archaeology of hearing and seeing by technical means. MIT Press, Cambridge

Sketching Symbiosis: Towards the Development of Relational Systems

Sougwen Chung

Abstract My approach to drawing using machine learning, data and robotics started as a way to evolve my drawing ability by designing artificial intelligence systems to explore collaboration with a robotic unit. For the past several years, my focus has been on the possibilities of robots as a medium of collective collaboration, communication and connection. The practice investigates the interactions between mark-made-by-hand and the mark-made-by-machine as an approach to understanding the dynamics of humans and systems. I believe interdisciplinary engagement lies at the foundation of speculative constructions of art and technology. In the work, I engage with interactive media like AI, robotics and augmented reality as a space of, and for, doubt and uncertainty rather than narrowly defined ends, creating works that address separation, merging and how we inhabit the relations between human, machine and non-human others. The following chapter is an outline, tracing a practice interested in the contours of where AI ends and the "I", the individual human subject begins, using technology of the present-day to ask questions of authorship on a space of a canvas and a performance over time. For me, this question of agency within systems is a microcosm of the wider implications of technological governance and its entangled relationship to the human subject.

Keywords Creativity · Robotics · Art · Collaboration · Authorship · Human–robotic interaction

1 Introduction

> The end of the human is not so much about the hypothesis that machines will completely replace human beings, because this may take longer from now than the extinction of the human species, but rather that machine intelligence will transform humans to an extent that

S. Chung (✉)
Unit 8, 37–42 Charlotte Road, London EC2A 3PG, UK
e-mail: hello@sougwen.com

C. Vear and F. Poltronieri (eds.), *The Language of Creative AI*, Springer Series on Cultural Computing, https://doi.org/10.1007/978-3-031-10960-7_15

is beyond their own imagination. We are in a flux of metaphysical force, which is in the process of carrying humans to an unknown destination.

—Yuk Hui *Yuk Hui, On the Limit of Artificial Intelligence*[1]

Were it not for shadows, there would be no beauty.

—*Jun'ichirō Tanizaki, In Praise of Shadows*[2]

Creative practitioners occupy a unique space in defining the intersection of cultural and technical production, inventing languages of artificial intelligence for artistic exploration. More broadly, the term artificial intelligence (AI) has come to describe an increasingly interwoven suite of technologies including robotics, machine learning, synthetic sensing and natural language processing. Indeed, the landscape of AI is vast and its role in society is pervasive and increasing, shaping industries across a multitude of fields. However, "artificial intelligence" as a term itself necessitates more questions than it does propose answers.

For instance, what do we mean by "artificial", and in what ways does the term presuppose conceptions of the "natural", of the resulting relation between human and machine subjects? In this sense, it could be argued that all artistic media made with the suite of technologies that comprise an AI system implies specific ontological positions in regards to these definitions. These positions, conveyed through installation, artefacts and performance, extend to shape contemporary ideas regarding the artificial and natural, the relation of human-machine-environment, and the shifting nature of intelligence brought upon by digitisation and the post-industrial revolution.

As such, the work of art and research is an ongoing process of inquiry and invention. As a practitioner, I develop operations, as seen in Fig. 1, using emerging technologies to explore phenomenological constructions, interrelations and alternative configurations of human and machine. The operations produce artefacts, vestiges of presence in the form of paintings, research studies, sculpture, installation and performance. I investigate the computable and uncomputable, interrogating the promises and pitfalls of meaning-making metaphors for understanding complex systems; considering sight as a metaphor for computer vision algorithms, embodiment as a metaphor for multi-robotic systems, and memory and learning as metaphors for data-driven machine models.

Throughout all these operations, the role of the human agent oscillates between relayer and receiver, architect and performer, observer and amplifier, designer and steward. At its core, the work takes a speculative approach using non-speculative research: the tools of the day. It asks:

What are the sensory mixes of the future? Where does "AI" end and "we" begin?

[1] Hui, Yuk. "On the Limit of Artificial Intelligence." *Philosophy Today*, vol. 65, no. 2, 2021, pp. 339–357., https://doi.org/10.5840/philtoday202149392.

[2] Tanizaki Jun'ichirō, et al. *In Praise of Shadows*. Vintage Books, 2011.

Fig. 1 Sougwen Chung, *Drawing Operations* performance with industrial robotic arm, 2017

2 Towards a Relational Intelligence

2.1 A Note on Language

One can make the argument that there is no such thing as Artificial Intelligence because there is no such thing as natural intelligence. As Donna Haraway[3] and others have said, we have been cyborgs for a long time now; that is, the intelligence we might attribute to that of the human, or the natural, has always been entangled with technology and machines. Yet, one presupposition that concerns me about the current discourse around AI is how it reveals the shortcomings of language in describing a suite of emerging, exciting and fallible technological systems. The language suggests a false binary between the artificial and the natural disproven by cognitive science and developmental psychology.

Within the presuppositions of the term "artificial" lie a range of ideas and beliefs, such as the secular atheist accelerationalist's belief in the artificial as transcendent, as seen in popular proponents of the technological singularity; something other-than-human and inscrutably above it. Perhaps these beliefs are founded on the conflating of resemblance as equivalence. While these systems of artificial intelligence seem to resemble competence in some traditionally-regarded-as-human tasks, to suggest its equivalence to human intelligence misses the mark.

It is along these same striations that a belief in the neutrality of AI systems also takes root. However, while the biases in AI systems have long since been discussed,

[3] Haraway, Donna Jeanne. *Simians, Cyborgs, and Women: The Reinvention of Nature*. Routledge, 1991.

the potentiality of foregrounding bias as a feature of AI systems is largely underutilised. One can begin by understanding the field of AI as a system of interconnected disciplines ranging from computer vision, robotics, data science and engineering, philosophy, art and so on. I include art and philosophy in the discipline of technological development as both are shaped by the cultural condition in which technologies manifest. I believe the construction of AI as "other" is a missed opportunity for a reinvention of the relations that shape the personal, the technological and the ecological. Within the premises of AI exists a provocation for the transformation of being as connected to tool and world and not separate from it.

2.2 *On Fallibility as Ground*

The most interesting thing I find in working with AI is its fallibility. Fallibility interests me as it is a shared trait across machines, human beings and the environment in which we are a part.

Recognising and exploring fallibility is a means of constructing relation through common ground. Doing so involves the development of relational AI systems trained on hyper-local, personal, individual datasets. It is my hope that this development offsets the misguided notion of a singular artificial intelligence towards a plurality of diverse, responsive, systems of play and governance.

The following projects follow a multi-generational approach to constructing relation as artistic inquiry, technological invention, and philosophical research and seek to offer constructions for moving beyond an ethno-human-centric view.

3 Practice-Based Research

> What is ownership? What is work?… How can we move from a production system in which human labour is merely a disposable means … to a process that depends on and expands connective relationships, mutual respect, the dignity of work, the fullest possible development of the human subject?
>
> —*Adrienne Rich*[4]

The following sections outline a brief overview of insights resulting from almost 10 years of explorations in human and machine co-creation. They stem from an approach that foreground hybridity of practice, including but not limited to research as art and art as research. A continuation of the antidisciplinary thinking made popular by the MIT Media Lab in the early 2000s, and the outmoded polarities of the humanities and the sciences, respectively.

---

[4] Rich, Adrienne. *Arts of the Possible: Essays and Conversations.* W. W. Norton, 2001.

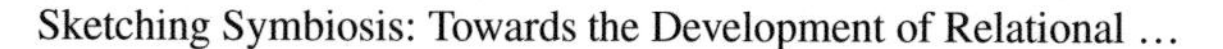

	GENERATION 1	GENERATION 2	GENERATION 3	GENERATION 4
	MIMICRY	MEMORY	MULTIPLICITY	SPECTRALITY
YEAR	2014 - 2016	2016 - 2018	2018 - 2019	2019- Ongoing
VISION:	Colour Tracking	Infra Red	Optical Flow	Form Tracking
RELATION:	Simultaneous, Mimicry	Input / Output, Memory	Multi-agent	Electro-physiological, EEG
TEMPORALITY:	Immediate	Data Archives	Site Interpretive	Neural Interpretive
LOCATION:	Site Specific	Site Non-Specific	New York City	Basel / London
ROBOTIC FORM:	4 axis Robotic Arm	4 axis Robotic Arm	Swarm Robotics (20)	6 axis Robotic Arm
TECHNIQUES:	Camera Jitter, Max MSP	Camera, Custom Software Recurrent Neural Net-	C++, Custom Robotic Modelling, Fabrication, Optical Flow, Custom Global / Local Positioning Software	Python, Custom End Effector, Custom Software, Biometric Headset

Fig. 2 Sougwen Chung, drawing operations unit: generation 1–5 diagram, 2020

In practice-based research, science transforms its languages; poetry invents its tongues.[5] The below perform as prompts and provocations for merging, the continued reimagining of collaborative systems, as well as signposts along the development of a practice informed by emerging concepts in the humanities and the development of technologies of the day.

Introduction to Drawing Operations Unit: Generation 1–4

Drawing Operations Unit: Generation 1–4 (*also known as D.O.U.G._1–4*) is an ongoing project exploring human and machine collaboration. The work is presented as code, drawing artefacts, narrative, sculpture, installation and live performance. *D.O.U.G._1-4* has utilised techniques in computer vision, deep learning on a dataset of two decades of my drawings, custom robotics, AR/VR and bio-sensors explore relation; to catalyse embodied configurations in human and machine creativity. You can think of them as responsive systems linked to my body, movements and biology. Each *D.O.U.G.* explores mimicry, memory, collectivity and spectrality as speculative prompts.

The artefacts are traces—traces of artistic speculation and investigative research. They mark a process of making that challenges, transmutes and distorts my own artistic agency as drawer, performer and programmer. The drawing operations take place in real time in the studio and in front of an audience, reflecting the processes, possibilities and paranoias of the time in which they were created.

D.O.U.G._1 began as a prototype of mimicry. *D.O.U.G._2* expanded on the notion of memory. *D.O.U.G._3* constructs collectivity. *D.O.U.G._4* explores spectrality and teases at symbiosis through *Flora Rearing Agricultural Network (F.R.A.N.)* (see Fig. 2).

---

[5] Glissant Édouard, and Betsy Wing. *Poetics of Relation*. The University of Michigan Press, 2010.

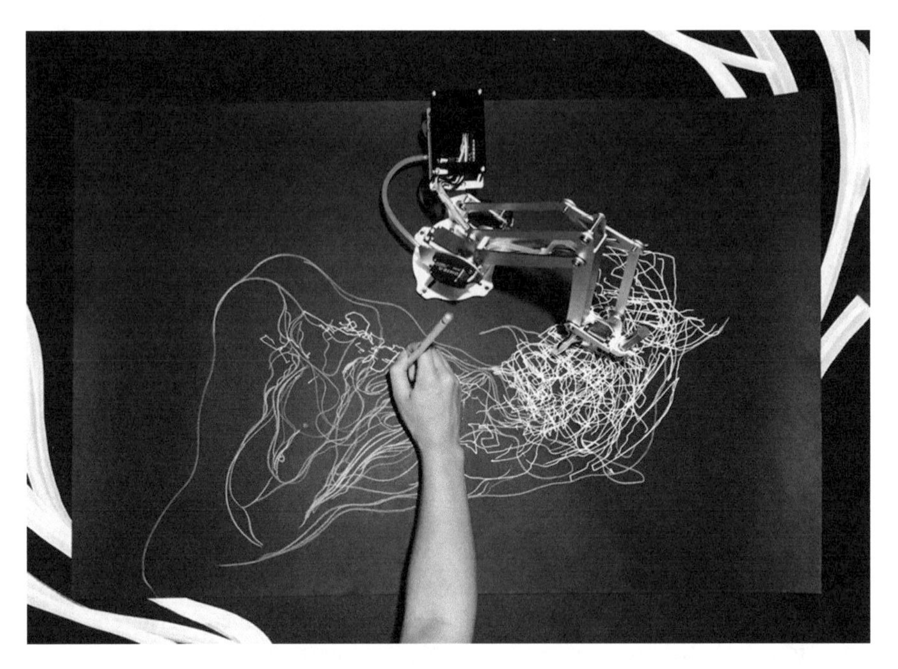

Fig. 3 Sougwen Chung, mimicry debut with D.O.U.G._1, New Museum Sky Room Performance in New York, 2014–2015

3.1 Mimicry: An Exercise in Behavioural Empathy

> Perhaps it belongs with the avant-garde abstract expressionist movement, under the context of Clement Greenberg's theory on medium specificity in which contingency is accepted as aesthetic sensibility. Rather, this collaborative performance between the body and a robot arm is musical.
>
> —*Naohiro Ukawa*[6]

Drawing Operations Unit: Generation 1 is the 1st stage of human and robotic interaction as an artistic collaboration. The robot mimics my drawing gesture via a system that gathers the data of my drawing movement through an overhead camera and analyses the position through computer vision software. The result is a synchronicity of movement between artist and machine in real time, resulting in an interpretive performance and captured as a drawing artefact, as shown in Fig. 3.

Shared Fallibilities

The performance is a process of slowing down, paying attention and communicating entirely through gesture. The robot mimics my movement like a partner in an improvisational singing performance. It is a robotic system that embraces every glitch, bug

[6] UKAWA, Naohiro. "Excellence Award—Drawing Operations Unit: Generation 1: Award: Entertainment Division: 2015 [19th]." *Japan Media Arts Festival Archive*, http://archive.j-mediaarts.jp/en/festival/2015/entertainment/works/19e_drawing_operations_unit_generation_1/.

and error. The drawing session, without pre-established harmony, frees itself from aesthetic constraints, while also examining the essence and phenomenon of beauty at the same time.

The artefacts are white ink on black paper and trace the limitations of the robotic positional translation and my own adaptations to drawing with a robotic unit for the first time.

Beyond Automation

The project came at a time when interactive art and media in an installation format created an interaction model in which the visitor in the space acted as a human catalyst for generative machine responses. *Drawing Operations Unit: Generation 1* sought to explore an inverted position of human and machine with machine as creative catalyst for a collaboratively composed composition of an image.

Drawing Operations Unit: Generation 1 references and re-imagines the premise of a predecessor in Harold Cohen's *A.A.R.O.N.* project in which Cohen as a painter extends his visual language through a flatbed plotter generating sequences of his own brushstrokes executed by mechanical and computational principles.[7]

This work has theoretical ties to Lawrence Shapiro's research on embodied cognition which describes the role of gesture and spatial reasoning in the experience and development of human cognition.[8]

In this first phase, the notion of collaboration is suggested through the creation of an interactive model beyond mere extension. While simplistic in its execution, the interaction model demonstrated in the D.O.U.G._1 configuration proposes a shift from automation to relation. Automation is the existing, hegemonic relation of human and machine paradigm derived from Industrial Revolution to relation.

3.2 Memory: Where Does "AI" End and "We" Begin?

Generation 2 (Memory 2015–2016) explored memory with deep learning and recurrent neural networks. It is an initial exploration into the machine learning of the drawing style of the artist's hand. The robotic arm's movement is generated from neural nets trained on two decades of my drawing archives, the basis of a gestural feedback loop based on my own style.

The white and blue artefacts show a hybrid human and machine drawing, in a sense I'm collaborating with two decades of my drawing as remembered by a machine, as shown in Fig. 4.

On a poetic level, the robotic arm has learned from the visual style of the artist's previous drawings and outputs a machine interpretation during the human/robot drawing duet. Gestures from previous drawings are collected and saved, existing as a

[7] Tate. "Harold Cohen 1928–2016." *Tate*, 1 Jan. 1966, https://www.tate.org.uk/art/artists/harold-cohen-925.

[8] Shapiro, Lawrence A. *Embodied Cognition*. Routledge, 2019.

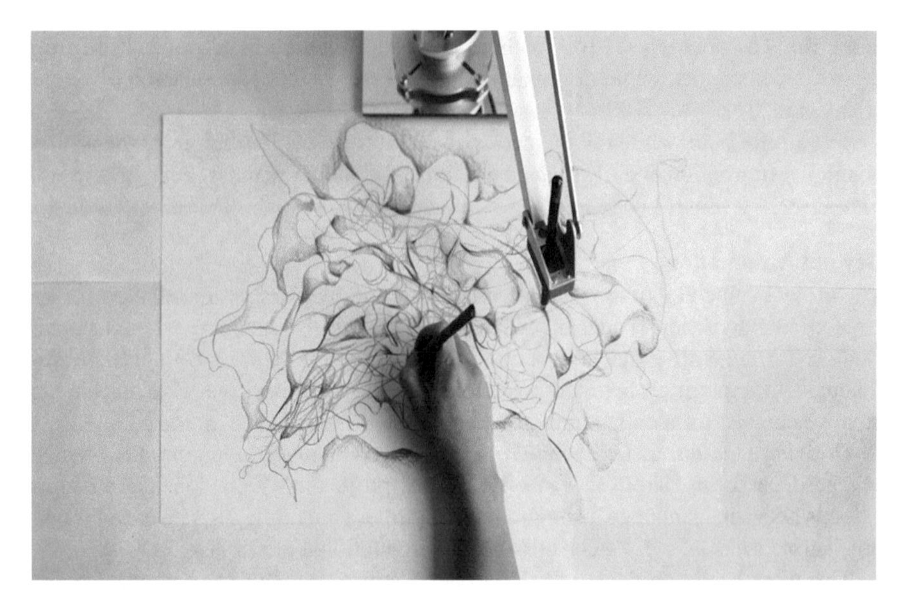

Fig. 4 Sougwen Chung, memory artefact with *D.O.U.G._2*, National Art Gallery in Tokyo, 2016

memory bank for *D.O.U.G._2*. Analysis of visual style of historic artists to translate into gesture as well as colour palette as a collective memory bank from which robotic arm will be able to select.

Foregrounding Artistic Bias

The work foregrounds the subjectivity of classifiers within a neural network, implementing human bias as an artistic style. By focusing on robotic memory using a recurrent neural network and point-based 2D path extraction. This project speculates at the beauty of a non-human move using two generations of an artist's drawing data as training as source material for a training model.

Art as Research Artefact

The project is based on Sketch-RNN, part of the research and development of Google Brain researcher David Ha.[9] The project uses custom software to convert two generations, two decades of drawing compositions on paper and various styles into machine readable paths. The project utilises an interactive system, an overhead camera in which the artist's pen position within the canvas is inputted to dog two and the robotic unit responds based on a statistical approximation of previous drawn line work of an artist, as shown in Fig. 5.

[9] Ha, David, and Douglas Eck. "A Neural Representation of Sketch Drawings." *ArXiv.org*, 19 May 2017, https://arxiv.org/abs/1704.03477.

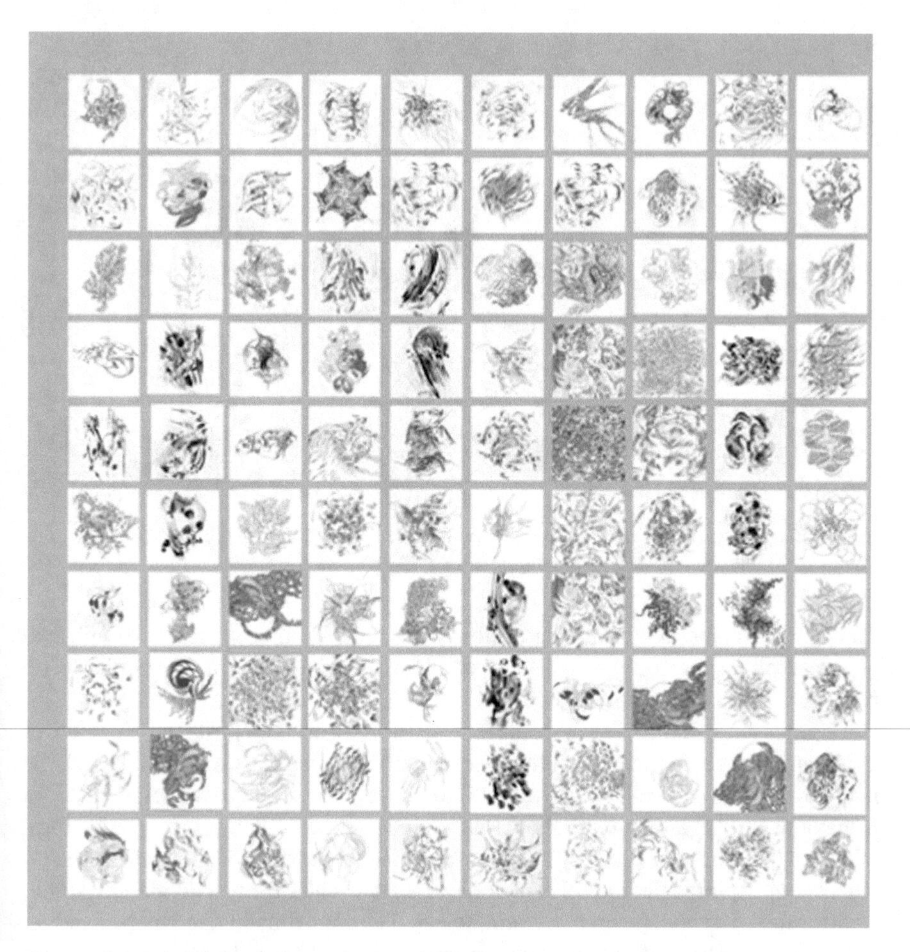

Fig. 5 Sougwen Chung, memory data from *D.O.U.G._2*, V & A permanent collection, 2016

3.3 *Collectivity: How Can Computational Ways of Seeing Reframe a Collective Imagination?*

Drawing Operations Unit: Generation 3 is a multi-robotic system linked to the flow of a city. In this project, I developed custom robotics and a computer vision system in collaboration with Andy Cavatorta and Larry Gorman, from Nokia Bell Labs to create a co-creative performance that traced the movements of New York City in various locations and times of day. The computer vision system was inputted with public camera feeds from New York City.

Its debut performance *Omnia per Omnia* reimagines the tradition of landscape painting as a collaboration between an artist, a robotic swarm, and the dynamic flow of a city, as shown in Fig. 8. The work explores the poetics of various modes of sensing: human and machine, organic and synthetic, and improvisational and computational.

Fig. 6 Sougwen Chung, Omnia per Omnia performance with D.O.U.G._3, Bell Labs and New Museum Exhibition at Mana Contemporary, 2018

Through a collaborative drawing performance between myself and a 20 unit, the project explores the composite agency of a human and machine as a speculation on new pluralities as shown in Fig. 6.

In Gutai art, "Matter never compromises itself with the spirit; the spirit never dominates matter. To make the fullest use of matter is to make use of the spirit".[10] The matter the artists of the Gutai tradition explored were paint, metal, canvas, with Akira Kanayama utilising primitive robotics as a remote painter. Today, matter in the form of digital data, biometrics and robotics has become responsive and interactive. *Omnia per Omnia* explores the interplay of today's matter with Spirit as defined by Gutai, a performance of human agency communicated through drawing and mark-making.

Omnia per Omnia found inspiration in the Situated Knowledges research from theorist Donna Haraway. Situated Knowledge establishes a view of knowledge that is embedded in, and thus affected by, the concrete historical, cultural, linguistic and value context of the knowing person.[11] In machine parlance, each perspective of each camera embedded in urban space comprising the "machine gaze" is determined by the material of the camera itself, the specific modality of the algorithm that constructs visual meaning from the data retrieved.

(Bio)Metrics of the City

The city of New York is a conductor for multi-robotic choreography. Motion vectors extracted from public cameras are linked to *D.O.U.G._3's* collective behaviour in the painting duet. I paint with this system of machines in the creation of an ephemeral portrait of a city in perpetual transition, in constant flux.

---

[10] *Gutai: Splendid Playground*. Guggenheim, 2013.

[11] Haraway, D. (1988). Situated Knowledges: The Science Question in Feminism and the Privilege of Partial Perspective. Feminist Studies, 14(3), 575–599. https://doi.org/10.2307/3178066

Surveillance Apertures
What do public cameras see? How do they see us? The positional data for the robots foregrounds the different states of the city via publicly available camera feeds. The optical flow algorithm categorises states of collective movement as density, dwell, direction and velocity, as shown in Fig. 7.

Ways of Seeing
The philosophical underpinnings of the computer vision algorithm captures the optical flow of a scene as opposed to the single object; it privileges the action of the collective (the behaviour of the crowd) over individual surveillance (face tracking and recognition), as shown in Fig. 6. Do computer vision systems that view a scene as a composition of discrete objects shape a sense of modern isolation?

Foregrounding a panopticon of eyes within an urban environment the states of the city the computer vision system was derived from a computer vision technique called optical flow which extracts the states within a scene and not an individual object. This is part of O'Gorman's ongoing research on creating "kinder" cameras which do not delineate the individuals of a scene but instead extract collective states of collective flow for interpretation and research.[12]

This approach to computer vision in the terrain of public cameras is in the direction of the development of privacy-centric deployment of vision systems in the public domain.

3.4 Spectrality: What is a "Body"?

> This is not a passive participation. Passivity plays no part in Relation. Every time an individual or community attempts to define its place in it, even if this place is disputed, it helps blow the usual way of thinking off course, driving out the now weary rules of former classicisms, making new "follow throughs" to chaos-monde possible.
>
> —*Eduoard Glissant, Poetics of Relation*[13]

Generation 4 (Mutations of Presence 2019–2021) explored biofeedback and was developed in isolation during the onset of the pandemic. I focused on internal flows of meditation captured through biometric recording with an EEG headset. I translated those states to the robotic unit as a physical expression of my meditative states during lockdown, as shown in Fig. 8.

Drawing Operations Unit: Generation 4 is a project exploring spectrality, the electrical signals, currents and pulses of human and machine cognition. Utilising EEG and biofeedback technologies measure, analyse and to catalyse alternative marks made by machine using human subject as a conduit. The work complicates the determinative notion of the drawn mark as linked to human intention, exploring a

[12] O'Gorman, Lawrence. "Putting a Kinder Face on Public Cameras." *Computer*, vol. 46, no. 8, 2013, pp. 79–81., https://doi.org/10.1109/mc.2013.286.

[13] Glissant Édouard, and Betsy Wing. *Poetics of Relation.* The University of Michigan Press, 2010.

Fig. 7 Sougwen Chung, *Omnia* film featuring optical flow with *D.O.U.G._3*, screening at Mana Contemporary in New Jersey, 2018

Fig. 8 Sougwen Chung, mutations of presence performance with *D.O.U.G._4*, Laurenz Haus Residency in Basel, 2020

configuration between human and machine operating at a sub-level of experience—that of the electrical signal. The work speculates on a relational recursion in which the human subject learns to draw through a muscle memory of the brain as an organ that can be trained through meditative and robotic practice.

Signal versus Noise

To draw parallels to computational development, the low level of the human system is accessed in Generation 4, the level of the signal versus the processor, the artistic intention of mark-making. This shifts the role of the human as agent to conduit—the human mind itself as a conduit for artistic co-creation.

Embodied Feedback Loops

D.O.U.G. 4's incubation, conceptualisation and development took place in lockdown during the beginning of the COVID-19 pandemic in 2020. As a response to prolonged

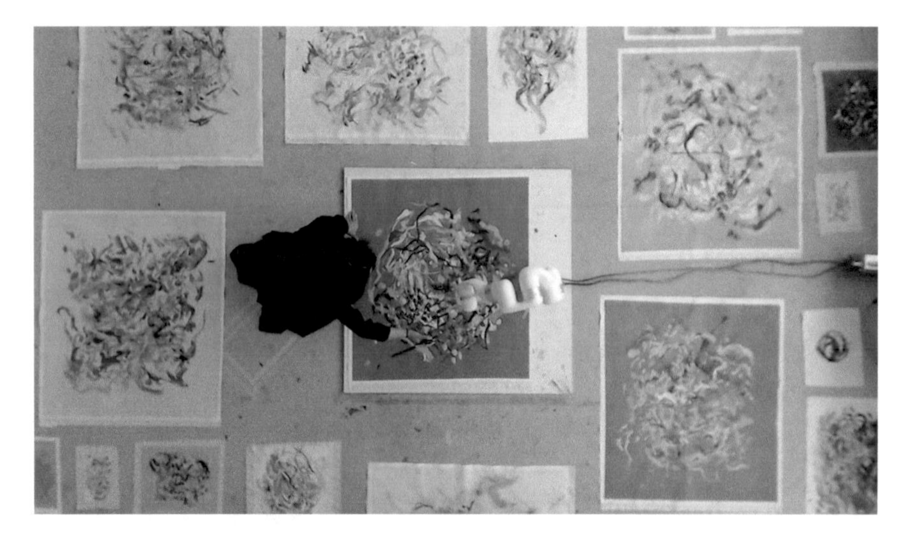

Fig. 9 Sougwen Chung, mutations of presence paintings with *D.O.U.G._4*, Laurenz Haus Residency in Basel, 2020

periods of isolation during the Laurenz Haus residency in Basel, Switzerland, I recorded my brain waves during experiments in meditation techniques ranging from vipassana (mantra), aural (sound), qi gong and visualisation approaches, as shown in Fig. 8. My biofeedback was recorded daily during 20 min meditation sessions. By measuring EEG fluctuations across the spectrum of alpha, beta, delta, gamma, theta, the extracted data is converted into kinematic positions resulting in a visual representation of the brain's electrical signals, as shown in Fig. 9.

Speculating Aura

D.O.U.G. 4 challenges the notion of the drawn line as an intentional, artistic gesture and posits a conceptualisation of drawing as inextricably linked electrical signals in the brain, the biology of the cognising human subject further moving away from the human as a control subject in the human/machine configuration.

An expansive and evolving relation between human and machine that generates movement and sound from the electrical signals generated by the human and robotic body.

The Electric Body: A Language of Unreason, But Which Carries a New Reason

> The electric body—at all scales, atmospheric, subatomic, molecular, organismic—is a quantum phenomenon generating new imaginaries, new lines of research, new possibilities.
>
> — Karen Barad
>
> Trans*/Matter/Realities and Queer Political Imaginings[14]

[14] Barad, Karen. "Trans Materialities." *GLQ: A Journal of Lesbian and Gay Studies*, vol. 21, no. 2–3, 2015, pp. 387–422., https://doi.org/10.1215/10642684-2843239.

Fig. 10 Sougwen Chung, *Exquisite Corpus* performance with *D.O.U.G._4*, Greek National Opera in Athens, 2020

> The earth flows. We are now aware of the deep historical coproduction, or "sympoiesis," of all kinds of material flows that we used to study separately. Flows of rock, flows of water, flows of air, flows of life, and even vast cosmic flows of matter are profoundly interdependent (relational) processes. What if we retold the history of the earth from this perspective?
>
> —*Thomas Nail, Theory of the Earth*[15]

Exquisite Corpus is an immersive performance piece that explores human and machine and ecological bodies. The project links my biofeedback to the mechanical unit and the visual immersive environment. My biofeedback is recorded through an EEG headset and streamed real time to the robotic unit and the immersive environment as shown in Fig. 10.

[15] Nail, Thomas. *Theory of the Earth.* Stanford University Press, 2021.

Fig. 11 Sougwen Chung, *Exquisite Corpus* EEG Visualization performance with *D.O.U.G._4*, Greek National Opera in Athens, 2020

The project exists in four chapters. Chapter 1 engages the audience in a sound exploration, neuro-audio orchestration derived from the artist's electroencephalogram and the robotic unit's electromagnetic frequencies in real time. Chapter 2 focuses on the shared gestural feedback loop between human and machine, the machine gestures derived from recorded and processed brainwave data. Chapter 3 integrates the human and machine performance alongside an immersive environment of real-time satellite feedback layered with the artist's visual interpretation of brainwave data. Chapter 4 combines these elements to speculate at a new co-naturality between the human subject, machine and ecology, as shown in Fig. 11.

3.5 *Sympoiesis*

> This is the same motion as that of the paintings—from dull to brilliant, and then back to dull, and then back to brilliant. Ecological pulses come from and enable the experience of ancestral power. Indeed, for power to come forth, it must recede. For shimmer to capture the eye, there must be absence of shimmer. To understand how absence brings forth, it must be

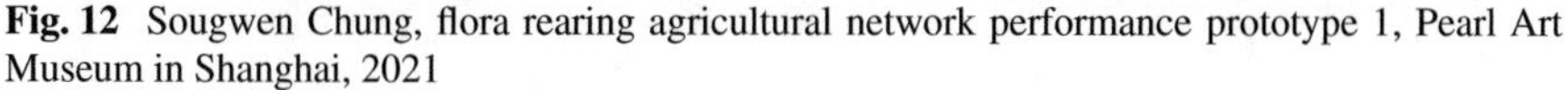

Fig. 12 Sougwen Chung, flora rearing agricultural network performance prototype 1, Pearl Art Museum in Shanghai, 2021

> understood not as lack but as potential. This is where one grasps, afresh, the awful disaster of extinction cascades: not only life and life's shimmer but many of its manifold potentials are eroding.
>
> —*Anna Tsing, Arts of Living on a Damaged Planet*[16]

Flora Rearing Agricultural Network (F.R.A.N) is a project exploring human, plant and machine co-naturality. The work speculates at an interdependent ecosystem of human, machine and flora. Explores linkages between machine and ecology through the development of a networked robotic system stewarding nature.

The first phase of the prototype is a sketch for a bio-mimetic machine. It lays out a conceptual blueprint for a network of custom robotic machines powered by microbial cell batteries exploring unconventional approaches to generating sustainable energy. The robotic units in the piece will be designed to steward the surrounding flora. What are the synthetic plant and machine hybrids of the future? How can we exist as part of a symbiotic feedback loop of caretaking machines attending to nature and humans as gardeners for both.

The process conceived in *F.R.A.N. 1* foregrounds regenerative systems and suggests alternative ways of conceptualising the dynamic configuration of human, machine and nature, as shown in Fig. 12. A configuration not as a productive force or creative force in service of the artistic artefact but as caretaking machines co-contributing to a renewed engagement with plant matter, entangled ecosystems and

[16] Tsing, Anna Lowenhaupt. *Arts of Living on a Damaged Planet: Ghosts of the Anthropocene.* University of Minnesota Press, 2017.

natural ecologies. *F.R.A.N.* addresses a shift in these views as a response to the extractive aspects of technological systems and its impact on planetary resources.

4 Summary

By pursuing a long-form process-driven work of this nature, the project posits that perhaps the future of human (and non-human) creativity isn't in what it makes, but how it comes together to explore new ways of making. Developing new approaches to embodiment, memory and improvisation is what excites me about technology and it's why I program my own creative systems. It's why art and tech intersections with robotics, AI and virtual reality matter—its about exploring new ways of creating and becoming-with machines.

In the work of building relational models, the realms of mimicry, memory, collectivity and spectrality are drivers for system-making enacted through the practice of artistic exploration. Relational AI couples technosocial research and art science practices as wayfinding towards new cosmotechnical pluralities and emerging conceptions of human and machine.[17]

[17] Hui, Yuan. *Art and Cosmotechnics*. University of Minnesota Press, 2020.

Printed in the United States
by Baker & Taylor Publisher Services